Specification B

Geography

John Belfield, Alan Brown, Jane Ferretti, Paul Guinness, Andy Leeder, Sue Lomas, Fred Martin, Ian Matthews, Garrett Nagle, David Payne and Ruth Totterdell

Series Editor: Tom Miller

www.heinemann.co.uk

✓ Free online support
✓ Useful weblinks
✓ 24 hour online ordering

01865 888080

Official Publisher Partnership

OCR AND HEINEMANN ARE WORKING ... T FOR YOU

Heinemann is an imprint of Pearson Education Limited, a company incorporated in England and Wales, having its registered office at Edinburgh Gate, Harlow, Essex, CM20 2JE. Registered company number: 872828

www.heinemann.co.uk

Heinemann is a registered trademark of Pearson Education Limited

First published 2009

13 12 11 10
10 9 8 7 6 5 4

British Library Cataloguing in Publication Data
A catalogue record for this book is available from the British Library

ISBN 978 0 43535370 4

Edited by Huw Jones
Proofread by Ann Kingdom
Project managed and typeset by Wearset Ltd, Boldon, Tyne and Wear
Original illustrations © Pearson Education Limited 2009
Illustrated by Wearset Ltd, Boldon, Tyne and Wear and HL Studios
Picture research by Q2AMedia
Cover photo © Masterfile/Rich Fischer
Printed in China (SWTC/04)

Websites
There are references to relevant websites throughout this book. In order to ensure that the references are up to date, that the sites are functional, and that the sites are not inadvertently linked to sites that could be considered offensive, we have made links available on the Heinemann website at www.heinemann.co.uk/hotlinks. When you access the site, you will need to enter the express code 3704.

Acknowledgements

The authors and publisher would like to thank the following individuals and organisations for permission to reproduce copyright material:

Photographs

pp2–3 Background Geoff Caddick/EPA; p4 Fig. 1.2 Pablo Aneli/Associated Press; p8 Fig. 1.7 Chris Howes/Wild Places Photography/Alamy; p9 Fig. 1.8 Images of Africa Photobank/Alamy; p10 Fig. 1.10 Image Source/Image Source Limited; p12 Fig. 1.15 Ingo Jezierski/PhotoDisc 1997; p13 Fig. 1.18 Leslie Garland Picture Library/Alamy. Fig. 1.19 Planetobserver/Science Photo Library; p15 Fig. 1.22 Ashley Cooper/PICIMPACT/Corbis; p16 Fig. 1.23 Arco Images GmbH/Alamy; p17 Fig. 1.24 Ashley Cooper/Alamy. Fig. 1.25 ICP/Alamy; p18 Fig. 1.26 Worldwide Picture Library/Alamy; p19 Fig. 1.28 Martin Bennett/Reuters; p20 Fig. 1.29 Stringer/Getty Images News/Getty Images Europe/Getty Images; p22 Fig. 1.32 Digital Vision; p25 Fig. 1.35 Manish Swarup/Associated Press; p28 Fig. 1.41 Lee Sanders UK and Ireland Out/EPA; p29 Fig. 1.43 Peter Marshall; p30 Fig. 1.45 Brook Fabian/Alamy. Fig. 1.46 Frank Blackburn/Alamy; p31 Fig. 1.47 Jacky Parker/Alamy. Fig. 1.48 Simon Grosset/Alamy; p32 Fig. 1.49 Bruce W. Heinemann/PhotoDisc.1996; p33 Fig. 1.52 Nick & Suzanne Geary/Alamy. Fig. 1.53 Michael Busselle/Corbis; p35 Fig. 1.56 Hervé Hughes/Hemis/Corbis. Fig. 1.57 Brandon Lennon; p36 Fig. 1.59 Anglia Images/Alamy; p37 Fig. 1.61 www.mike-page.co.uk. Fig. 1.62 Jim Whiteside; p38 Fig. 1.65 Mark Goble/Photographers Direct; p39 Fig. 1.66 Martin O'Neill, New Forest National Park; p41 Fig. 1.69 ASP Norfolk Images/Alamy. Fig. 1.70 Istockphoto; p42 Fig. 1.72 Horizon International Images Limited/Alamy; p43 Fig. 1.74 Sipa Press/Rex Features; p45 Fig. 1.78 Claire Dunstan; p47 Fig. 1.80 Courtesy of Pevensey Coastal Defence Ltd. Fig. 1.81 Courtesy of Pevensey Coastal Defence Ltd. Fig. 1.82 Manor Photography/Alamy; p50 Fig. 1.85 Darrell Young/Alamy; p51 Fig. 1.88 CircumerroStock; p54 Inset Gregg Newton/Corbis; pp54–55 Background Reuters/Paulo Whitaker; p57 Fig. 2.2 Jordi Cami/age fotostock/Photolibrary; p59 Fig. 2.3 Jean-Marc Giboux/Contributor/Getty Images News/Getty Images Europe/Getty Images; p62 Fig. 2.6 Dinodia Photo Library; p64 Fig. 2.9 Gideon Mendel/Corbis; p65 Fig. 2.10 Tim Graham/Alamy; p66 Fig. 2.11 Yoshikatsu Tsuno/AFP; p68 Fig. 2.13 Owen Franken/Corbis; p69 Fig. 2.14 Cheryl Ravelo (Philippines)/Reuters; p71 Fig. 2.17 Sakchai Lalit/Associated Press; p73 Fig. 2.18 Tata Motors; p74 Fig. 2.19 Patrick Robert/Corbis; p75 Fig. 2.20 Rebecca Blackwell/Associated Press; p76 Fig. 2.23 Jose Fuste Raga/Corbis; p77 Fig. 2.24 Paul Almasy/Corbis; p78 Fig. 2.25 Alicia Clarke/Alamy; p79 Fig. 2.26 Sajjad Hussain/AFP; p80 Fig. 2.27 Fred Martin. Fig. 2.28 Fred Martin; p81 Fig. 2.29 Tom Biebrach; p82 Fig. 2.30 Christopher Furlong/Staff/Getty Images News/Getty Images Europe/Getty Images; p83 Fig. 2.31 Ashley Cooper/Corbis; p84 Fig. 2.32 Jim West/Alamy; p85 Fig. 2.33 67photo/Alamy; p87 Fig. 2.36 Eddie Mulholland. Fig. 2.38 Alastair Grant/Associated Press; p89 Fig. 2.40 Robert Stainforth/Alamy; p90 Fig. 2.41 RichMcD Photography. Fig. 2.42 David Cheshire/Alamy; p91 Fig. 2.44 http://www.jimwestphoto.com/; p92 Fig. 2.45 www.thermalcities.com; p94 Fig. 2.49 Nick David/Alamy; p96 Fig. 2.51 Christopher Nicholson/Alamy; p97 Fig. 2.53 Jonathan Banks/Rex Features; p98 Fig. 2.54 Art Directors; p100 Fig. 2.55 kyscan/Corbis; p101 Fig. 2.57 Courtesy of Cambridge Newspapers Ltd; p103 Fig. 2.59 Robert A Dickinson. Fig. 2.60 Rob Atherton/Photographers Direct. Fig. 2.61 Bob Harrison/Photographers Direct. Fig. 2.62 London Aerial Photo Library; p105 Fig. 2.63 Stephen Coyne/Sylvia Cordaiy Photo Library Ltd/Alamy; pp106–107 Background China Photos/Getty Images; p108 Fig. 3.1 Marco Garcia/Contributor/WireImage/Getty Images. Fig. 3.2 Head of the Geotechnical Engineering Office and the Director of the Civil Engineering and Development, the Government of the Hong Kong Special Administrative Region; p111 Fig. 3.5 Jacques Descloitres, MODIS Rapid Response Team, NASA/GSFC; p114 Fig. 3.11 Adam Eastland/Alamy. Fig. 3.12 David Harlow/US Geological Survey/Time Life Pictures/Getty Images; p115 Fig. 3.13 Jim Sugar/Corbis; p117 Fig. 3.17 Carlos Angel/Contributor/Getty Images News/Getty Images North America/Getty Images; p118 Fig. 3.19 Bettmann/Corbis. Fig. 3.21 NASA; p119 Fig. 3.22 Marco Busoni/Photographers Direct. Fig. 3.23 Stephen Bardens/Alamy; p120 Fig. 3.24 Roger Ressmeyer/Corbis; p121 Fig. 3.26 Mauro Fermariello/Science Photo Library; p124 Fig. 3.29 Roger Ressmeyer/Corbis; p125 Fig. 3.30 R1/Alamy. Fig. 3.31 Visions of America, LLC/Alamy; p126 Fig. 3.33 Bobby Yip/Reuters; p127 Fig. 3.34 Associated Press. Fig. 3.35 ChinaFotoPress/Contributor/Getty Images News/Getty Images AsiaPac/Getty Images; p129 Fig. 3.38 www.CartoonStock.com; p131 Fig. 3.41 Neon1974; p132 Fig. 3.42 Reuters/Luis Enrique Ascui; p133 Fig. 3.45 Reuters/Mian Khursheed; p136 Fig. 3.49 Charles Sykes/Rex Features; p137 Fig. 3.50 Bill Haber/Associated Press; p139 Fig. 3.52 NASA. Fig. 3.53 NASA; p140 Fig. 3.54 Martin Bennett/RTR1R09V; p141 Fig. 3.55 Alex Brandon/Associated Press; p143 Fig. 3.59 Marion Kaplan/Alamy; p147 Fig. 3.63 AP Photo/Anthony Mitchell. Fig. 3.65 Les Stone/Sygma/Corbis; p150 Fig. 3.68 Warren Millar/Fotolia; p151 Fig. 3.69 Ami Vitale/Getty Images; p153 Fig. 3.72 Herbie Knott/Rex Features; p156 Inset, left Lester Lefkowitz/Getty Images. Inset, right Andre Jenny/Alamy; pp156–157 Background Charlotte Thege/Alamy; p158 Fig. 4.1 Gary Cook/Alamy; p160 Fig. 4.4 Kaleff/Dreamstime. Fig. 4.5 Robert Harding/Robert Harding Travel/Photolibrary; p163 Fig. 4.11 2008 Alain Evrard; p164 Fig. 4.12 Ho New/Reuters; p166 Fig. 4.16 WaterAid; p167 Fig. 4.19 WaterAid; p168 Fig. 4.20 www.oxfam.org.uk; p169 Fig. 4.25 Computeraid; p170 Fig. 4.26 Schlegelmilch/Corbis; p172 Fig. 4.28 Natalie Behring-Chisholm/Getty Images; p172 Think About It box 123RF; p174 Fig. 4.30 Francis Frith Collection/Photolibrary; p175 Fig. 4.32 Reuters; p176 Fig. 4.33 BlueMoon Stock/Alamy; p177 Fig. 4.35 Paul A. Souders/Corbis; p179 Fig. 4.37 Sean Gallup/Getty Images; p180 Fig. 4.39 Guy Vanderelst/Getty Images; p181 Fig. 4.40 Sam Toren/Alamy. Fig. 4.41 Chris Pearsall/Alamy; p183 Fig. 4.43 Cisco. Fig. 4.44 Getty Images/Stuart O'Sullivan; p185 Fig. 4.48 AP Photo/Douglas Engle. Fig. 4.49 Carlos Cazalis/1599/Corbis; p189 Fig. 4.53 Chris Rank/Sygma/Corbis; p190 Fig. 4.54 Richard Vogel/Associated Press; p191 Fig. 4.56 Stu Smucker/Getty Images; p192 Fig. 4.57 Stan Gamester/Alamy; p193 Fig. 4.58 Colin Garratt/Milepost 92½/Corbis; p194 Fig. 4.61 Godfrey Tsetse; p195 Fig. 4.63 Johnson Matthey; p196 Fig. 4.65 Wayne Lawler/Ecoscene; p197 Fig. 4.67 Tim Boyle/Staff/Getty Images News/Getty Images North America/Getty Images; p201 Fig. 4.72 David Roos/Shutterstock; p202 Fig. 4.74 NASA; p203 Think About It box Bruno Fert/Corbis; p204 Fig. 4.75 Mark Lennihan/Associated Press; p205 Fig. 4.76 Stephen Finn/Dreamstime; p210 Exam café NASA; pp212–213 Background Paul Prescott/Alamy; p215 Fig. 5.1 Pearson Education Ltd, Ian Wedgewood; p216 Fig. 5.4 Gideon Mendel for The Global Fund/Corbis; p219 Fig. 5.8 Homer W Sykes/Alamy; p220 Fig. 5.9 Sean Sprague/Sprague Photo Stock; p221 Fig. 5.12 Greenshoots Communications/Alamy; p223 Fig. 5.15 MARKA/Alamy; p224 Fig. 5.16 Penny Tweedie/Alamy; p226 Fig. 5.19 Fairtrade Foundation; p227 Fig. 5.21 Mark Boulton/Alamy; p230 Fig. 5.25 Robert Taylor; p231 Fig. 5.26 Denis Sinyakov (Russia)/Reuters; p232 Fig. 5.28 Ajay Verma/Reuters; p233 Fig. 5.29 Edward Parker/Alamy. Fig. 5.30 Forest Stewardship Council; p234 Fig. 5.31 Getty Images/Contributor/Getty Images Sport/Getty Images Europe; p235 Fig. 5.33 Clive Mason/Getty Images; p236 Fig. 5.34 Stuart Clarke/Rex Features; p237 Fig. 5.35 The Beijing Organizing Committee for the Games of the XXIX Olympiad. Fig. 5.36 Dave Ellison/Alamy; p238 Fig. 5.37 Guiziou Franck/Hemis/Photolibrary; p239 Fig. 5.38 Jon Sparks/Alamy. Fig. 5.39 Ashley Cooper/Alamy; p241 Fig. 5.42 Juan Jose Pascual/Photolibrary; p243 Fig. 5.44 Gary Bell/Getty Images; p245 Fig. 5.47 Nik Wheeler/Corbis; p247 Fig. 5.50 Reuters; p248 Fig. 5.52 Skyscan/Corbis; p250 Fig. 5.53 Steven Gillis HD9 Imaging/Alamy. Fig. 5.54 Cro Magnon/Alamy.

Artwork and text

p15 Fig. 1.21 *Geofactsheet no 189*, Curriculum Press (Jan 2006); p16 'Planning authorities across Britain' Environment Agency; p17 'The following statements' Dudley Council; p19 Fig. 1.27 Adapted from Crown copyright data supplied by the MET Office; p21 Fig. 1.30 Environment Agency; p24 'UN fears malnutrition' Copyright Guardian News & Media Ltd 2004; 'Aftermath of the 2004' WaterAid, WaterAid's mission is to overcome poverty by enabling the world's poorest people to gain access to safe water, sanitation and hygiene education (Charity reg numbers: 288701 (England and Wales) and SC039479 (Scotland)); p25 Fig. 1.36 Asian Development Bank, 2002; p26 Fig. 1.38 Environment Agency; p27 Fact file Environment Agency; Fig. 1.39 Environment Agency; p28 'A Summer to Remember!' Crown copyright, 2008; p29 'Britain needs a million new homes' Jill Sherman, *The Times*, 2005; p42 'Tourist resort development' Tourism Concern; p45 Fig. 1.77 West Dorset Council/Hilary Bosworth; p49 Fig. 1.85 Environment Agency; p56 Table 2.1 United Nations, accessed September 2008; Table 2.2 United Nations, accessed September 2008; Fig. 2.1 US Census Bureau; p58 Table 2.3 UK Statistics Authority website, Crown copyright; p62 Table 2.4 NationMaster website; p63 Fact file Audit Commission; Fig. 2.7 US Census Bureau; p64 Fig. 2.8 *Africa: Atlas of Our Changing Environment*, 2008, the United Nations Environment Programme; Table 2.5 NationMaster website; p65 Fact file Unicef; p68 Fig. 2.12 US Census Bureau; p71 Fig. 2.16 US Census Bureau; p72 Table 2.8 Food and Agricultural Organization of the United Nations/Thomson Reuters; p73 'Tata Nano' Ashling O'Connor, *The Times*, 2008; p75 Fig. 2.22 Hein de Haas, International Migration Institute, University of Oxford; p76 Fact file *Migration News*; p77 Fact file Global Land Tool Network (GLTN) as facilitated by UN-HABITAT; p80 Table 2.9 UK Statistics Authority website, Crown copyright; Table 2.10 UK Statistics Authority website, Crown copyright; p82 Table 2.11 National Statistics website; Table 2.12 National statistics website; p83 'Government statement on developing the regions, 2003' *A Modern Regional Policy for the United Kingdom*, Crown copyright, 2003; p84 Table 2.13 Optimum Population Trust, an environmental think tank; Fact file *Migration News*; p85 Fact file UK Statistics Authority website, Crown copyright; p86 Fig. 2.35 Office for National Statistics, Crown copyright, 2007; p87 'My life in England' *Daily Telegraph*, 2008; p93 'I think eco-towns' Tom Peterkin, *Daily Telegraph*, 2008; p95 Fig. 2.50 Greenwich Millennium Village; speech bubbles Comments from members of the public to the Sustainable Development Commission; p98 Fact file *High Street Britain* (government report) House of Commons All-Party Parliamentary Small Shops Group; p101 'Comments about a proposal' *Daily Telegraph*, 2008; p116 Fig. 3.15 Ingeominas, Instituto Columbiano de Geologíy Minería; p126 Fig. 3.32 Map provided courtesy of the ReliefWeb Map Centre, UN Office for the Coordination of Humanitarian Affairs; p127 'Gao Li Qiang – Writing from Sichaun' Gao Li Qiang, Tibetan Trekking, Threshold website; p128 Fig. 3.36 GFZ, German Research Centre for Geosciences; p129 Fig. 3.37 US Geological Survey; p133 Fig. 3.44 Earthquake Reconstruction and Rehabilitation Authority; p137 'Looters take advantage' Associated Press; p140 'Jumpstarting livelihoods in a cyclone-hit village' United Nations Development Program; p141; 'Katrina floods wipe out years of research' Associated Press; Fig. 3.56 US Census Bureau; p143 Fig. 3.58 Adapted from *New York Times*; p148 Fig. 3.66 Jacaranda Project, John Wiley and Sons; p149 Fig. 3.67 Ryebuck Media Pty Ltd, Australia; p151 Development Compass Rose, Activities Development Education Centre, Tide Global Learning; p152 'Expert guidance' Hugh Possingham; p159 Fig. 4.3 *Geofactsheet no 228*, Curriculum Press; p165 Fig. 4.14 Reproduced with the permission of Nelson Thornes Ltd from *The New Wider World* by David Waugh, ISBN 978-0-7487-7376-3 first published in 1998; p167 Fig. 4.18 WaterAid, WaterAid's mission is to overcome poverty by enabling the world's poorest people to gain access to safe water, sanitation and hygiene education (Charity reg numbers: 288701 (England and Wales) and SC039479 (Scotland)); p168 'With these goats' Oxfam GB; p169 Fig. 4.23 Simon Birch, *Independent*, 2006; p170 Table 4.3 Office for National Statistics, Crown copyright, 2006; p176 Fig. 4.34 *North America in Focus*, Hodder Education, 1990; p177 Fact file Washington Grain Alliance; p183 Fig. 4.45 OCRI; p184 Table 4.6 Census 2000, US Census Bureau; p186 Table 4.7 *Fortune Global 500* © 2007 Time Inc. All rights reserved; Fig. 4.51 *Geofactsheet no 147*, Curriculum Press; p188 Fig. 4.52 Paul Guinness and Garrett Nagle, *Advanced Geography: Concepts and Cases*, Hodder Education, 1999; p195 'Extracts from the report' ActionAid; p198 Fig. 4.68 *Financial Times*; p200 Fig. 4.70 Scripps Institution of Oceanography, UC San Diego; Fig. 4.71 Reproduced with the permission of Nelson Thornes Ltd from *Geography: An Integrated Approach* by David Waugh ISBN 978-0-7487-7376-3 first published in 2000; p201 Fig. 4.73 Human Development Report, 2007/2008, Palgrave Macmillan; p204 Table 4.10 *Financial Times*, 5/6 May 2007; p216 Fig. 5.3 *UC Atlas of Global Inequality*; p218 Fig. 5.5 HMSO; Fig. 5.6 Health Development Agency, *The smoking epidemic in England*, 2004, HDA-London, reproduced with permission; p219 Fig. 5.7 NHS Across Sheffield; p221 Fig. 5.11 © Copyright 2006 SASI Group (University of Sheffield) and Mark Newman (University of Michigan); p239 'The Les Arcs ski resort is committed' Les Arcs resort website; p240 Fig. 5.40 Reprinted by kind permission of *New Internationalist*. Copyright New Internationalist; Fig. 5.41 bmibaby; p241 Fact file Tourism Concern/Reprinted by kind permission of *New Internationalist*. Copyright New Internationalist/*Earth Report*; p242 Leeds Development Education Centre; p246 *BP Statistical Review of World Energy 2008*, BP plc; Fig. 5.49 *The Oil Drum: Europe*, Crown copyright; p249 'Britain cannot do without nuclear power' Professor Ian Fells; 'Research shows that even 10 new reactors' Greenpeace UK; 'There is no need for nuclear power' Friends of the Earth (England, Wales and N. Ireland); p251 Fact file Greenpeace UK/Energy Saving Trust.

Every effort has been made to contact copyright holders of material reproduced in this book. Any omissions will be rectified in subsequent printings if notice is given to the publishers.

Contents

Chapter 4 Economic development

Chapter 5 Investigating geographical issues

Welcome to OCR GCSE Geography B

This book has been written specifically to support you during the OCR GCSE Geography B course. Resources for this course include a Student Book with ActiveBook CD-ROM, a Teacher Guide and an ActiveTeach CD-ROM.

About your course

The OCR GCSE Geography B course has been developed to be exciting, relevant and interesting. The issues it deals with are the most important challenges facing the world today, and it will give you inside information on the causes of these different challenges, the impact they have on us all and the environment we live in, and what we can do to help overcome the problems we all face.

The OCR course is divided into three units:

- Unit B561: Sustainable Decision Making
- Unit B562: Geographical Enquiry
- Unit B563: Key Geographical Themes.

Unit B561: Sustainable Decision Making

This is a one-hour written exam which will assess knowledge and understanding of a sustainable development issue from one of the four themes. This part of the course is worth 25 per cent of your GCSE grade.

This is an exciting, different style of examination. It focuses on a sustainable development issue from one of the four themes: Rivers and coasts, Population and settlement, Natural hazards and Economic development.

Three weeks before your exam your school or college will receive copies of the OCR Resource Booklet, which contains different geographical resources to help you develop your knowledge and understanding of the sustainable development issue. These could be maps, graphs, diagrams, data tables, cartoons, photographs, points of view, satellite images – many different things. You will use these resources in your lessons running up to the exam, and you can take them into your exam too. Then in the exam you will receive another booklet with the questions in and spaces for you to write your answers.

This Student Book contains Decision Making Exercises relating to the first four chapters. These exercises will give you practice in the types of skills you will need for the sustainable decision making exam. They are each supported by additional resources in the ActiveBook CD-ROM.

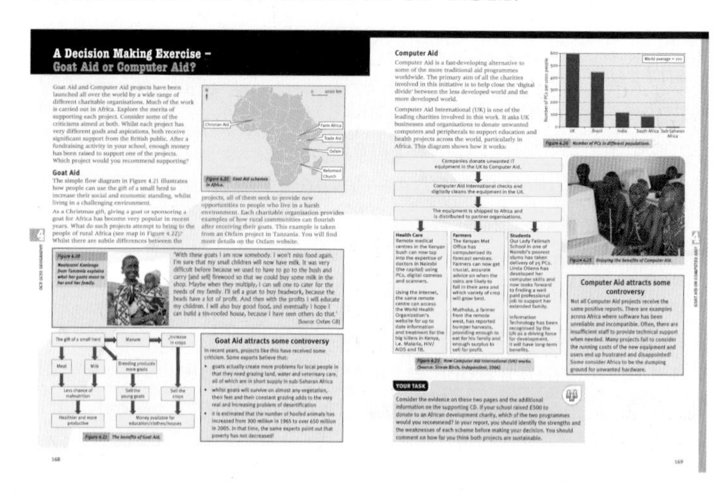

Unit B562: Geographical Enquiry

This makes up 25 per cent of your GCSE grade and is based on two pieces of Controlled Assessment.

- Fieldwork Focus: based on the collection and analysis of students' own geographical information and data.
- Geographical Investigation: an issue-based investigation researching and using secondary data and information.

Your teacher will tell you more about what you need to do for this part of the course, because it will be individual to your school or college. The fieldwork focus is also based on one of the four themes that make up the first four chapters of this book.

This Student Book gives you some guidance on geographical issues and geographical investigations in Chapter 5, which is all about this part of the course. There are nine possible geographical issues for the Geographical Enquiry unit: this book gives you some introductory material on six of them. The full list of issues is:

- Disease
- Trade
- Ecosystems
- Sport
- Fashion
- Energy
- New technologies
- Crime
- Tourism.

This book covers: disease, trade, ecosystems, sport, tourism and energy.

Unit B563: Key Geographical Themes

The sustainable decision making exam is on one theme – what about the other three? These will be covered in the exam: key geographical themes. This is a more traditional type of exam that will ask you a series of questions. There will also be a Resource Booklet for this, but you will not see it until you are in the exam.

The final question in each section of the exam will involve you discussing a case study you have learned about. As well as covering all the content required for each theme, this Student Book also provides you with case studies throughout.

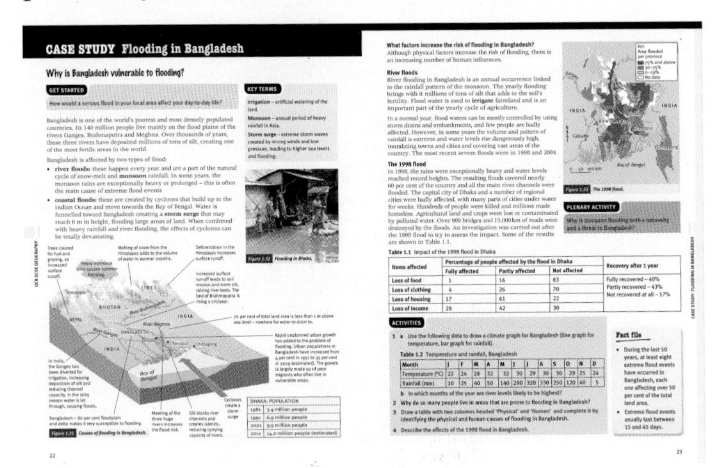
CASE STUDY Flooding in Bangladesh

Why is Bangladesh vulnerable to flooding?

Other features of this textbook

There are other features in this book, including activities to get the topic started and ones to summarise what you have learned, plus questions to get you discussing issues as a class.

Activities

The activities have been designed to help you understand the specification content and develop your geographical skills.

Key words

Definitions of new words can be found near where they occur in bold in the text. All key words can also be found in the Glossary.

Research link

Research activities help you take your learning further.

Fact file

Fact files give you extra information about a topic.

Grade Studio

Grade Studio is designed for you to improve your chances of achieving the best possible grades. You will find Grade Studio activities in the Student Book and on the ActiveBook CD-ROM.

Exam Café

Exam Café is to be used when revising and preparing for exams. Exam Café could be used in revision classes after school or in revision lessons.

Exam Café will help you prepare for the final exam. Like Grade Studio, Exam Café is in the Student Book with additional resources on the CD-ROM. From the CD-ROM you will be able to access a number of useful resources which will help you to organise your revision, practise exam questions, access sample mark schemes and locate extra resources to stretch yourself.

Websites

Throughout the book there are references to websites that will provide additional information on the topics covered in the text. We recommend that you access these through the Heinemann website, www.heinemann.co.uk/hotlinks, entering the express code 3704 when asked to do so.

ActiveBook CD-ROM

In the back of this Student Book you will find your FREE ActiveBook CD-ROM. This is for your individual use and includes a digital copy of the book on screen, plus the unique Grade Studio and Exam Café resources.

We hope that you will enjoy your course!

Chapter 1

Rivers and coasts

Rivers and coasts are dynamic systems that act like natural conveyor belts, removing, transporting and depositing material. The rate at which this happens is often determined by the weather. Heavy rainfall will increase the flow in a river. Strong winds will create large waves, which can then act as a powerful force when they break against cliffs. Increasing numbers of people live near rivers and coasts because they have economic and environmental value. Consequently there is a need to understand these areas so that they can be managed sustainably.

Consider this

As extreme weather events become more common in the UK, it is likely that an increasing number of us will be affected by river and coastal flooding. It is not possible to protect every place from the threat of flooding. Decisions have to be taken about which places to protect, and which to leave unprotected. In the future, if river and coastal areas are to be sustainable there will be an increasing need to work with the environment in order to reduce the risk of flooding.

QUESTIONS FOR INVESTIGATION

- **a** How do systems ideas help us to understand physical processes?
- **b** How does river flooding illustrate the interaction between natural processes and human activity?
- **c** What processes and factors are responsible for distinctive landforms within a river basin?
- **d** What processes and factors are responsible for distinctive coastal landforms?
- **e** Why is the management of coastlines increasingly important?

Why are rivers seen as part of a wider system?

GET STARTED

Discuss with a partner why water is perhaps the most recycled natural resource.

Hydrology is the study of water. The hydrological cycle (or water cycle) describes the continuous transfer of water vapour from stores such as oceans or lakes into the atmosphere, then back to the land as **precipitation**, finally returning to the original stores (Figure 1.1).

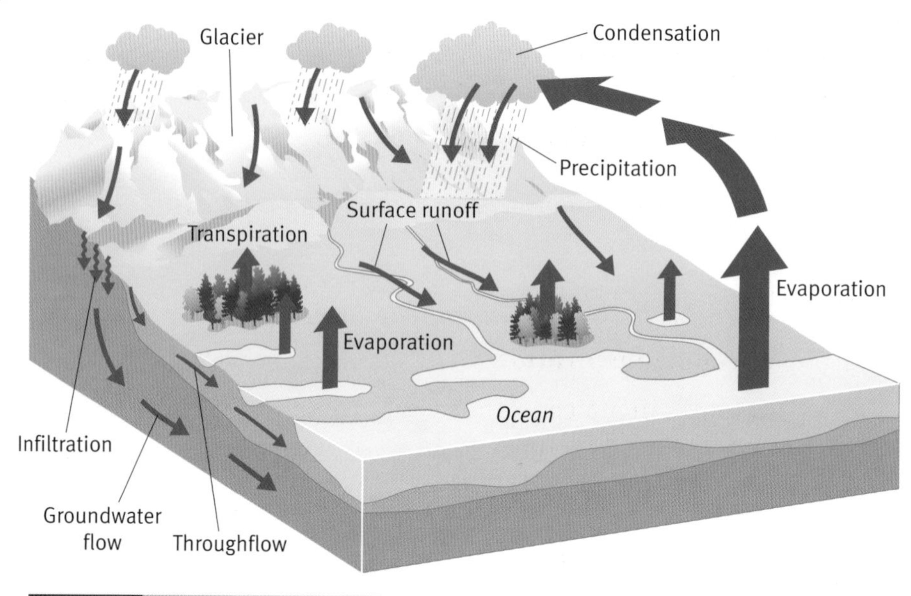

Figure 1.1 The hydrological cycle.

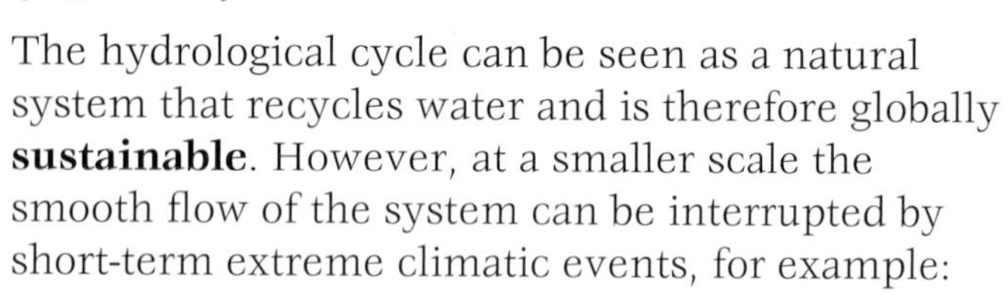

The hydrological cycle can be seen as a natural system that recycles water and is therefore globally **sustainable**. However, at a smaller scale the smooth flow of the system can be interrupted by short-term extreme climatic events, for example:

- periods of prolonged rainfall can create a surplus of water on the land that may lead to flooding
- periods of drought can cause water shortages that may in turn lead to food shortages in some parts of the world (Figure 1.2).

The hydrological cycle is frequently adapted for human use, and increasingly parts of it are modified after extreme climatic events to try to reduce the risk of flooding and drought in the future.

Figure 1.2 The effects of drought.

Fact file

- If nature did not recycle water, at the rate the world uses it humanity would run out of water in about three weeks!
- During the 2008 Olympic Games, storm clouds were artificially seeded to make it rain in order to disperse potential rainstorms before they reached Beijing.

The river system

The river system is the part of the hydrological cycle operating on land. It is made up of four key parts:

1 **Inputs:** water entering the system through precipitation.

2 **Stores:** water stored in lakes, rocks, soil or vegetation. Storage can be temporary and is linked to the amount of rainfall.

3 **Transfers:** processes that move water through the system such as **surface runoff**, **infiltration** and underground flow.

4 **Outputs:** where water is lost to the system as rivers reach the sea or through **evapotranspiration**.

The river system is not a constant system – variables are changed by the amount of precipitation. Because there are variations in the amount of inputs and outputs, the river system is called an open system (see Figure 1.3).

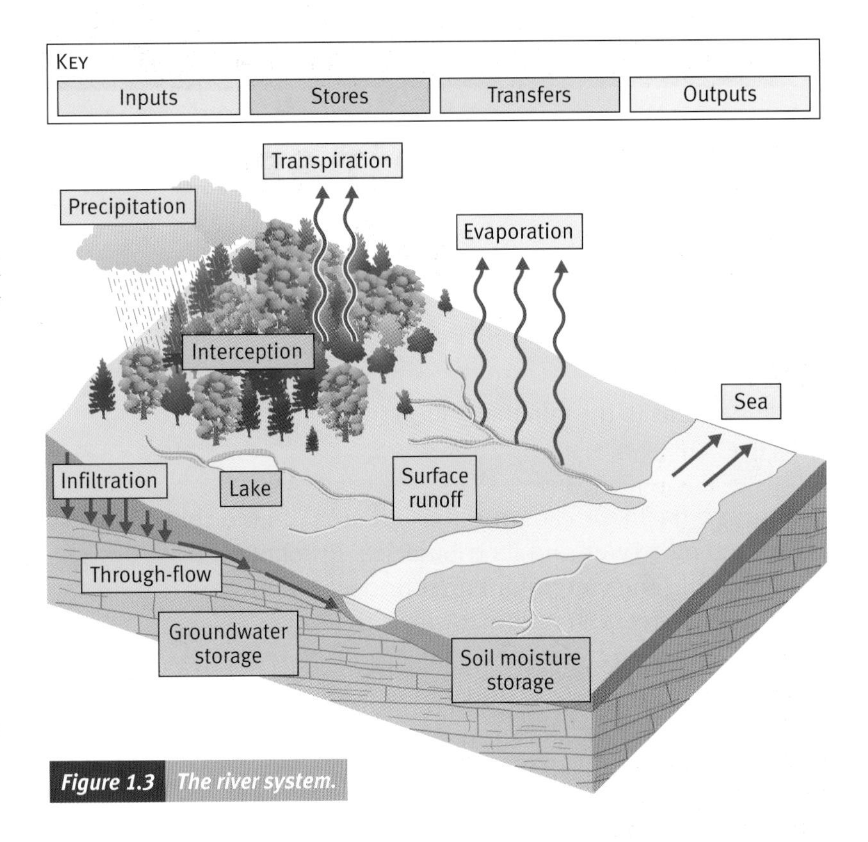

Figure 1.3 The river system.

KEY TERMS

Evaporation – water turning into water vapour.

Evapotranspiration – the sum of evaporation from the Earth's surface together with the transpiration of plants.

Groundwater flow – movement of water underground through rocks.

Infiltration – seeping of water into soil.

Interception – collection of water by vegetation.

Precipitation – moisture that falls from the atmosphere in any form.

Surface runoff – all water flowing on the Earth's surface.

Sustainable – capable of existing in the long term.

Through-flow – movement of water through the soil.

Transpiration – loss of moisture from plants.

Water table – the upper level of underground water.

ACTIVITIES

1 Why is the hydrological system described as a 'natural system'?

2 All vegetation will intercept rainfall. The following information shows average rates of interception (per cent) for different types of vegetation:

- pine forest – 70 per cent
- rainforest – 60 per cent
- deciduous forest – 20 per cent
- crops (wheat, etc.) – 15 per cent
- grass – 30 per cent.

  a Present these figures as a clearly labelled pictogram.

  b Suggest reasons for the differences shown.

3 Describe and explain what might happen to a river system after:

  a a period of heavy rainfall

  b a period of drought.

PLENARY ACTIVITY

How might climate change affect the hydrological cycle?

What is a drainage basin?

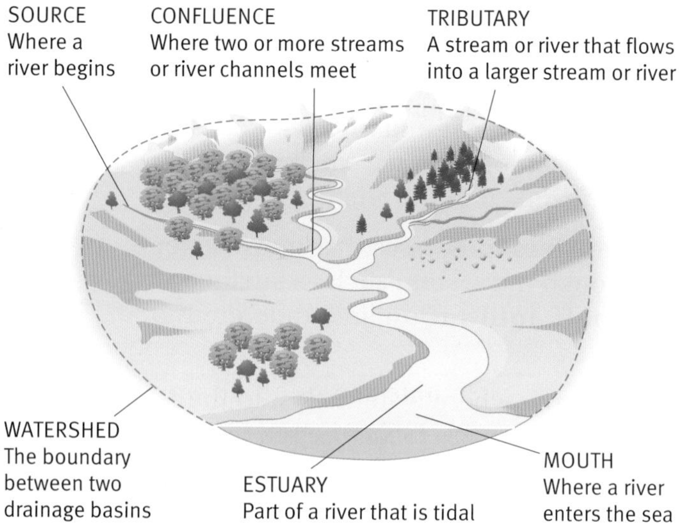

GET STARTED

What happens to the rain that falls in the area where you live?

A drainage basin, or catchment, is an area of land that is drained by a river and its tributaries. Precipitation falling within the catchment area finds its way into streams and rivers that flow towards the sea.

The edge of a drainage basin is called a watershed and is often a ridge of higher land. Precipitation falling on the other side of the high land flows into a neighbouring drainage basin. The main features of a drainage basin can be seen in Figure 1.4.

Fact file

The world's largest drainage basins

1. Amazon – 6,150,000 km^2
2. Congo – 3,730,000 km^2
3. Nile – 3,255,000 km^2
4. Mississippi – 3,200,000 km^2
5. Ob – 2,950.000 km^2

(Figures are approximate; for comparison the total area of the UK is 243,820 km^2)

SOURCE
Where a river begins
CONFLUENCE
Where two or more streams or river channels meet
TRIBUTARY
A stream or river that flows into a larger stream or river
WATERSHED
The boundary between two drainage basins
ESTUARY
Part of a river that is tidal
MOUTH
Where a river enters the sea

Figure 1.4 *Main features of a drainage basin.*

THINK ABOUT IT

Why do you think drainage basins are also called catchment areas? What are they catching?

RESEARCH LINK

Use the internet to investigate how one other large river has been adapted for human use.

How are drainage basins adapted for human use?

Water is one of the world's most important natural resources for human survival and economic development. Consequently many drainage basins

The source of the River Tees is Cross Fell, the highest point on the Pennine moorland at 893 m above sea level. The area receives over 2000 mm of precipitation a year. Infiltration rates are low, resulting in a wide area of saturated peat bog. Water drains out of the bog creating small streams.
The Tees has a wide tidal estuary with sand banks and mud flats. Many of the natural features of the estuary have been adapted to suit the needs of heavy industry and shipping.
Cross Fell
Steep-sided valley with rapids and waterfall.
North Sea
Middleton
Watershed
Middlesbrough
At its mouth the River Tees enters the North Sea.
Barnard Castle
Darlington
Yarm
Croft
KEY
Height (metres)
Over 600
301–600
121–300
61–120
0–60
N
0
10 km
The River Tees is joined by many tributaries throughout its course.
The river meanders through a wide flood plain before reaching the estuary, where building development has taken place alongside the river embankment to reduce the risk of flooding.

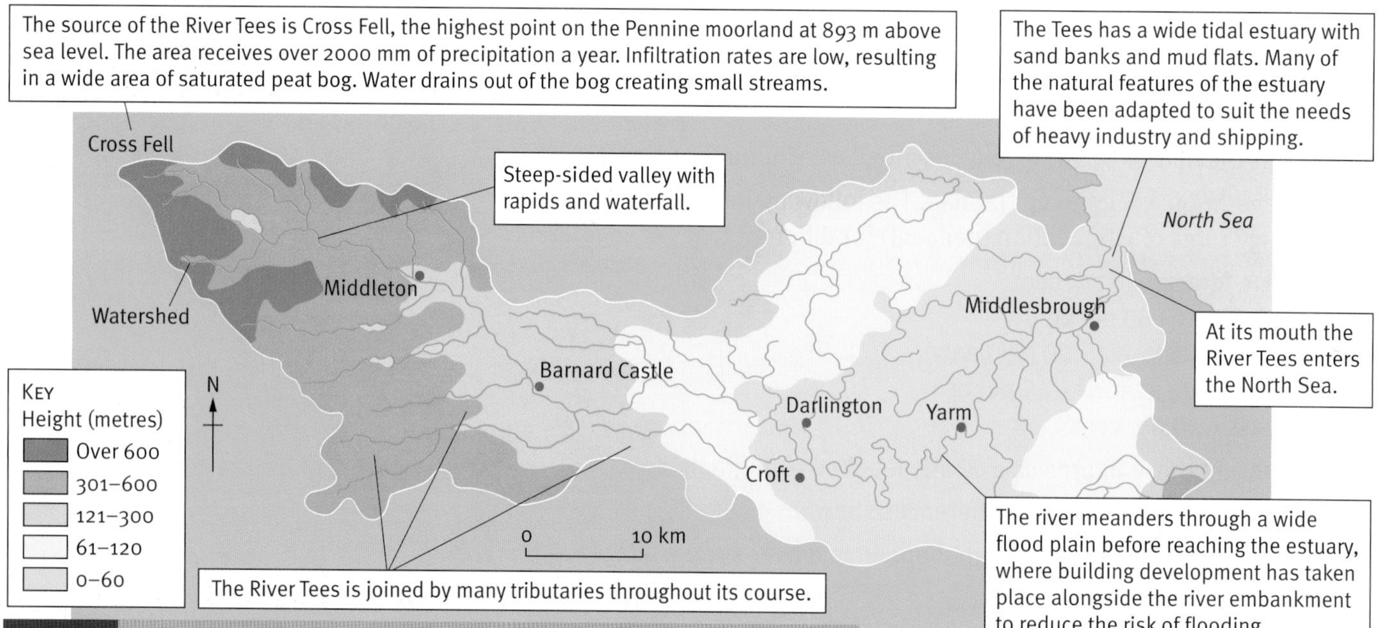

Figure 1.5 *Example of a river basin and its features: the Tees Basin, north-east England.*

are used as a source of water and a means of dealing with waste water, especially in towns and cities. Drainage basins are adapted in many ways, including:

- building dams and reservoirs
- irrigation schemes for agriculture
- water abstraction for industry and power stations
- widening or straightening river channels for navigation
- constructing flood protection walls and barriers.

Changes to any part of a drainage basin may affect the natural flow of rivers within it, and changing one part of a river system may have significant effects on other parts.

The following example shows how drainage basins can be adapted for human use.

River basin management – The Colorado River, USA

The Colorado River and its tributaries flow through seven American states before reaching the Gulf of California (Figure 1.6). Much of the catchment is a semi-desert landscape, making the rivers a vital source of water for over 15 million people.

It has been said that the Colorado River Basin is one of the most managed systems in the world and that very little of the system is in its natural state. Water flow is totally controlled, and so much water is taken out of the river that during some months of the year it enters the Gulf of California as a mere trickle.

How much has the river basin been adapted?

- The Hoover Dam was constructed in 1935 to control the rising water levels in spring and summer, and to reduce the risk of flooding. A further 18 dams have been constructed throughout the river basin, creating a number of massive artificial lakes. A number of these dams are used to generate hydro-electricity.

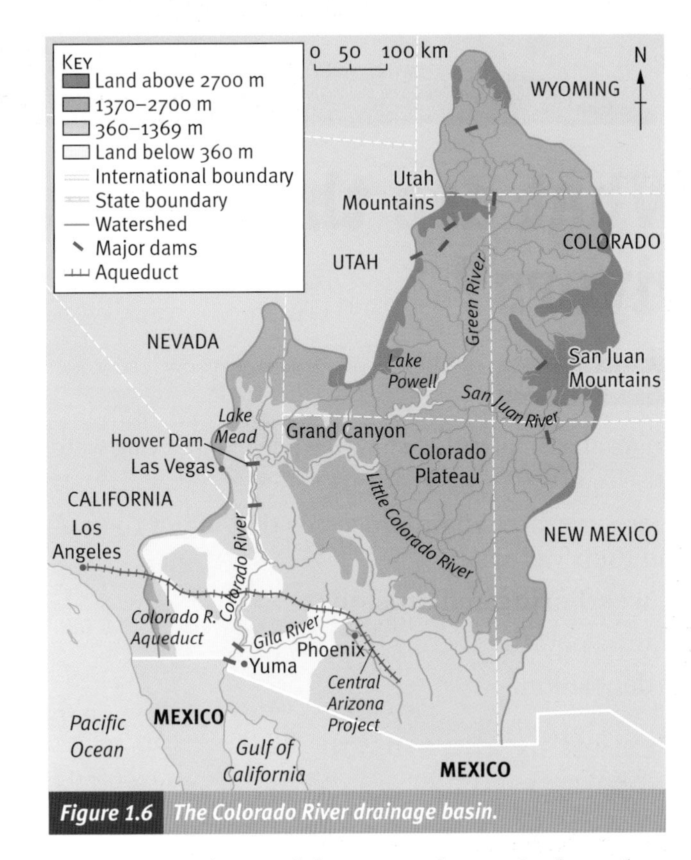

Figure 1.6 The Colorado River drainage basin.

- Increasing demand for water for agriculture has seen the development of a number of canals and aqueducts that transport water hundreds of kilometres to places such as the Central Valley in California. Large-scale sprinkler systems have been developed to allow the semi-desert landscape to be used for agriculture.
- Population growth in the western USA has increased demand for water for both residential and industrial use. Water is transported to the western coastal cities as well as being used in the growing urban areas within the river basin itself. Phoenix, to the south of the river basin, has become the USA's seventh biggest urban area with over 6 million people. The development of high-tech industry has attracted thousands of people to the area. Large developments of new houses with swimming pools, large gardens and community golf courses have been built in this desert environment.
- The growth of tourism has led to the development of hotels and leisure facilities in many areas – Las Vegas alone attracts over 8 million visitors a year, and millions more visit the spectacular scenery of the Grand Canyon. Tourists demand a full range of leisure facilities including water sports and golf courses.

ACTIVITIES

1. Why is a water catchment area called a basin?
2. Briefly describe the characteristics of the River Tees:
   a. from Cross Fell to Middleton
   b. from Croft to Yarm
   c. from Middlesbrough to the North Sea.
3. a. Describe and explain the changes that have been made to the Colorado River Basin.
   b. Suggest any problems that these changes may create.

PLENARY ACTIVITY

Why are large scale river management schemes often found in highly developed countries?

What are the processes that affect rivers?

GET STARTED

Think of examples that show the power of moving water.

KEY TERMS

Bedload – larger particles moved along a river bed.

Erosion – the wearing away and removal of rocks by the action of water, wind or ice.

Freeze–thaw – the continued freezing and thawing of moisture in rocks that will eventually cause them to break.

Weathering – breaking up of rocks by the action of weather, plants, animals and chemical processes.

There are three main processes that affect rivers and their valleys:

- **weathering** and **erosion**
- transportation
- deposition.

Weathering and erosion

Weathering breaks down rocks and makes it easier for the processes of erosion and transportation to operate. In upland areas, the process of **freeze–thaw** weathering can shatter rocks on valley sides. Rock fragments fall into the river channels and are shaped and moved by processes of erosion and transportation.

Erosion

There are four main processes of erosion.

- **Abrasion/corrasion:** fragments of rock carried by a river wear away the bed and banks of the river (Figure 1.7). When a river is in flood it can carry tons of material and the effect of abrasion will be greater. Abrasion is the major type of erosion responsible for deepening river valleys in upland areas.
- **Attrition:** pebbles and rocks collide with each other, reducing their size and making them increasingly smooth.
- **Hydraulic action:** the power of moving water being forced against river banks causes them to collapse and be washed away (Figure 1.7).
- **Corrosion/solution:** chemical reaction occurs when slightly acid water dissolves calcium, breaking down rock such as limestone.

Figure 1.7 River bank erosion.

What factors affect the rate of erosion?

The ability of a river to erode is determined by the amount of energy it has. This is variable because it depends on the volume of water in the river. After heavy rainfall, the volume of water in a river will increase and its level of energy will also increase, giving it the power to move significant amounts of material, including large rocks and boulders (Figure 1.8). After a prolonged period of dry weather the volume of water in the river will decline, reducing its ability to erode and transport material. In some cases rivers may totally dry up until the next spell of wetter weather.

Other factors affecting the rates of erosion include:

- **gradient:** the natural process of a river is to flow downhill until it reaches its base level (usually sea level). If a river has a steep gradient the amount of energy available will be greater, increasing the rate of erosion. The gradient of a river can be seen by looking at its long profile (Figure 1.9)
- **rock type:** weaker, softer rocks such as sand and gravel are more easily eroded than harder, solid rock structures such as limestone and granite
- **bedload:** as the amount of bedload increases more abrasion will take place
- **human factors:** cutting down trees and changing the landscape may decrease the time it takes for rainfall to reach a river. This will increase the volume of water in the river.

Figure 1.8 A river in flood.

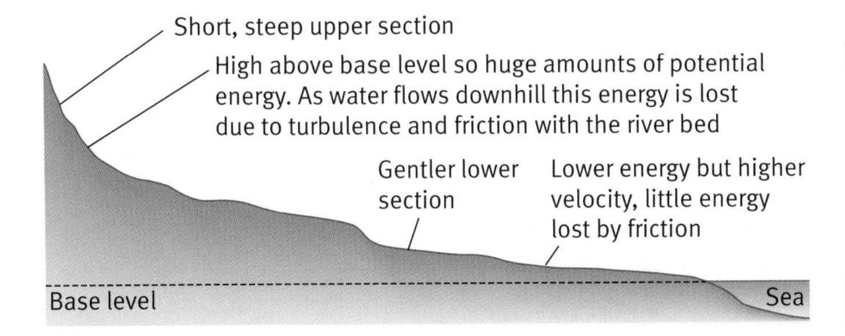

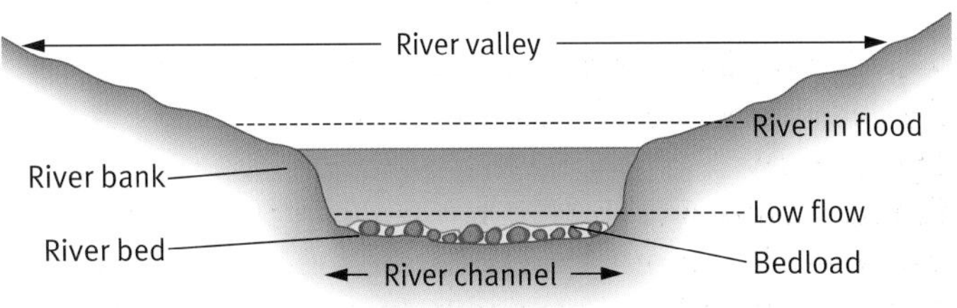

Figure 1.9 River profiles.

Transportation

A river transports material in the following three ways:

- **as bedload:** larger fragments rolled along the river bed (traction) or bounced along the river bed (saltation)
- **as suspended load:** smaller fragments carried in the flow of the river
- **in solution:** dissolved minerals carried in the water.

Deposition

Deposition occurs when a river does not have enough energy to carry material any further.

This can happen because of:

- a reduction in gradient
- a reduction in the volume of water
- increasing friction with the river bed and banks, causing the river to slow down
- adaption of the river for human use.

ACTIVITIES

1 Why are rates of erosion greater after heavy rainfall?
2 Use an annotated sketch to explain how abrasion and hydraulic action can erode a river bank.
3 Write a brief paragraph under each of the following sub-headings:
   a natural factors affecting erosion
   b human factors affecting erosion.
4 Describe the major processes that affect the bed of a river;
   a near the source
   b near the mouth.

PLENARY ACTIVITY

Why does the bedload of a river become smaller and less angular as it moves towards the sea?

RESEARCH LINK

How do people change the flow of a river? Search the internet for dams, wing dykes and diversions.

What features are associated with the upper course of a river?

GET STARTED

What spectacular river features might you expect to find in upland areas? In what type of landscape in the UK are the upper courses of our rivers? Describe the landscape to someone who has not been to the UK before.

In the upper course of a river, erosion is the dominant force because the gradient of the river is steeper and it has a considerable amount of energy. Abrasion causes both **lateral** and **vertical** erosion of the sides and bed of a river. The valley sides are also affected by weathering. After heavy rainfall, loose soil and rock will slump down the sides of the valley into the river and be carried downstream. This continuing process causes the valley sides to become steeper and have an increasingly V-shaped profile (Figure 1.11).

Figure 1.10 Interlocking spurs.

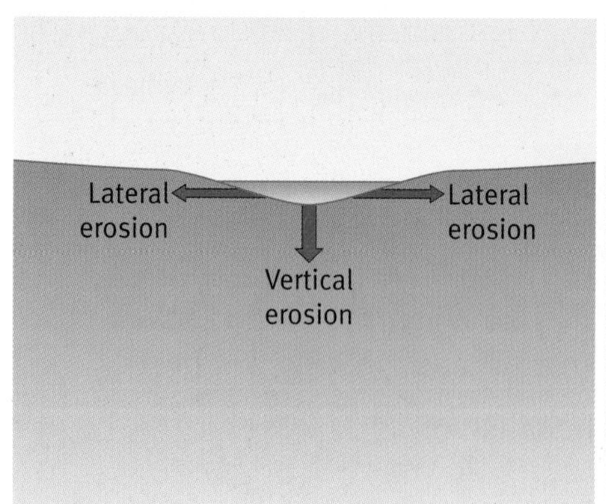

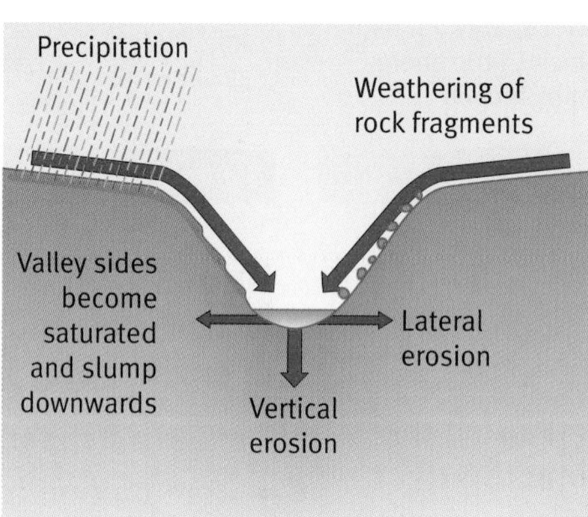

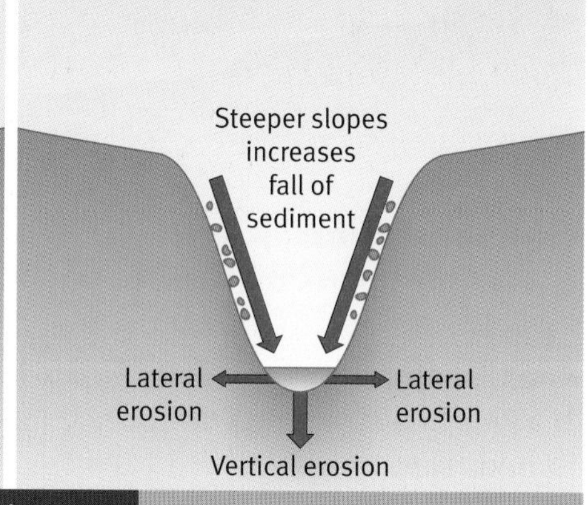

Figure 1.11 The formation of a V-shaped valley.

Near to their source most rivers follow a fairly straight course, but then begin to wind around obstructions in their path, forming a series of interlocking spurs that jut out across the valley (Figure 1.10).

The formation of waterfalls and gorges

The gradient of a river becomes less steep as it flows downstream from its source. Where harder rock lies across a river's course, this slows down the rate of erosion and reduces the gradient. As the river reaches softer rock it begins to cut downwards rapidly forming a series of rapids or a waterfall (Figure 1.12). The turbulent water at the base of the waterfall erodes the softer rock underneath until the lip of the harder rock is unsupported. Fragments of the harder rock break away and fall, leaving a boulder-strewn river bed. In this way the waterfall gradually moves back upstream or retreats, leaving a steep-sided gorge downstream (Figure 1.13).

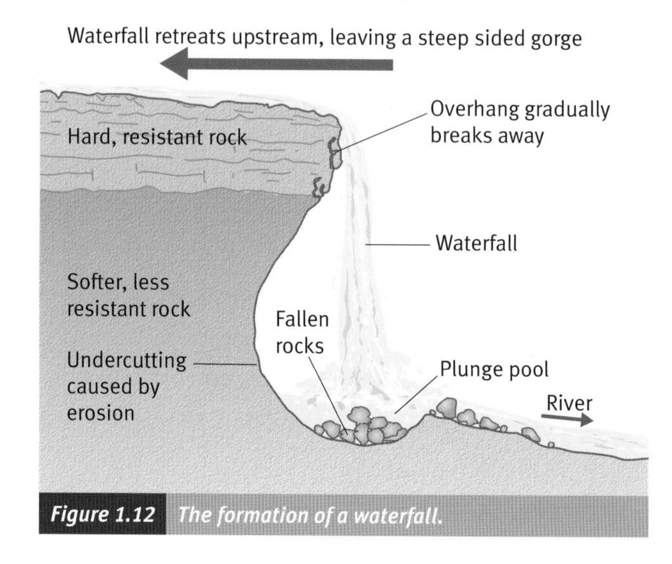

Figure 1.12 The formation of a waterfall.

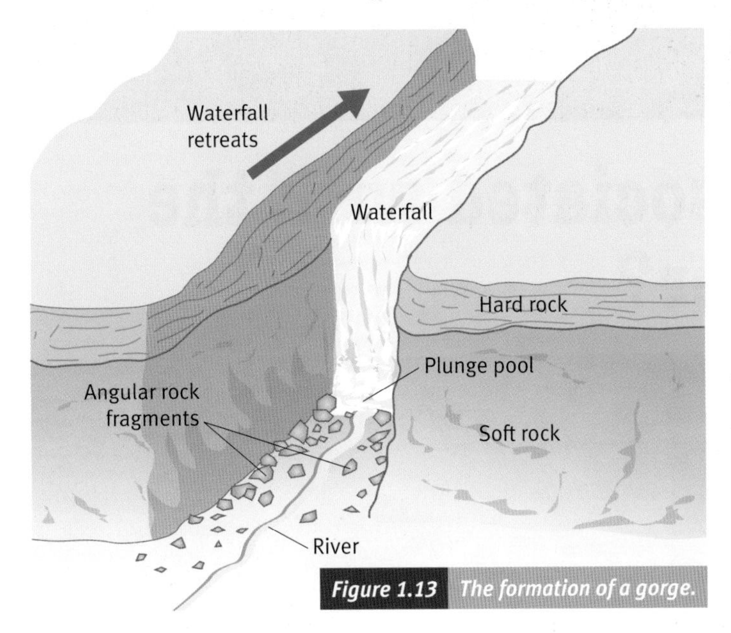

Figure 1.13 The formation of a gorge.

KEY TERMS

Hydro-electricity – electricity produced by flowing water.

Lateral erosion – erosion of the sides of a valley.

Vertical erosion – downward erosion of a river bed.

Waterfalls and human activity

Waterfalls provide some of the world's most spectacular scenery, and are seen as a major tourist attraction. For example, the Victoria Falls in Zimbabwe has developed a tourist industry based entirely on the environment of the River Zambezi and its dramatic waterfalls. Nearly a million people a year visit the area to admire the scenery and the local wildlife. The river and its rapids are used for activity holidays, including white-water rafting. The following example looks at Niagara Falls – perhaps the most famous waterfall in the world.

Niagara Falls

Niagara Falls, on the border betweeen the USA and Canada, is made up of three waterfalls: Horseshoe Falls, American Falls and the small Bridal Veil Falls. Horseshoe Falls is one of the widest waterfalls in the world, with an overall width of nearly 1 km and a fall of over 50 m. The American Falls are over 300 m wide, with a fall of only 20 m because of a build-up of massive rocks at the base of the waterfall. At its peak, the flow of water over Niagara Falls reaches 6000 cubic metres per second (cumecs) although **hydro-electric** plants draw some of the water into their reservoirs before it reaches the falls.

Niagara Falls was visited by over 20 million people in 2008 and is a major tourist attraction and money earner for the area. Tourism creates thousands of jobs and is an important part of the local economy. There are many hotels in the area; a number have been built with the rooms overlooking the waterfalls. More recently, casinos and other tourist attractions have been built, the Falls increasingly acting as magnets in attracting a wide range of visitors. To observe the Falls, a series of walkways have been built and tunnels have also been made near the bottom of Horseshoe Falls on the Canadian side of the border so that the waterfall can be observed from underneath the lip. For an even more spectacular view, the Falls can be observed from helicopter or hot-air balloon. Visit the Niagara Falls tourism website to find out more.

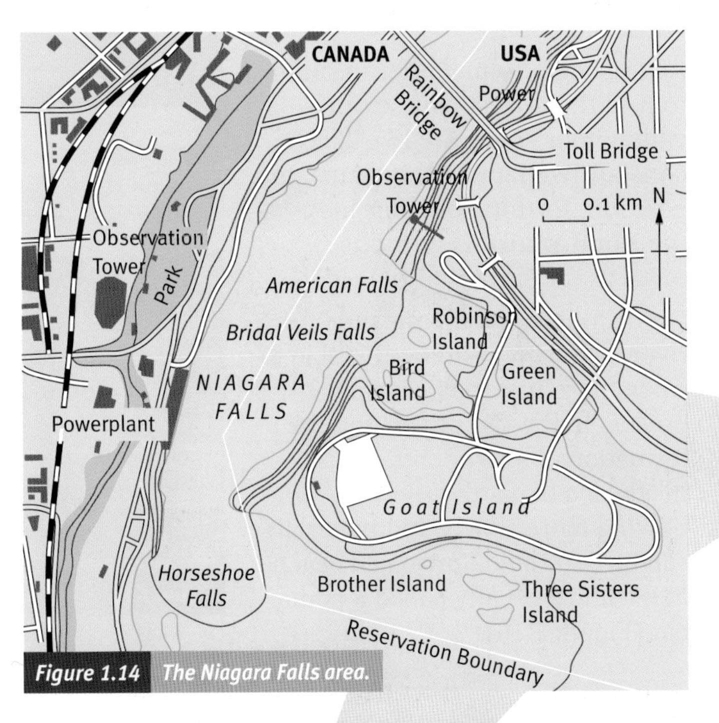

Figure 1.14 The Niagara Falls area.

ACTIVITIES

1. Explain the difference between vertical and lateral erosion.
2. Use an annotated sketch to explain the formation of a waterfall and a gorge.
3. Describe the characteristics of Niagara Falls.
4. Niagara Falls is an important part of the local economy – explain this statement.

RESEARCH LINK

Use the internet to investigate the Iguazu Falls (a national park and a world heritage site) on the borders of Argentina, Brazil and Paraguay.

PLENARY ACTIVITY

Why are rivers an important economic resource?

What features are associated with the lower course of a river?

GET STARTED

What happens to the volume and speed of water in a river as it flows towards the sea? What are the effects of this?

Meanders

Meanders are bends in the course of a river. They begin to appear as the valley floor becomes flatter and lateral erosion becomes a more significant force than vertical erosion.

Rivers begin to meander in order to find the lowest course on a valley floor (Figure 1.15). They constantly change their shape and position, forming broader loops as the **flood plain** widens out towards the mouth of the river.

On the outside of the meander, the river is flowing faster and is able to carry more material, causing erosion to the river bank and the formation of a river cliff. On the inside of a bend, the river is flowing more slowly and material can be deposited, forming a beach-like feature usually called a point bar (Figure 1.16).

Figure 1.15 A meandering river.

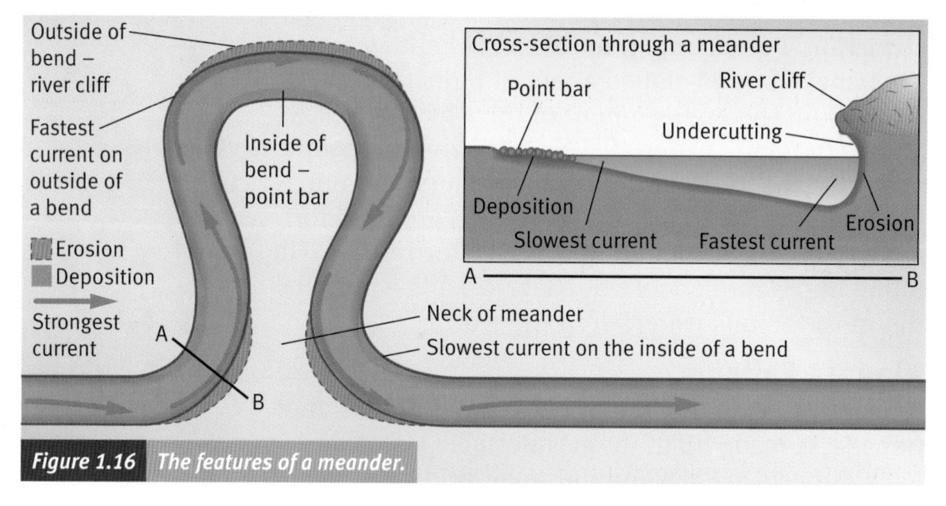

Figure 1.16 The features of a meander.

Ox-bow lakes

Continued erosion of the neck of a meander will cause the river to break through. When this happens, the river will take the straight course. Deposition will eventually block off the old meander, leaving an ox-bow lake (Figure 1.17). Gradually the ox-bow lake will dry up, forming a meander scar.

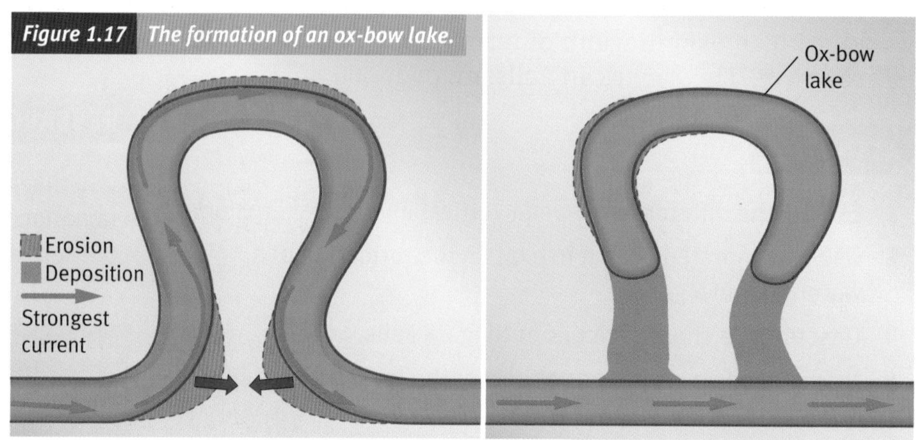

Figure 1.17 The formation of an ox-bow lake.

The mouth of a river

As the river reaches the sea it flows into an **estuary** or a **delta**.

Estuaries

Most rivers flow into tidal estuaries as they reach the sea. At this point the land is at sea level, so the river will use its energy to widen the river channel. Estuaries are usually the widest part of a river and they become wider as they reach the sea. Because of the volume and velocity of water, the river is able to carry large amounts of silt. Much of this will be deposited as fine sand and mud, forming large areas of salt marsh on either side of the river channel, an important habitat for wildlife.

Estuaries are often adapted for human use by draining the wetlands at the sides of the main river channel. This provides an excellent location for sheltered ports and the development of heavy industry. This can be seen in the River Tees estuary in North-East England (Figure 1.18). It is the second largest port in Britain and a major centre of heavy industry, much of it built on **reclaimed** mudflats.

Figure 1.18 The River Tees estuary.

Deltas

As a river enters the sea, its speed will suddenly decrease and the river will deposit most of the silt it has been carrying. If the rate of deposition is greater than the rate of sediment removal by the sea, a delta will develop.

The Nile Delta in Egypt is a fan-shaped delta covering over 25,000 km². As the river reaches the flat delta landscape it divides into hundreds of distributaries before entering the Mediterranean Sea. Ocean currents can remove some of the deposited material or smooth the edge of the delta (Figure 1.19). The Nile Delta is one of the most productive areas of farmland in Africa. The soil is up to 20 m deep, and the high temperatures and irrigation from the River Nile make it ideal for growing a variety of crops. Over 30 million people live in the delta region, relying on the River Nile for water supply in an area that often gets less than 100 mm of rainfall a year.

Figure 1.19 The Nile Delta.

KEY TERMS

Delta – where a river breaks into many distributaries before it reaches the sea.

Estuary – tidal part of a river mouth.

Flood plain – flat area next to a river that is liable to flood.

Meanders – large bends in a river.

Reclaimed land – land that has been drained so that it can be used for development.

ACTIVITIES

1. Use diagrams to describe and explain how both erosion and deposition are important processes in the formation of oxbow lakes.
2. Why is heavy industry often found on estuaries?
3. Describe and explain the formation of the River Nile.

RESEARCH LINK

Use the internet to find out more about;
- the types of land use in the Tees estuary
- the crops grown on the Nile delta.

PLENARY ACTIVITY

What evidence suggests that deposition becomes an increasingly dominant process as a river gets nearer the sea?

How are precipitation and runoff linked?

GET STARTED

What data are needed to calculate the volume of water at a particular point in a river?

When it rains the water will either:

- be held in storage in lakes or the soil
- be lost through evapotranspiration
- make its way into rivers, either on the surface or underground.

All precipitation that reaches a river is called runoff.

What is a hydrograph?

A hydrograph is used to show how the **discharge** of a river changes over time at a particular point on the river. A flood (or storm) hydrograph is usually drawn for a particular period of time when rainfall is unusually high. It shows how river discharge responds to short-term storm conditions (Figure 1.20).

Why are storm hydrographs important?

Storm hydrographs can help to show how a river will react to heavy rainfall, and can be used to predict floods and plan for future flood events. Some of the points on the hydrograph that are particularly useful are:

- **the rising limb:** if it is very steep, it shows that the surface run-off is high, which may add to flood risks
- **the lag time:** if it is short, it means that the river levels will rise quickly, adding to the risk of flooding
- **the falling limb:** if it is gently sloping, it suggests that the river takes quite a long time to get back to its normal flow pattern. If during this time there is heavy rainfall, the risks of flooding will be increased
- **the peak discharge:** local authorities will have an understanding of how high the discharge of a river can rise before it floods. If it appears that the discharge may go above this figure, people living nearby can be warned.

Fact file

Measuring river discharge

River discharge is calculated at a particular point on a river by measuring:

- the velocity (speed) of the river
- the cross-sectional area of the river.

Discharge = velocity × cross-sectional area expressed in cumecs (cubic metres per second).

KEY TERMS

Discharge – the volume of water in a river passing a point in a given time, measured in cumecs (cubic metres per second).

THINK ABOUT IT

The runoff of a river will always be less than the precipitation. Why?

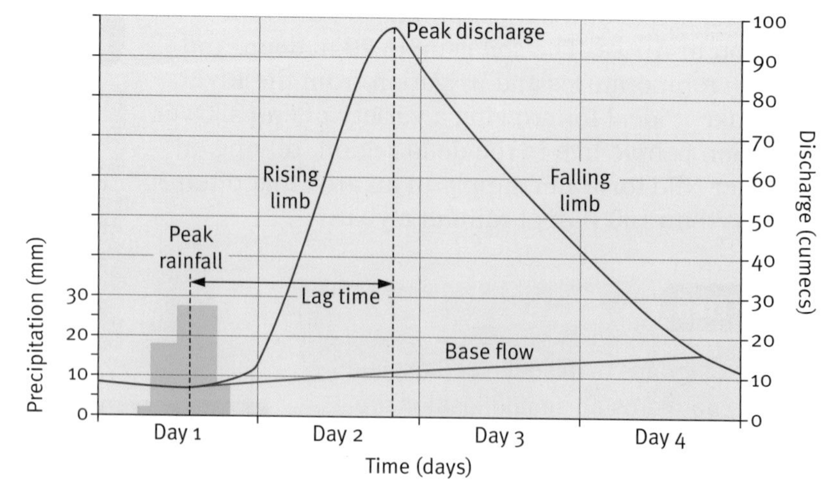

Base flow – expected discharge for the time of year
Rising limb – increasing discharge as rainfall finds its way into the river
Falling limb – decreasing discharge as the river carries storm rainfall away
Lag time – time between the highest rainfall and the highest (peak) discharge

Figure 1.20 A storm hydrograph.

The following example shows how unusually high levels of rainfall caused a rapid increase in river discharge and subsequent flooding in the city of Carlisle.

Figure 1.22 *The Carlisle flood, 2005.*

Carlisle flood, January 2005

In early January, a severe storm brought heavy rainfall to the north-west of England. River levels rose quickly, and within hours the River Eden flooded in the city of Carlisle.

Timeline for the flood

- **December 2004:** a month of above-average rainfall; the ground became increasingly saturated.
- **5–7 January 2005:** nearly 20 per cent of the expected annual rainfall fell in just 3 days, including 200 mm in 36 hours in the Carlisle area. Accompanying the rain were gales with winds of over 80 mph that brought down trees, blocking drains and smaller streams.
- **8 January 2005:** with drains blocked, water levels quickly rose and Carlisle began to flood in the early morning. By midday, the city was totally cut off, the River Eden was a torrent of water of nearly 1500 cumecs (1500 tonnes per second) (see hydrograph: Figure 1.21).
- **10 January 2005:** flood waters had begun to fall and within a week the river levels were back to normal.

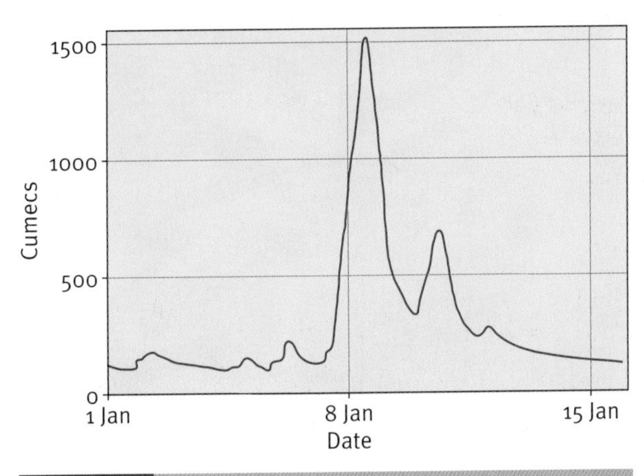

Figure 1.21 *Hydrograph for the River Eden, January 2005.* (Source: *Geofactsheet no 189*, Curriculum Press (Jan 2006))

The effect of the flood

- Three people were killed.
- Over 1500 homes were flooded.
- Over 5000 people had to be rehoused.
- Thousands of people were without electricity for days.
- Roads and railways disrupted for several days.
- Local social services (schools, police, fire service) disrupted.
- The total cost of the flood was estimated at over £500 million.

ACTIVITIES

1 The following hydrographs in an urban area and a rural area were recorded after a heavy rainstorm.

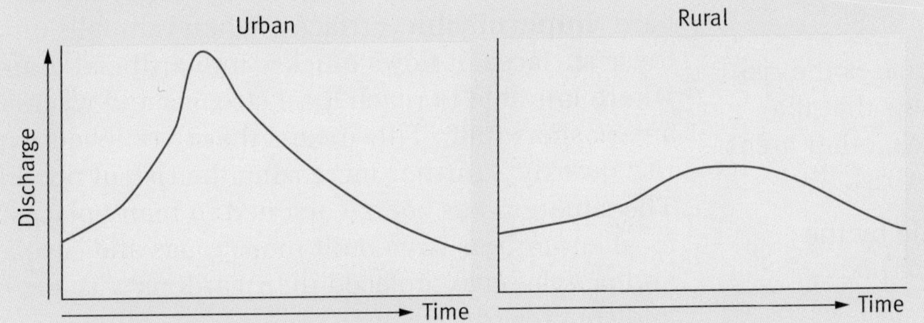

For each of the hydrographs:

a describe the pattern of discharge

b suggest reasons for the pattern

2 In Figure 1.21 for the River Eden, explain what might have happened to the storm hydrograph had there been heavy rain on 9–10 January.

3 Suggest how factors other than the January rainfall may have added to the flood risk in Carlisle.

RESEARCH LINK

Find out more about the Carlisle floods: look at the BBC website for a good starting place.

PLENARY ACTIVITY

How might the study of past storm hydrographs be useful for future flood risk planning?

Managing an increasing flood risk

GET STARTED

Why are so many areas alongside rivers becoming developed with new housing?

KEY TERMS

Impermeable – a surface that does not allow water to pass through it.

Permeable – a surface that allows water to pass through it.

Why is the flood hazard increasing?

In recent years, the risk for people of being affected by flooding has increased in many areas because of:

- increasing building development on flood plains
- the growth of hard surfaces in urban areas.

Building on flood plains

Flood plains act as a natural defence against flooding. After heavy rainfall, river levels rise to expand temporarily across flood plains. Eventually river levels fall back to their natural level and flood plains return to their previous state. Historically, flood plains were left undeveloped, often used for grazing animals when water levels were lower.

In recent years, the increasing demand for building land has meant that many towns and cities have developed right up to the sides of rivers – often encasing the river channel in concrete banks to prevent it from flooding. Riverside areas are seen as increasingly desirable places to live and often command high property prices.

Building on flood plains not only increases the risk of flooding (Figure 1.23) but also means that if a river is not able to expand in one place, other areas further downstream may be at greater risk.

The following observations were made by the Environment Agency, the government body responsible for flood management:

Planning authorities across Britain are ignoring the Environment Agency's advice and allowing developers to build on flood plains – open areas that soak up water as rivers rise. The agency predicts that the amount of building on flood plains will increase the number of people at risk from flooding to 6 million by 2020.

Figure 1.23 Urban flooding.

The increasing amount of hard surfaces in urban areas

Urban areas contain an increasing proportion of hard, **impermeable** surfaces. When rain falls on these surfaces, it flows quickly into artificial drains where it is able to reach local streams and rivers in a very short time. This means that river levels can rise quickly – further increasing the risk of flooding. The situation has been worsened in many places as local authorities have built more roads and car parks and people have replaced their gardens with paving and concrete driveways.

Fact file

- Around 5 million people in 2 million properties are at risk from flooding in England and Wales.
- Changes in climate such as wetter winters and more storms will increase the risk of flooding.

How can the risk of flooding be reduced in urban areas?

The risk of flooding in urban areas can be reduced by:

- restricting development on flood plains
- increasing the amount of **permeable** surfaces.

Managing development on flood plains

In the early 2000s the UK suffered a number of serious flood events. As a result of this, in 2006 changes were made to planning rules and local authorities were told to 'consider carefully before allowing development in flood risk areas'. However, a number of planning decisions had already been taken before 2006 and it has been estimated that up to a third of new houses built by 2020 may be on flood plains.

A number of local authorities have taken note of the increasing risk of flooding and have put 'flood plain management strategies' in place. The following statements are taken from a planning document for Dudley Council in the West Midlands:

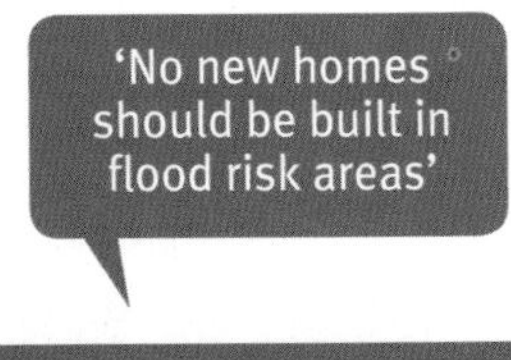

'No new homes should be built in flood risk areas'

'The Environment Agency flood risk mapping service should be used before building is allowed'

'Any new homes built in flood risk areas should be designed to withstand a flood and built with flood resistant materials'

'Any homes built on flood plains should also have flood defences'

THINK ABOUT IT

Why might insurance companies refuse to provide house insurance for people living near rivers?

RESEARCH LINK

Find out if your home is at risk from flooding at the Environment Agency website.

ACTIVITIES

1. Suggest why riverside areas are seen as attractive places to live.
2. Explain how the development of hard surfaces adds to the flood risk in urban areas.
3. Use the photographs (Figures 1.24 and 1.25) to explain how increasing the amount of green spaces in cities can reduce the risks of flooding.

Figure 1.24 Creating 'green roofs'

Increasing the amount of 'green space' in urban areas

Increasing the amount of permeable or 'green' surfaces in urban areas helps to absorb rainfall and also increases the time it takes rainfall to reach nearby rivers. Both of these factors slow down the rise in river levels after rainfall and reduce the risk of flooding.

How can the amount of green space be increased in urban areas?

- By developing parks and landscaped open spaces.
- By planting vegetation alongside roads, walkways and rivers.
- By developing 'green roofs' on buildings (Figure 1.24).

Encouraging more green spaces in urban areas not only reduces the risk of flooding, it also improves urban areas by:

- providing open areas for meeting people and relaxation (Figure 1.25)
- reducing noise pollution and creating a more peaceful environment
- reducing excessive heat produced by hard surfaces in the summer.

Figure 1.25 Green space in an urban area.

PLENARY ACTIVITY

Why is the flood risk increasing in many urban areas?

CASE STUDY The Boscastle flood 2004

Flash flooding in North Cornwall

GET STARTED

What impression does the term 'flash flood' suggest?

There are three main types of flooding event:

- **slow onset floods:** develop over a period of days and last a week or more
- **rapid onset floods:** occur more quickly, often in highland areas that feed large river catchments. This type of flood does not always last long, but can be destructive
- **flash floods:** an immediate response to short periods of intense rainfall. River levels rise without warning, often reaching a peak within minutes or hours. This type of flooding is very destructive and can cause a significant danger to life.

Fact file

The risk of flash flooding is increased by:

- building on flood plains
- **canalisation** of rivers, which increases flow rates
- development of catchments – especially on steep slopes
- removing trees and vegetation
- not maintaining drains and gullies.

Boscastle 2004 – a flash flood

The Boscastle flood of 2004, in North Cornwall, is an example of a flash flood – a rare event in the UK, where slow onset floods are more usual. The following article describes the day and the flood in Boscastle.

THINK ABOUT IT

In the UK freak storms are more likely to occur in June, July and August than in any other month of the year. Why do you think that is so?

Cornish village devastated by flash floods

The morning of 16 August started quietly for the residents of Boscastle, a village built at the confluence of the Valency and Jordan Rivers. Holiday-makers enjoyed the early morning sunshine as they sat and admired the beautiful scenery, or took time to shop for souvenirs in the local gift shop.

Within hours, the village was a disaster zone. At lunchtime, clouds began to gather and torrential rain fell on the surrounding hillsides. Within minutes, river levels rose and the three rivers that flow through Boscastle burst their banks – flooding the whole village.

It was estimated that over 500 mm of rain fell in 4 hours – creating a tide of water that swept through the village at nearly 40 mph, destroying everything in its path.

Figure 1.26 *The village of Boscastle.*

What were the causes of the Boscastle flood?

The natural characteristics of the area make the village of Boscastle vulnerable to flooding. The catchment is small (about 23 km^2) and includes the relatively impermeable upland area of Bodmin Moor (Figure 1.27). Steep-sided valleys converge as they run towards the sea, funnelling water towards Boscastle. After heavy rainfall, surface run-off quickly reaches the rivers, increasing the likelihood of flash flooding.

During the summer of 2004, a number of unusual factors occurred. It had been an extremely wet summer, and by August the ground was saturated. On 16 August thundery clouds developed, the remnants of Hurricane Alex that had moved across the Atlantic Ocean. These clouds remained stationary over North Cornwall, because of converging winds. Throughout the afternoon an unprecedented amount of rain fell, estimated at over 1.400 million litres in just 2 hours – nearly 200,000 litres a second! The rainfall quickly made its way into rivers that rose at an alarming rate as they flowed towards Boscastle (Figure 1.28).

The flood risk in Boscastle had been increased by the amount of building alongside the river and also the construction of small bridges across it. These trapped material being washed downstream, creating a 'dam-like' effect.

Figure 1.28 ***Boscastle during the 2004 flood.***

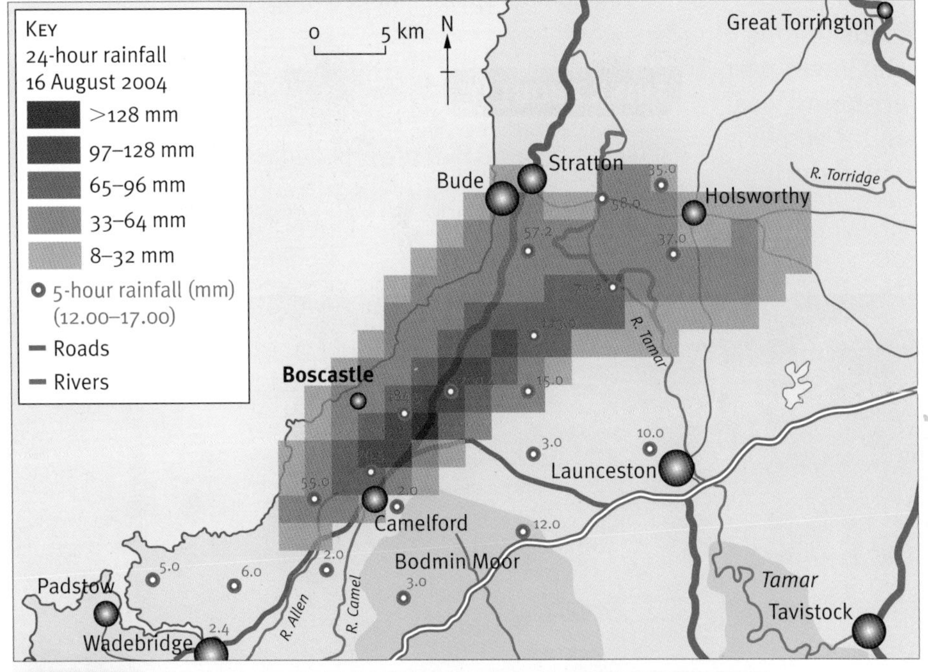

Figure 1.27 ***Rainfall map: Boscastle, 16 August 2004.*** **(Source: Adapted from Crown Copyright data supplied by the MET Office)**

Diary of events on 16 August 2004

- 12.15 Dry in Boscastle – dark clouds visible in Camelford.
- 12.30 Heavy rain begins to fall.
- 12.39 Flood-watch issued.
- 15.00 First of many power cuts caused by lightning.
- 15.30 River valency begins to flood.
- 15.46 Reported rise of 2 m in rivers in one hour.
- 15.53 Fire brigade mobilised.
- 16.00 All access roads closed.
- 16.30 A 3-m wall of water flows through Boscastle at 64 kph.
- 17.00 Floods reach peak level, cars swept away and buildings destroyed. Rain is so heavy that it is difficult to see.
- 17.10 Major incident declared – RAF search and rescue alerted.

ACTIVITIES

1. Why are flash floods the most dangerous type of flood?
2. Describe the village of Boscastle before the flood.
3. **a** Explain how both physical and human factors make Boscastle vulnerable to flooding.

   b What particular events in August 2004 triggered the flash flood?

RESEARCH LINK

Find out more about the Boscastle flood at the Met Office website.

PLENARY ACTIVITY

Could the Boscastle flood have been predicted? Would prediction have reduced the impact of the flood?

KEY TERMS

Canalisation – making a river more like an artificially built canal.

Reducing the impact of future floods

GET STARTED

View the video footage of the 2004 Boscastle flood on the internet. What evidence shows the power of the flood?

Boscastle had been affected by minor flooding events before 2004, but was unprepared for a flood of the scale that it experienced in August 2004.

The flood destroyed homes and businesses and swept away a hundred vehicles. Bridges and roads were damaged, and rescue services had to mount the largest peacetime rescue operation in Britain's history.

When the flood waters receded, thousands of tons of mud and debris were left throughout the lower part of the village. The effect on the local economy was devastating – the area relies on tourism for 90 per cent of its income, most of which is earned during the summer months.

Fact file

Effects of the Boscastle flood

- 25 business properties destroyed.
- 50 buildings flood damaged.
- 4 footbridges washed away.
- Visitor centre destroyed.
- Pavements and gardens damaged by the weight of the flood water.
- Stress and anxiety of local people.
- Insurance companies paid out an estimated £20 million to repair damaged property.

THINK ABOUT IT

It was estimated that the probability of a flood of the scale seen at Boscastle was 1 in 400 years!

Figure 1.29 ***The effects of the Boscastle flood.***

Responding to the flood – planning for the future

After the 2004 flood in Boscastle the Environment Agency investigated the causes of the flood and found a number of human factors had added to the flood risk. These included:

- the building of low bridges, which trapped boulders and trees that had been washed down the valley
- allowing trees to grow alongside the river; during the flood these were washed into the river, blocking channels
- artificially narrowing the river as it passed through Boscastle, reducing its carrying capacity
- building alongside the river and not allowing for its expansion during periods of high water flow
- removing vegetation from the sides of the valleys, which increased the rate of surface runoff and meant that rainfall reached the river very quickly.

In order to reduce the risks of future floods, a £4.6 million flood defence scheme was completed in 2008. The scheme aims to reduce the flood risk whilst preserving the character and amenities of the village. The main features of the flood defence scheme are shown in Figure 1.30.

RESEARCH LINK

Las Vegas (US) has been the victim of flash floods. Use the internet to investigate the causes and the impacts of flooding in Las Vegas.

Figure 1.30 ***Boscastle flood defence scheme.***
(Source: Environment Agency website, 2006)

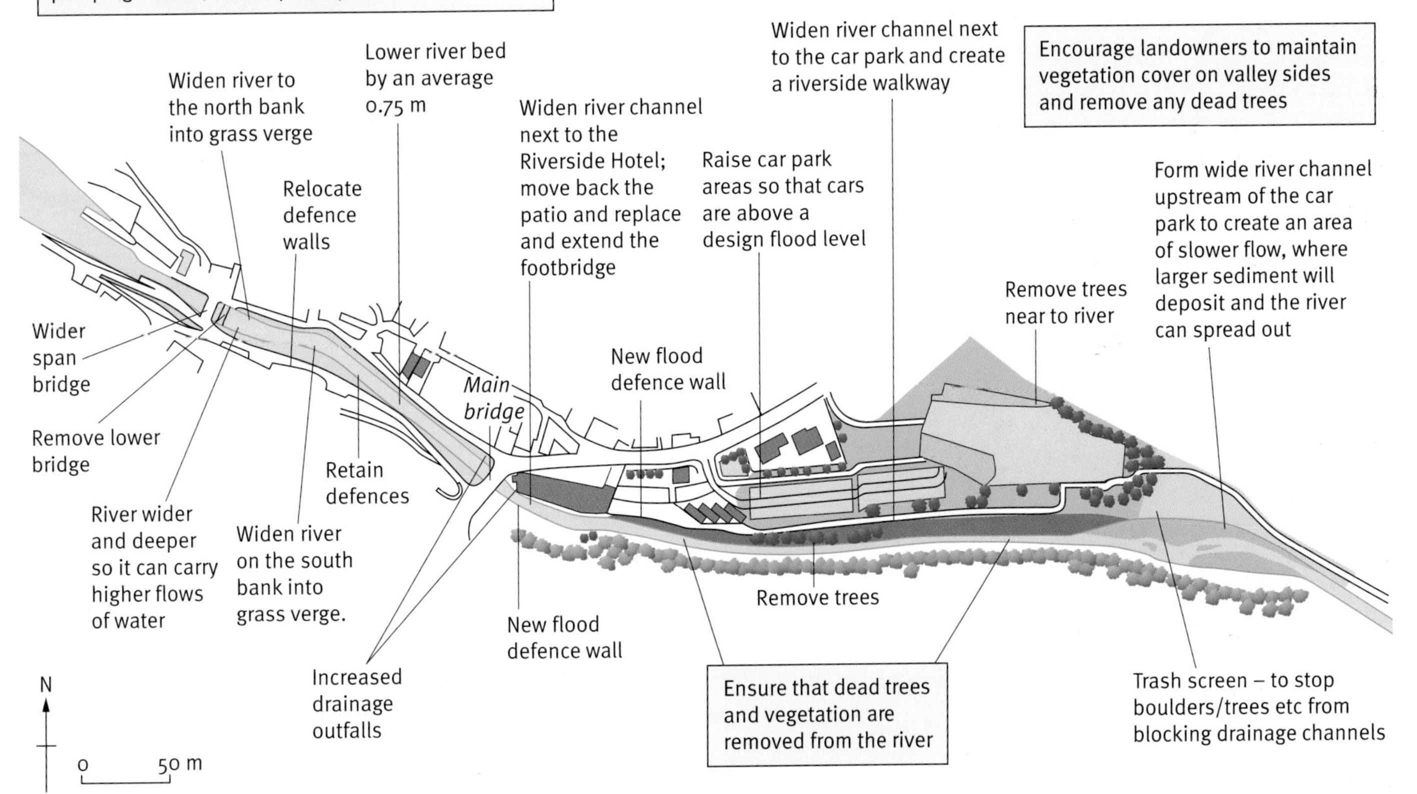

ACTIVITIES

1 a Using Figure 1.29, describe the effects of the flood on the village of Boscastle.

b Suggest how the flood affected the local economy.

2 Explain how the development of Boscastle increased the flood risk.

3 Explain how any five parts of the flood defence scheme might reduce the risk of future flooding.

PLENARY ACTIVITY

Explain the following statement: 'The flood defence scheme in Boscastle will not stop a flood event on the scale of 2004, but should reduce the impact'.

CASE STUDY Flooding in Bangladesh

Why is Bangladesh vulnerable to flooding?

GET STARTED

How would a serious flood in your local area affect your day-to-day life?

KEY TERMS

Irrigation – artificial watering of the land.

Monsoon – annual period of heavy rainfall in Asia.

Storm surge – extreme storm waves created by strong winds and low pressure, leading to higher sea levels and flooding.

Bangladesh is one of the world's poorest and most densely populated countries. Its 140 million people live mainly on the flood plains of the rivers Ganges, Brahmaputra and Meghna. Over thousands of years, these three rivers have deposited millions of tons of silt, creating one of the most fertile areas in the world.

Bangladesh is affected by two types of flood:

- **river floods:** these happen every year and are a part of the natural cycle of snow-melt and **monsoon** rainfall. In some years, the monsoon rains are exceptionally heavy or prolonged – this is often the main cause of extreme flood events
- **coastal floods:** these are created by cyclones that build up in the Indian Ocean and move towards the Bay of Bengal. Water is funnelled toward Bangladesh creating a **storm surge** that may reach 6 m in height, flooding large areas of land. When combined with heavy rainfall and river flooding, the effects of cyclones can be totally devastating.

Figure 1.32 **Flooding in Dhaka.**

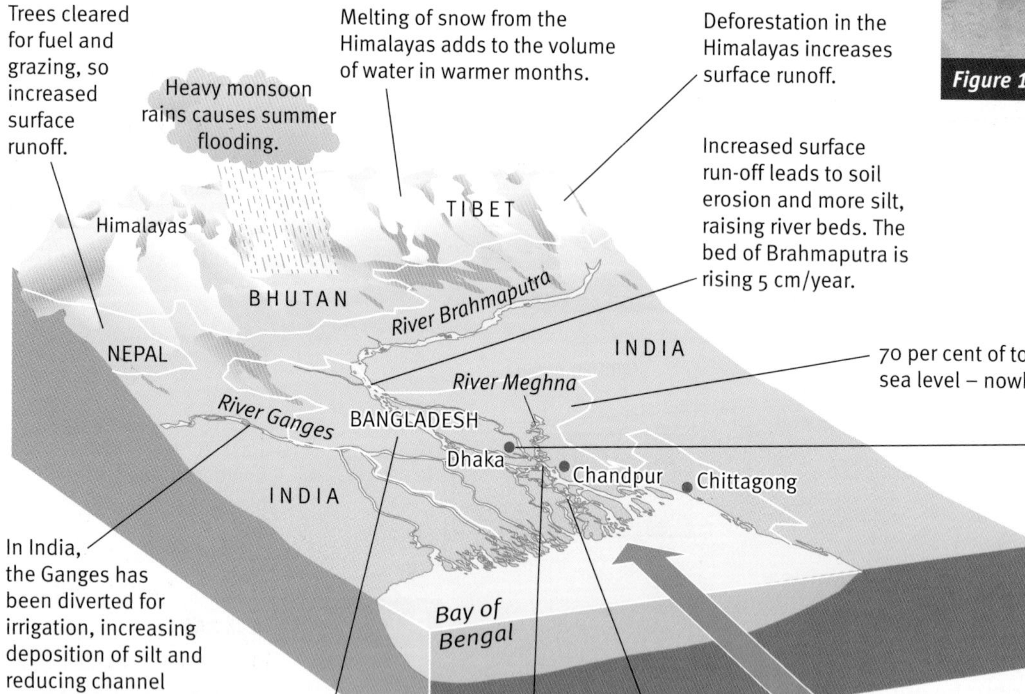

Trees cleared for fuel and grazing, so increased surface runoff.
Heavy monsoon rains causes summer flooding.
Melting of snow from the Himalayas adds to the volume of water in warmer months.
Deforestation in the Himalayas increases surface runoff.
Increased surface run-off leads to soil erosion and more silt, raising river beds. The bed of Brahmaputra is rising 5 cm/year.
Himalayas
TIBET
BHUTAN
NEPAL
River Brahmaputra
INDIA
River Meghna
River Ganges
BANGLADESH
Dhaka
Chandpur
Chittagong
INDIA
Bay of Bengal
70 per cent of total land area is less than 1 m above sea level – nowhere for water to drain to.
Rapid unplanned urban growth has added to the problem of flooding. Urban populations in Bangladesh have increased from 4 per cent in 1951 to 35 per cent in 2009 (estimated). The growth is largely made up of poor migrants who often live in vulnerable areas.
In India, the Ganges has been diverted for irrigation, increasing deposition of silt and reducing channel capacity. In the rainy season water is let through, causing floods.
Bangladesh – 80 per cent floodplain and delta makes it very susceptible to flooding.
Meeting of the three huge rivers increases the flood risk.
Silt blocks river channels and creates islands, reducing carrying capacity of rivers.
Cyclones create a storm surge

Figure 1.31 **Causes of flooding in Bangladesh.**

DHAKA: POPULATION	
1981	3.4 million people
1991	6.9 million people
2001	9.9 million people
2011	14.0 million people (estimated)

What factors increase the risk of flooding in Bangladesh?

Although physical factors increase the risk of flooding, there is an increasing number of human influences.

River floods

River flooding in Bangladesh is an annual occurrence linked to the rainfall pattern of the monsoon. The yearly flooding brings with it millions of tons of silt that adds to the soil's fertility. Flood water is used to **irrigate** farmland and is an important part of the yearly cycle of agriculture.

In a normal year, flood waters can be mostly controlled by using storm drains and embankments, and few people are badly affected. However, in some years the volume and pattern of rainfall is extreme and water levels rise dangerously high, inundating towns and cities and covering vast areas of the country. The most recent severe floods were in 1998 and 2004.

The 1998 flood

In 1998, the rains were exceptionally heavy and water levels reached record heights. The resulting floods covered nearly 60 per cent of the country and all the main river channels were flooded. The capital city of Dhaka and a number of regional cities were badly affected, with many parts of cities under water for weeks. Hundreds of people were killed and millions made homeless. Agricultural land and crops were lost or contaminated by polluted water. Over 900 bridges and 15,000 km of roads were destroyed by the floods. An investigation was carried out after the 1998 flood to try to assess the impact. Some of the results are shown in Table 1.1.

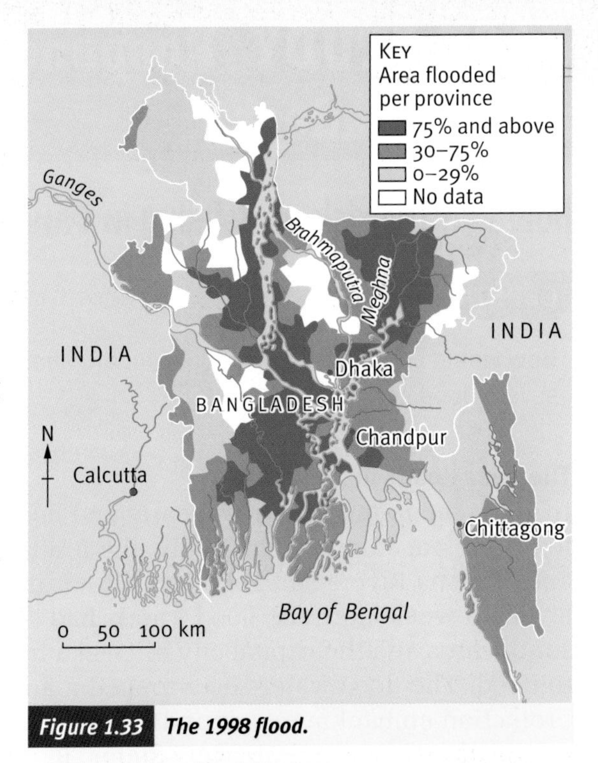

Figure 1.33 *The 1998 flood.*

PLENARY ACTIVITY

Why is monsoon flooding both a necessity and a threat to Bangladesh?

Table 1.1 Impact of the 1998 flood in Dhaka

Items affected	Percentage of people affected by the flood in Dhaka			Recovery after 1 year
	Fully affected	Partly affected	Not affected	
Loss of food	1	16	83	Fully recovered – 40% Partly recovered – 43% Not recovered at all – 17%
Loss of clothing	4	26	70	
Loss of housing	17	61	22	
Loss of income	28	42	30	

ACTIVITIES

1 a Use the following data to draw a climate graph for Bangladesh (line graph for temperature, bar graph for rainfall).

Table 1.2 Temperature and rainfall, Bangladesh

Month	J	F	M	A	M	J	J	A	S	O	N	D
Temperature (°C)	22	24	28	32	32	30	29	30	30	29	25	24
Rainfall (mm)	10	25	40	50	140	290	320	330	250	120	40	5

b In which months of the year are river levels likely to be highest?

2 Why do so many people live in areas that are prone to flooding in Bangladesh?

3 Draw a table with two columns headed 'Physical' and 'Human' and complete it by identifying the physical and human causes of flooding in Bangladesh.

4 Describe the effects of the 1998 flood in Bangladesh.

Fact file

- During the last 50 years, at least eight extreme flood events have occurred in Bangladesh, each one affecting over 50 per cent of the total land area.
- Extreme flood events usually last between 15 and 45 days.

CASE STUDY Responding to the flood hazard in Bangladesh

How can the risk of flooding in Bangladesh be reduced?

GET STARTED

How would your life be affected if your area had a severe flood every year?

The river floods in 2004

In 2004, the monsoon arrived early and heavy rain fell from late June. By early July, the Brahmaputra and Meghna Rivers had risen above their danger levels. A week later, the flood waters had drained southwards and the capital city of Dhaka had begun to flood. The flood waters over-topped many flood protection embankments, inundating areas not provided with storm drainage systems. By mid-August, the flood waters had fallen in most areas, leaving contaminated mud and general destruction. Government figures estimated that the flood had affected 36 million people.

UN fears malnutrition after flooding in Bangladesh

Up to 1.5 million women and children are at risk of 'acute malnutrition' after Bangladesh's worst floods for six years, two UN agencies said yesterday.

Monsoon rains and floods inundated half the country in July and August, killing 766 people and affecting more than 30 million.

Without intervention, the number of 'acutely malnourished children' in the flooded areas could rise to more than a million within eight weeks, a joint statement by Unicef and the World Food Programme said.

It said that more than 500,000 pregnant women and breast-feeding mothers were likely to face 'serious malnutrition', and that babies born to malnourished women were more likely to become ill.

(Source: *Guardian* website, 2004)

Aftermath of the 2004 Bangladesh flood: WaterAid report

Dhaka, the capital of Bangladesh, was a city swimming in sewage. More than half the city was submerged when the floodwaters began pouring, first from the swollen rivers as they burst their banks, then from overburdened sewers disgorging back on to the streets.

(Source: WaterAid website)

THINK ABOUT IT

The word 'monsoon' comes from the Arabic word for 'season'. Looking back at Table 1.2, which months are the monsoon months in Bangladesh?

RESEARCH LINK

Use the internet to find out more about the causes and effects of the 2004 flood in Bangladesh.

Reducing the flood risk

Historical records show that the frequency and scale of flooding in Bangladesh has increased in the last 50 years, increasing the need for flood planning and preparation. However, despite more use of flood control measures, the cost of damage caused by flooding has steadily risen.

The following examples look at two approaches to flood management: Figure 1.34 describes part of the 'Preparedness Programme' being supported by Oxfam. Figure 1.36 shows part of the Dhaka Integrated Flood Protection Project, a $100 million government project that includes a number of **hard engineering** methods.

Fact file

Cities where over 60 per cent of the land is covered by roads and buildings are six times less likely to flood if they have a storm drain system.

KEY TERMS

Hard engineering – use of concrete barriers to control water.

- **Cluster villages:** a cluster village is a village that has been raised 2 m above water level. Each village houses between 25 and 30 families.
- **Raised homestead:** individual homes are raised 2 m above water level on earth banks. The earth banks are planted with grass to prevent erosion.
- **Flood shelter:** around 2 hectares of raised land where people can bring livestock. Each shelter has space for over 100 families and includes a community room and toilets.
- **Rescue boats:** rescue boats are located around the areas most at risk from flooding and near to flood shelters
- **Radios:** radios given to each 'preparedness committee'. Flood warnings can be issued and the preparedness plan put into action.

Figure 1.34 ***Part of the Preparedness Programme.***

ACTIVITIES

1. Describe the immediate and longer-term effects of the 2004 floods in Bangladesh.
2. Explain how the Dhaka Flood Protection Project will reduce the risks of flooding.
3. Describe and explain how the Flood Preparedness Programme will help people in rural areas.

PLENARY ACTIVITY

Why are developing countries often more badly affected by floods?

Figure 1.36 ***Dhaka Integrated Flood Protection Project.*** **(Source: Asian Development Bank, 2002)**

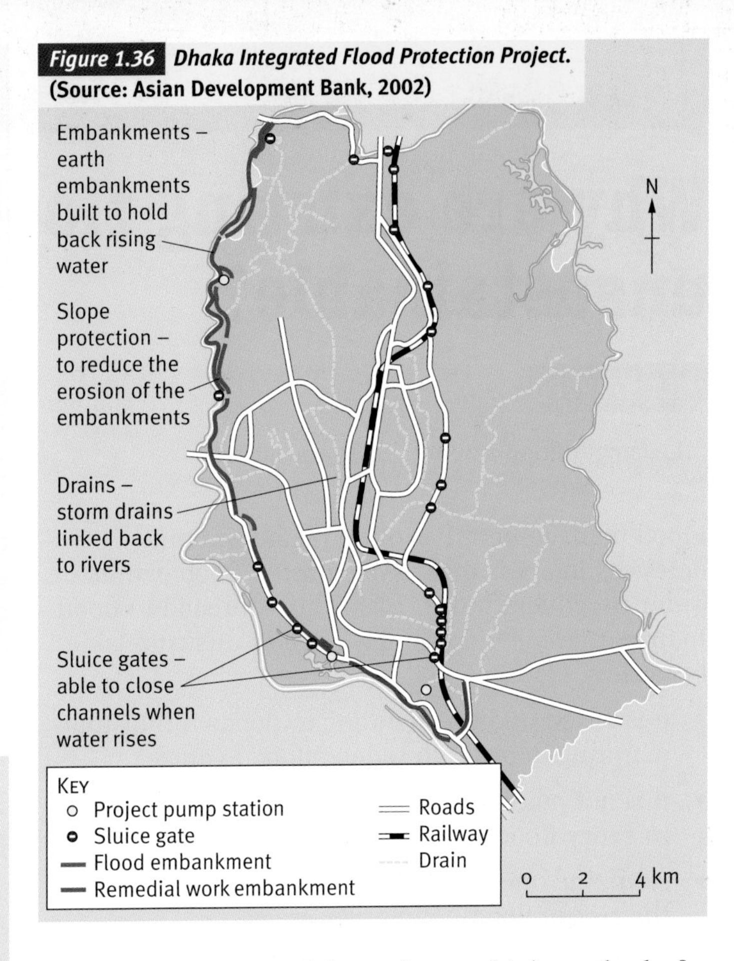

There is an ongoing debate about which method of flood protection should be used. Hard engineering can control large amounts of water, but it is expensive and can create environmental problems – for example, building embankments may reduce the flow of water to farming areas. Smaller-scale projects can be used across a wider area, but may have limited effect in large urban areas.

Figure 1.35 ***The Dhaka embankment.***

Why are existing flood defence strategies unsustainable?

GET STARTED

How might flood defences be sustainable?

Flood management has often meant building defences in areas that have suffered flood events in order to reduce future risks. This approach to flood management is increasingly seen as unsustainable because:

- it is based upon responding to floods rather than planning in order to reduce flood events
- it is not possible to build and maintain defences in every flood risk area
- changing one part of a river basin can increase the risk of flooding in other areas
- flood risks may increase in the future, so there is a need to manage the whole of the drainage basin and not just parts of it.

In order to plan for the future, **Catchment Flood Management Plans (CFMPs)** are being developed throughout England and Wales.

What are Catchment Flood Management Plans (CFMPs)?

CFMPs look at a complete river basin and set out different strategies in order to reduce flood risks over the next 50–100 years. The plans take into account building developments in the area and also consider the possible effects of climate change. This approach is more sustainable because it is not just about responding to floods, but about long-term planning and using the natural environment to reduce the risk of flooding.

The Thames Region Catchment Flood Management Plan

The Thames catchment covers the main part of the River Thames and its tributaries (Figure 1.37). It includes rural and densely populated urban areas, including London, Reading and Guildford. Although it covers only 10 per cent of the area of England and Wales, it is home to 25 per cent of the population. The region has suffered a number of flood events in recent years. It is estimated that over 100,000 properties are at risk of flooding in the area.

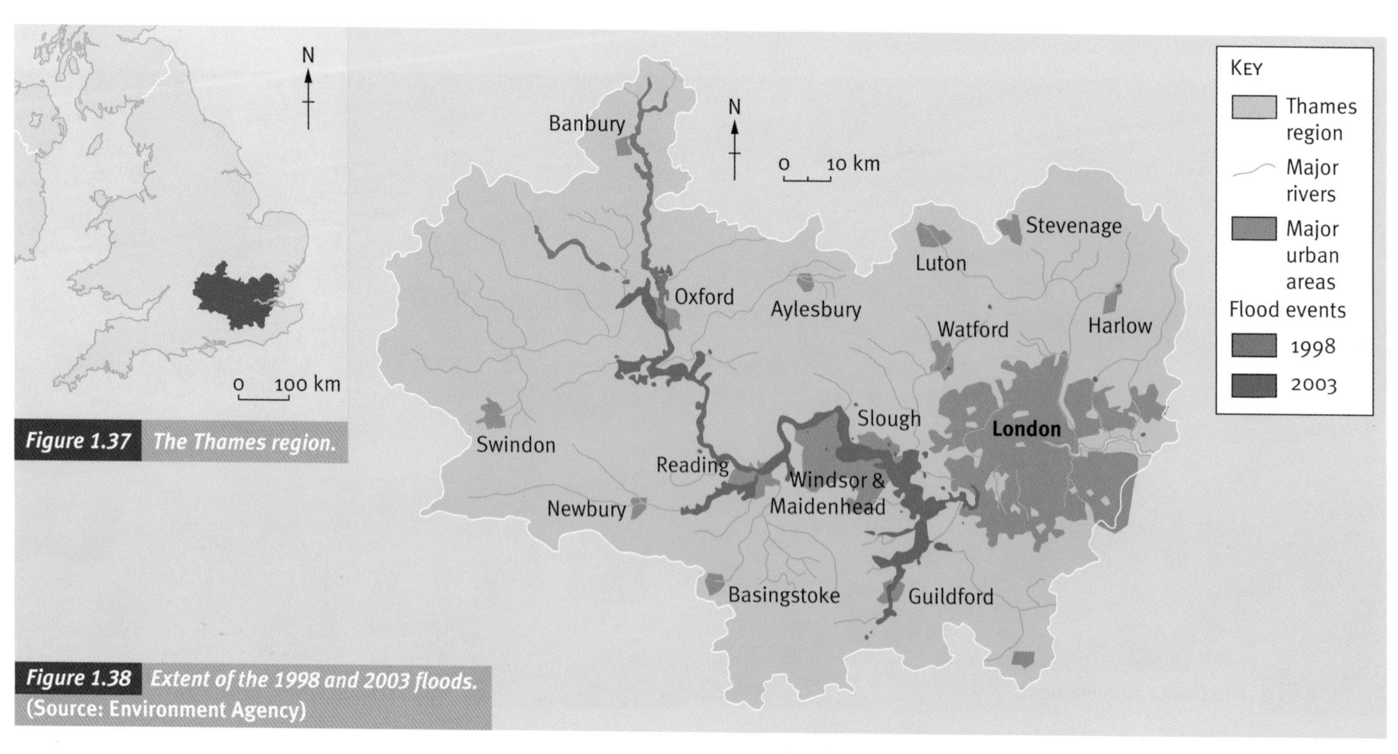

Figure 1.37 The Thames region.

Figure 1.38 Extent of the 1998 and 2003 floods. (Source: Environment Agency)

Flood management strategies in the Thames (Catchment Flood Management Plan) area

Maintaining river channels

- Weeds and silt cleared to ensure river flow is not interrupted.
- River banks maintained to reduce the risk of bank collapse.

Use of the natural flood plain

- Use of flood plain increased to store water during times of flood.
- Storage ponds built to collect storm water and prevent nearby towns from flooding.

Re-create river channels in urban areas

- Number of concrete channels reduced by restoring urban rivers.
- New river channels constructed with space to flood naturally.

KEY TERMS

Catchment Flood Management Plans (CFMPs) – plans setting out strategies to reduce flood risks in a river basin over 50–100 years.

Fact file

- In urban areas, 50–80 per cent of the flood plain is developed.
- In London, nearly 50 per cent of all water courses are artificial.
- On average, a flood affecting 50,000 properties is expected once every 100 years.

(Source: Environment Agency)

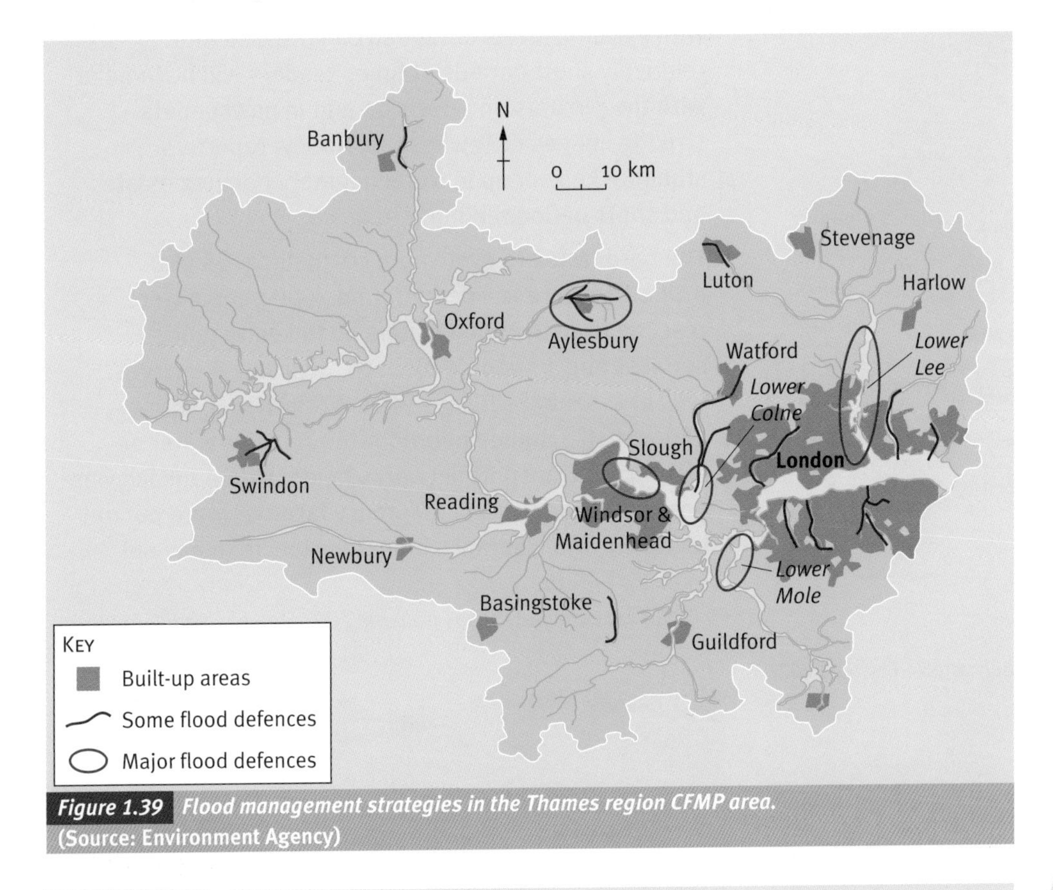

Figure 1.39 Flood management strategies in the Thames region CFMP area. (Source: Environment Agency)

ACTIVITIES

1 Why are CFMPs an important part of the sustainable management of rivers?

2 a Describe the human and physical characteristics of the Thames region.

 b Name the settlement that was affected by both the 1998 and 2003 floods.

 c Suggest reasons why this might have happened.

3 Describe and explain how the Thames region flood management strategies will reduce the risk of flooding.

PLENARY ACTIVITY

Why are CFMPs seen as a more sustainable type of river management?

A Decision Making Exercise – Should we continue to build on flood plains?

In the summer of 2007, much of the UK suffered from severe flooding. The worst-hit areas can be seen on the map below. The floods brought misery to thousands of people, some of whom were still waiting to return to their homes over a year after the event. Flood plains have always attracted housing, industry, roads and rail networks. Level ground, which is easy to build on, reduces construction costs. Many redundant brownfield sites awaiting redevelopment are found on flood plains. In this exercise, you will be asked to consider the costs and benefits of continuing to exploit such locations for much-needed additional housing.

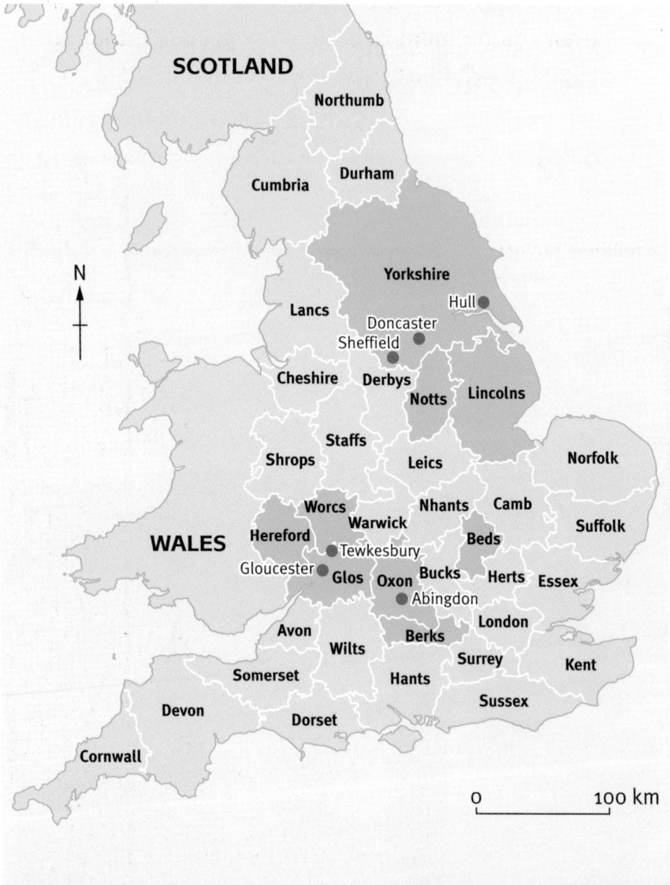

Figure 1.40 *English counties and towns badly affected by the 2007 summer floods.*

A Summer to Remember!

(Extracts from the official government report published 25 June 2008)

The floods of last year caused the country's largest peacetime emergency since World War II.

We witnessed the wettest summer since records began, with extreme levels of rainfall compressed into relatively short periods of time. Readers will be familiar with the pictures on television and in newspapers – striking images of Tewkesbury Abbey, reporters standing knee deep in water in empty housing estates and shots of flooded infrastructure.

The hard facts are even more compelling. 55,000 properties were flooded. Around 7000 people were rescued from the flood waters by the emergency services and 13 people died. We also saw the largest loss of essential services since World War II, with almost half a million people without mains water or electricity. Transport networks failed, a dam breach was narrowly averted and emergency facilities were put out of action. The insurance industry expects to pay out over £3 billion.

(Source: *Learning Lessons from the 2007 Floods: The Pitt Review*, Crown copyright)

MORE FLOODS PREDICTED

Experts suggest that we need to get used to events such as the 2007 floods. Global warming will be responsible for more than just rising temperatures! Changing weather patterns will result in an increased risk of significant summer rainstorms. Areas at particular risk are low-lying coastal areas. These face the additional problem of rising sea levels.

Figure 1.41 *Tewkesbury was badly affected.*

Britain needs a million new homes

Experts project an increase of 3.5 million households by 2021 or 175,000 per year. The explosion is due mainly to more people remaining single, rising divorce rates and older people living on their own rather than in nursing homes. Other facts include:

- increasing life expectancy (now typically 79 years)
- increasing immigration
- increasing migration into the south-east, putting additional pressure on the shortfall in available housing.

(Source: Jill Sherman, *The Times* website, 2005)

'Many submissions to the Review call for a complete end to building on the floodplain. This is not realistic. The country cannot end all development along the Thames, or bear the costs of building critical infrastructure, such as water treatment works or power stations, away from the water supplies they need to function.'

(Source: Sir Michael Pitt, Independent Chair of the 2008 Flood Review Report)

Environmentalist viewpoint: 'Councils will be forced into using brownfield sites if they maintain their commitment to preserving green-belt land, even though there is an inevitable flood risk.'

The Thames Gateway – a solution to the housing crisis?

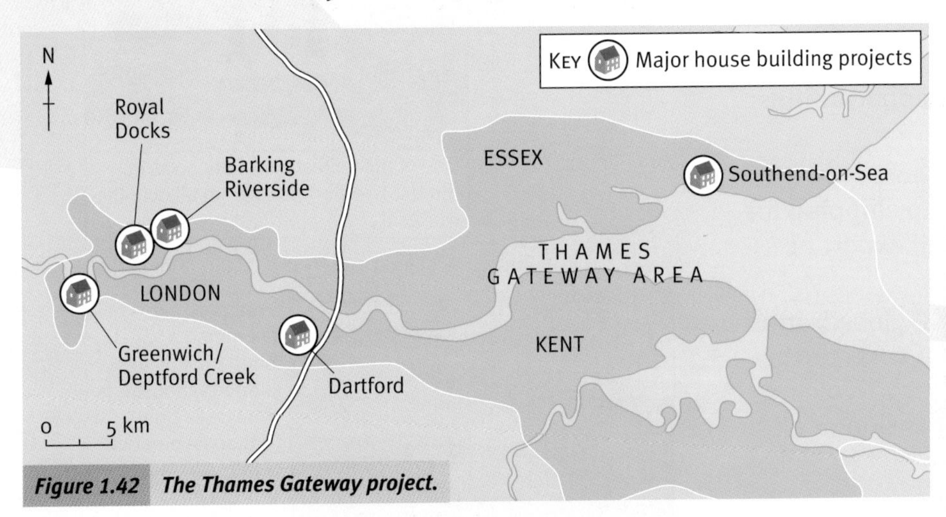

Figure 1.42 *The Thames Gateway project.*

Figure 1.43 *Proposed development at Barking Riverside.*

The Thames Gateway will help to solve the housing crisis in the south-east. Private developers have been encouraged by the government to submit housing and infrastructure plans for planning approval. The government will support the scheme to build 160,000 homes and create 225,000 jobs with £9 billion of public money.

One of the largest areas targeted is Barking Riverside, where over 26,000 homes are planned, along with schools, recreation areas and major transport improvements. The site is redundant and was formerly occupied by old industrial works.

YOUR TASK

Alex Nixon, lead environmental officer for Thames Gateway, is confident that flood-prone areas can be used if money is spent on flood-resistant buildings with a range of flood prevention measures in place. He believes that people must think in a holistic way.

Consider the evidence on these two pages and the additional information on the supporting CD. Should Alex Nixon be confident that developers will want to build here and that people will want to live here? Divide your group into two; each half should present evidence to suggest why the scheme at Barking Riverside should or should not go ahead.

Why is the coast called a multi-use resource?

GET STARTED

Think of a use for the coast for every letter of the alphabet.

Over 4 billion people live in coastal areas. This number is expected to grow rapidly in the next 50 years, making it increasingly important that the natural processes and economic demands that shape these areas are fully understood.

Coastal areas are a valuable resource because they provide the location for a number of opportunities, including:

- **industrial development:** coastal areas can offer large areas of flat land and access to shipping for the import of raw materials and the export of finished products
- **transportation:** growing industrial globalisation and trade has meant that industrial ports are increasingly important. Also, the growth of cruising holidays has seen the expansion of a number of ports
- **recreational activities:** increasing wealth and more leisure time have increased the demand for water sport activities
- **residential development:** coastal areas are seen as attractive places to live because of the scenery and recreational opportunities they offer
- **nature conservation:** salt marsh and sand dune habitats are important environments for rare plant and animal species.

The following example looks at some of the ways the coast is used in the area of the Solent, southern England.

THINK ABOUT IT

The population of some coastal areas doubles in the summer with the influx of holiday visitors. What impacts might this have on the lives of local residents?

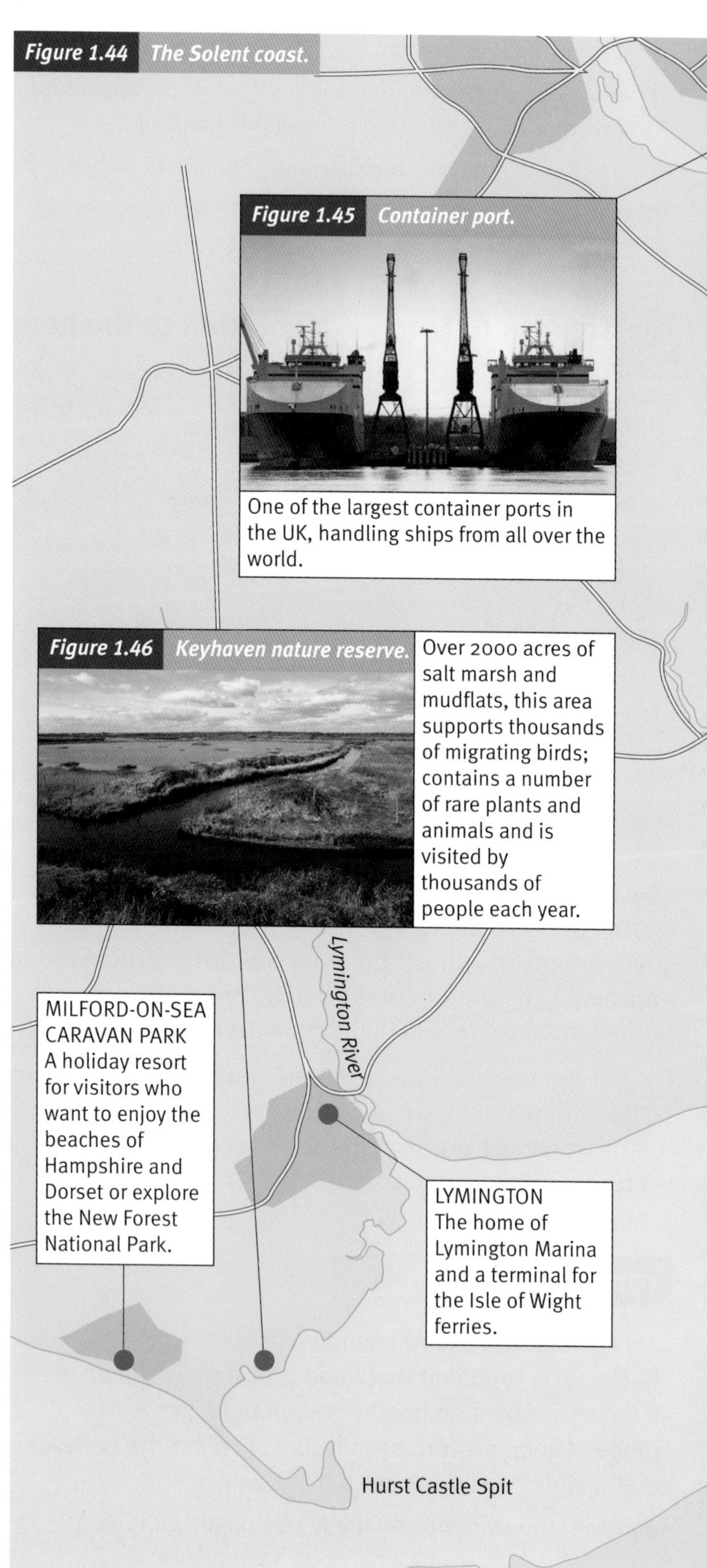

Figure 1.45 Container port.
One of the largest container ports in the UK, handling ships from all over the world.
Figure 1.46 Keyhaven nature reserve.
Over 2000 acres of salt marsh and mudflats, this area supports thousands of migrating birds; contains a number of rare plants and animals and is visited by thousands of people each year.
Lymington River
MILFORD-ON-SEA CARAVAN PARK
A holiday resort for visitors who want to enjoy the beaches of Hampshire and Dorset or explore the New Forest National Park.
LYMINGTON
The home of Lymington Marina and a terminal for the Isle of Wight ferries.
Hurst Castle Spit

Figure 1.44 The Solent coast.

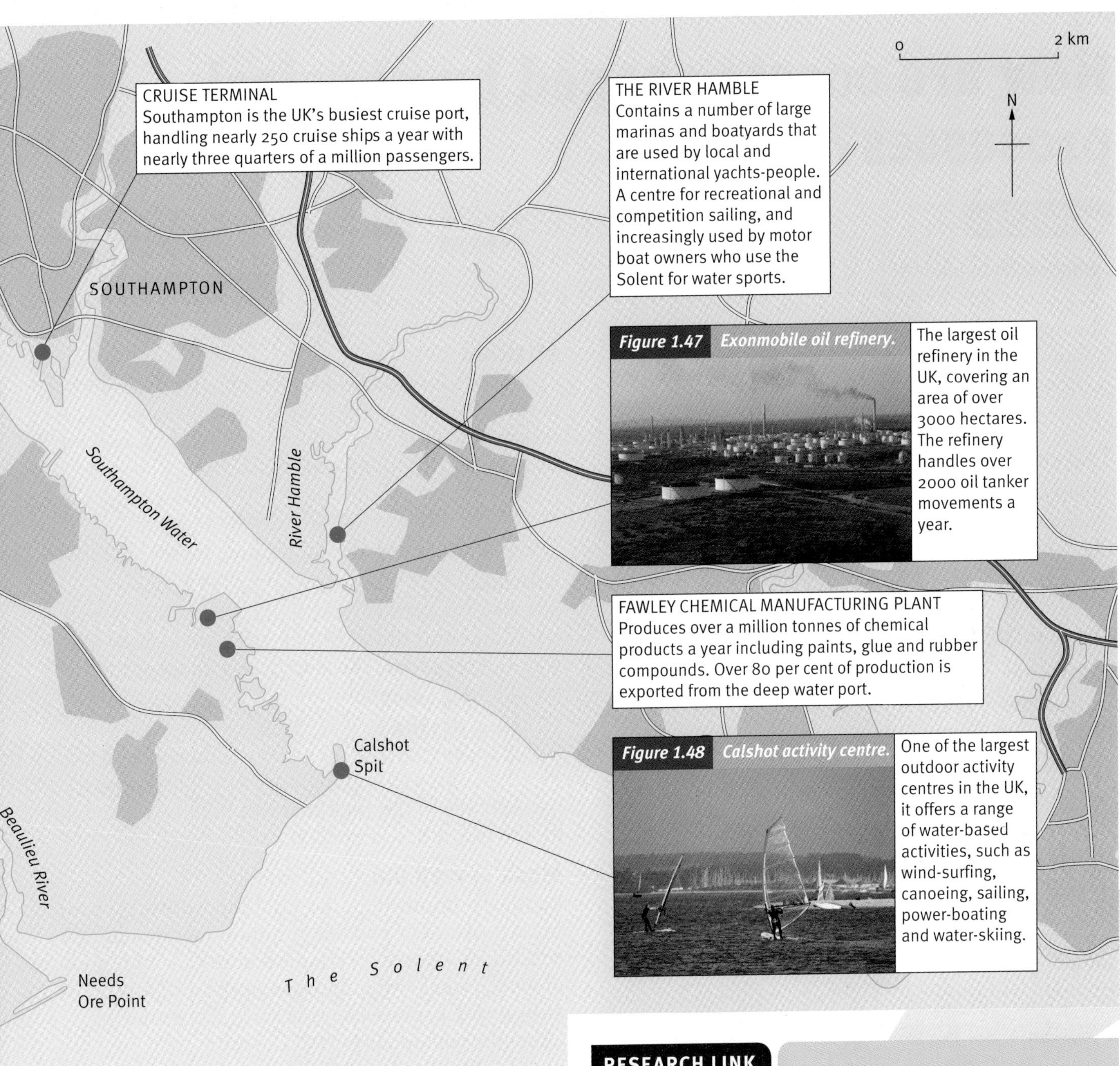

0
2 km
N
CRUISE TERMINAL
Southampton is the UK's busiest cruise port, handling nearly 250 cruise ships a year with nearly three quarters of a million passengers.
THE RIVER HAMBLE
Contains a number of large marinas and boatyards that are used by local and international yachts-people. A centre for recreational and competition sailing, and increasingly used by motor boat owners who use the Solent for water sports.
SOUTHAMPTON
Southampton Water
River Hamble
The largest oil refinery in the UK, covering an area of over 3000 hectares. The refinery handles over 2000 oil tanker movements a year.
FAWLEY CHEMICAL MANUFACTURING PLANT
Produces over a million tonnes of chemical products a year including paints, glue and rubber compounds. Over 80 per cent of production is exported from the deep water port.
Calshot Spit
One of the largest outdoor activity centres in the UK, it offers a range of water-based activities, such as wind-surfing, canoeing, sailing, power-boating and water-skiing.
Beaulieu River
Needs Ore Point
The Solent
ISLE OF WIGHT

Figure 1.47 *Exonmobile oil refinery.*

Figure 1.48 *Calshot activity centre.*

RESEARCH LINK

Find out more about the western Solent on the New Forest District Council website.

ACTIVITIES

1. Coastal areas are often called 'multi-use areas': why is this?
2. Explain why the Solent coastal area is an important economic and environmental resource.
3. Suggest what activities in and around the Solent may harm the marine environment.

PLENARY ACTIVITY

Explain why the demands of different users in the Solent area may cause conflict.

How are coasts shaped by physical processes?

GET STARTED

What evidence might show the power of storm waves?

The coast is the frontier between land and sea and is one of the most fragile environments on Earth. It is constantly under attack from both marine processes and weathering.

The coastal system

The coast acts as a giant conveyer belt; material is worn away from some places, moved by waves, and deposited in other places (Figure 1.50).

Marine erosion

The main types of marine erosion are;

Abrasion

During storm conditions, waves have the energy to pick up large quantities of sand and pebbles. These are hurled at the face of the cliffs as the wave breaks, creating a 'sand blasting' effect. This process causes the most rapid rates of erosion in the UK.

Hydraulic pressure

In some areas, waves break directly against the base of the cliffs. The sheer force of water hitting the cliffs will break off fragments of rock. Also, as waves hit a cliff face air is forced into cracks, blasting away fragments of rock.

Figure 1.49 The power of the sea.

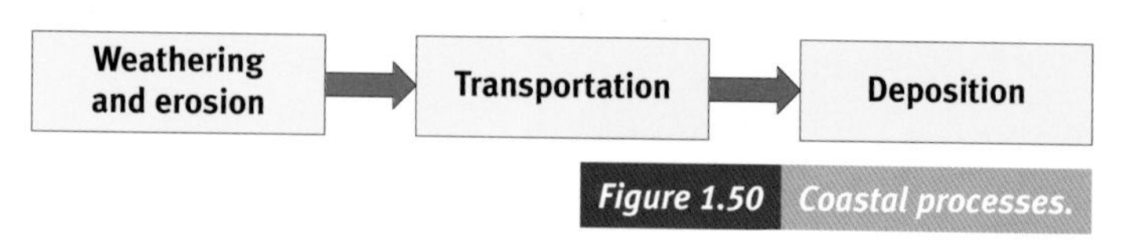

Figure 1.50 Coastal processes.

Attrition

Sand particles and pebbles are constantly colliding with each other as they are moved by waves. This action gradually wears the material away, making it increasingly rounder and smaller.

Weathering in coastal areas

The main types of coastal weathering are as follows.

Solution

Sea water is very corrosive and can slowly dissolve chalk and limestone. Salt crystals are formed as salt water evaporates. These crystals can grow in size, forcing rocks to break.

Wetting/drying

Softer rocks such as clay expand and contract as they become wet and then dry out. This causes weaknesses in the rock that can then be picked out by the processes of erosion.

Mass movement

Rock falls, mudslides and landslides are all types of mass movement and are common features of cliff coastlines, often occurring because of a combination of waves weakening the base of the cliff and **sub-aeriel processes** (erosion and weathering) attacking the upper part of the cliff.

In more resistant rocks such as chalk and limestone, erosion at the base of a cliff can lead to rock falls. Water soaking through softer rocks such as clay can weaken the structure of the rock and lead to landslides.

KEY TERMS

Geology – the nature and structure of rocks.

Marine erosion – the wearing away of rocks by the action of the sea.

Sub-aeriel processes – processes active on the face and top of cliffs.

The impact of geology on coastal landforms

Rock type and structure can have a significant influence on coastal landforms. More resistant rocks such as chalk and limestone are eroded more slowly and often produce spectacular cliffs and headland features. Weaker rocks, such as clay and sands, are frequently weakened by heavy rainfall and consequently eroded more easily, resulting in a lower cliff profile with cliff slumping and mudslides. An example of the effect of geology on coastal landforms can be seen on the Dorset coast (Figure 1.51).

Figure 1.52 Lulworth Cove, Dorset.

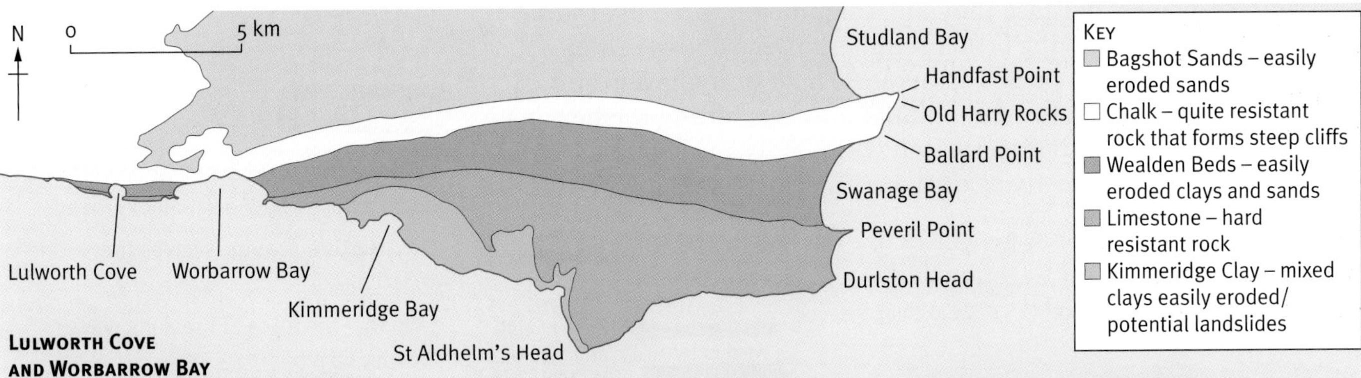

Lulworth Cove and Worbarrow Bay
The sea has broken through the narrow coastal strip of hard Portland limestone and been able to erode the softer Wealden beds until it reached the harder chalk. This resulted in the formation of the circular cove at Lulworth and a wider bay at Worbarrow.

Kimmeridge Bay to St Aldhelm's Head
The soft Kimmeridge Clay cliffs have been worn away by the effects of weathering and marine erosion, resulting in a slumped coastline with mudslides.

Chalk headland at Handfast Point
The area of chalk between Ballard Point and Handfast Point has been eroded more slowly than the Bagshot Sands at Studland Bay and the Wealden Beds at Swanage Bay. The result is the spectacular chalk headland, including 'Old Harry Rocks'.

Figure 1.51 The geology of the Dorset coast.

Figure 1.53 Chalk headlands – Old Harry Rocks.

ACTIVITIES

1. Explain what is meant by 'the coastal system'.
2. Construct a table with the heading 'Coastal processes'. List the processes of erosion and weathering down the left-hand side of your table and write a brief description of each.
3. When walking along a shingle beach in front of a cliff, what might you look for in order to identify the coastal processes taking place?
4. Using the Dorset coast as an example, explain how geology can affect the rates of erosion and weathering.

PLENARY ACTIVITY

Can you think of ways that human activity may increase the rate of coastal erosion?

What landforms are associated with destructive coastlines?

As waves approach a shoreline they pick up sediment. Breaking waves force water and sediment up a beach as **swash**. When the wave has reached its furthest point, it moves back down the beach as **backwash**.

Waves that remove sediment from a beach are referred to as **destructive waves**, because they are a significant force in eroding the coastline and removing beach sediment. Destructive waves break with tremendous power, generating a lot of hydraulic and abrasive erosion. The force and volume of water hitting the beach creates a large backwash effect, often completely removing a beach so that waves break directly against a cliff face (Figure 1.54).

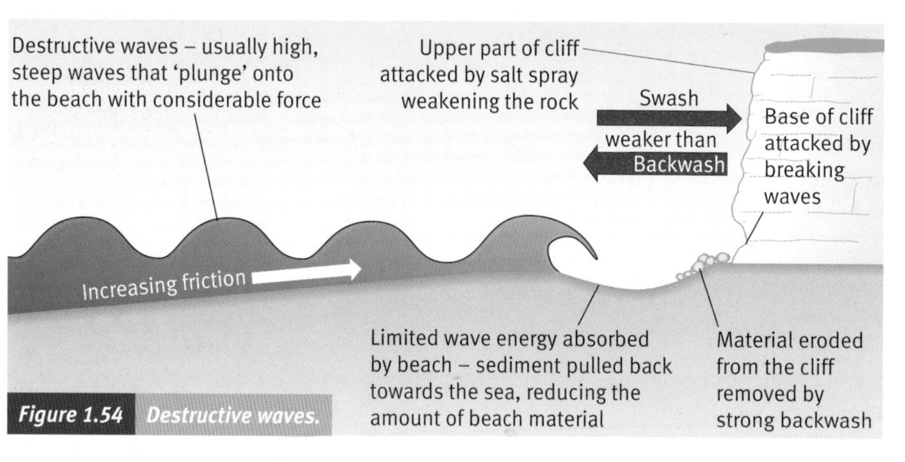

Destructive waves – usually high, steep waves that 'plunge' onto the beach with considerable force
Upper part of cliff attacked by salt spray weakening the rock
Swash
weaker than
Backwash
Base of cliff attacked by breaking waves
Increasing friction
Limited wave energy absorbed by beach – sediment pulled back towards the sea, reducing the amount of beach material
Material eroded from the cliff removed by strong backwash

Figure 1.54 *Destructive waves.*

GET STARTED

Discuss with a partner why some coastlines are described as 'destructive'.

KEY TERMS

Backwash – the movement of water down a beach by the action of gravity.

Destructive waves – waves that erode coastlines.

Swash – the force of breaking waves moving up a beach.

Destructive waves illustrate:

- the power of the waves to remove beach material and attack the base of cliffs
- the importance of a beach in absorbing wave energy and protecting cliffs from wave attack.

Landforms associated with resistant rocks – headlands

Where destructive waves attack more resistant coastlines, very distinctive features may be seen. This is especially noticeable on headlands that can be attacked by the sea on both sides. The waves gradually erode the base of the headland, picking out areas of weakness. Over a long period of time, a number of particular features may be formed (Figures 1.55 and 1.56).

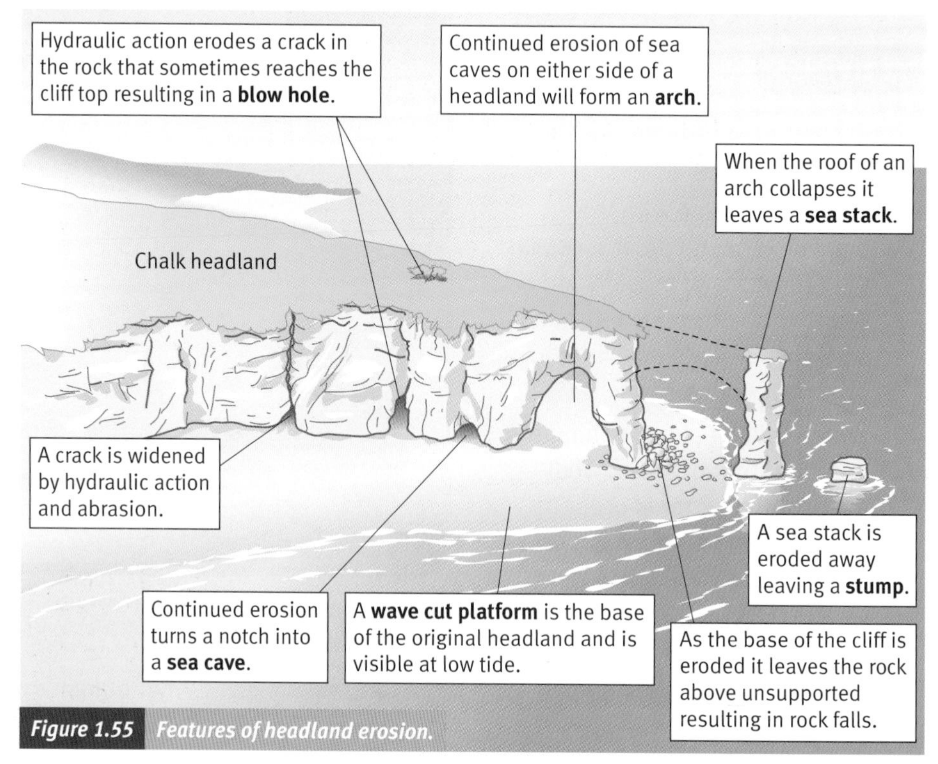

Hydraulic action erodes a crack in the rock that sometimes reaches the cliff top resulting in a **blow hole**.
Continued erosion of sea caves on either side of a headland will form an **arch**.
When the roof of an arch collapses it leaves a **sea stack**.
Chalk headland
A crack is widened by hydraulic action and abrasion.
A sea stack is eroded away leaving a **stump**.
Continued erosion turns a notch into a **sea cave**.
A **wave cut platform** is the base of the original headland and is visible at low tide.
As the base of the cliff is eroded it leaves the rock above unsupported resulting in rock falls.

Figure 1.55 *Features of headland erosion.*

Figure 1.56 Chalk headland.

Landforms associated with less resistant rock – landslides

All slopes are under stress because of the force of gravity. When the force acting on a slope (erosion/weathering) becomes greater than the strength of the rocks that it is made of, there will be a sudden movement, resulting in a landslide.

This often happens with softer, less resistant rocks such as sand and clay. Heavy rainfall may make less resistant rocks unstable, and where destructive waves erode the base of unstable cliffs a landslide may result.

The following newspaper article describes a recent landslide on part of the Dorset coast, one of the most unstable stretches of coastline in the UK.

Dorset coast crumbles after heavy rainfall

The weakening of the clay, sandstone and limestone cliffs by storm waves and heavy rain created one of the largest landslides in the recent history of the UK. The area of the landslide, to the east of Lyme Regis, is famous for landslides and the fossil-hunting opportunities they create. A local fossil hunter, looking up at the exposed cliff created by the latest landslide said, 'It is a great day for us, we could find all sorts of fossils and artefacts over the next few months.'

Other people are less happy; one local resident described the event, 'There was a loud rumbling, like a thunderstorm, and the next minute huge boulders the size of cars started rolling down the cliff. It was lucky it was high tide and no one was on the beach. Today, my house is a little nearer to the edge of the cliff.'

Local householders are worried that any further slides of this scale will put their homes in danger.

Figure 1.57 Crumbling Dorset coast.

ACTIVITIES

1. a Draw a sketch of the landscape in Figure 1.56.
   b Annotate your sketch with the names of four features of erosion.
2. Describe and explain the process of headland erosion.
3. Why may people have different views about whether coastlines ought to be protected from landslide events?

PLENARY ACTIVITY

'Landforms linked to hard rocks are caused by erosion, landforms linked to softer rocks are caused by weathering and erosion.' Explain this statement.

Why are soft coastlines vulnerable to rapid erosion?

GET STARTED

The risks of living near a soft coastline are well known to geographers. Why do people buy houses in vulnerable coastal areas?

Many parts of the UK have soft coastlines made up of clays, gravels or sand dunes. Clays and gravels are not well **consolidated** and become increasingly unstable when wet. Where waves break at the base of softer cliffs the combination of heavy rainfall and wave attack can cause rapid retreat of the cliffs, often putting houses and other buildings in danger (Figure 1.58).

Figure 1.58 *The retreat of soft coastlines.*

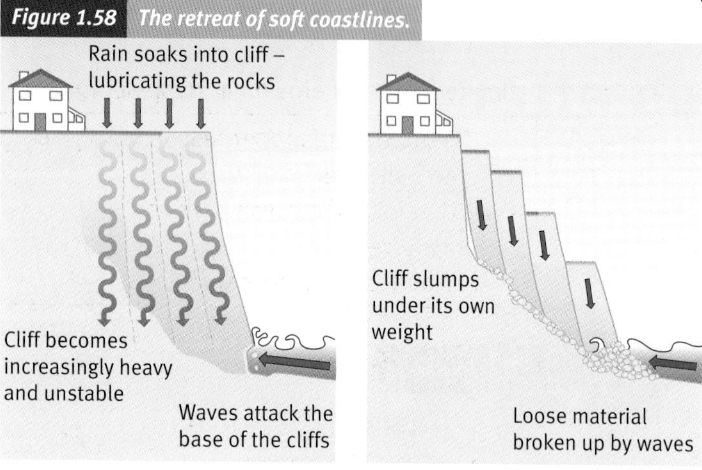

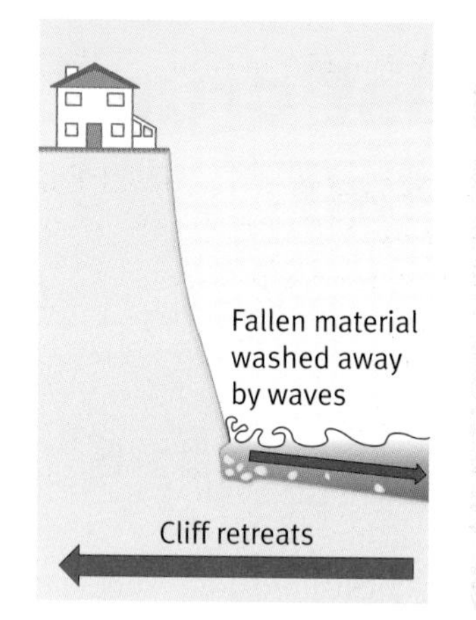

Fact file

- About 10 million tonnes of gravel are **dredged** each year from the seabed in the North Norfolk area. It is used in the construction industry.

The North Norfolk coast – a soft coastline

Much of the North Norfolk coast is made up of soft rocks that have very little strength, and are easily attacked by the storm waves generated by the strong, north-easterly winds.

Figure 1.59 *Storm waves attacking a soft coastline.*

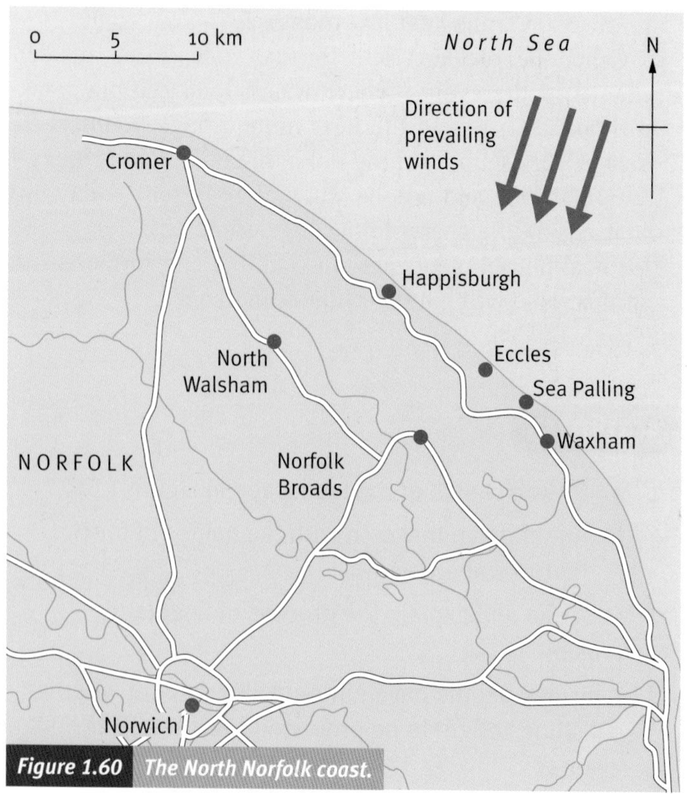

Figure 1.60 *The North Norfolk coast.*

Happisburgh: the problem of rapid erosion

The coastline at Happisburgh (Figure 1.61) is made up of soft clays and gravels. After heavy rainfall, the cliffs become saturated and very unstable. Waves attack the base of the cliffs, triggering landslides and rock falls. A single storm can erode hundreds of tonnes of cliff material. It is estimated that the cliffs are retreating by an average of 12 m a year; 30 properties have already been lost to the sea.

Timber defences were built in the 1950s and 1960s, but over the years they have been broken down by the powerful North Sea waves, and the sea now reaches the cliffs at high tide. In 2002, the North Norfolk District Council provided the money for 4000 tonnes of rock to be put to be put along the bottom of the cliff as a temporary measure against further erosion. This has been extended using £50,000 from the 'buy a rock for Happisburgh' local appeal.

Decisions about which coastal areas to protect from wave attack are often taken at national level. It has been decided that Happisburgh will not get any more money for coastal protection. The result of this decision will be further loss of land and property. This decision was taken because it was felt that Happisburgh was not economically valuable enough to justify the cost of coastal defences.

Figure 1.61 *Homes under threat at Happisburgh.*

Figure 1.62 *Old timber defences and recent rock erosion at Happisburgh.*

Table 1.3 Estimated effect of erosion at Happisburgh

By 2055	• Loss of 20 more properties, including 18 listed buildings • Further loss of caravan park and agricultural land • Loss of access to the beach

KEY TERMS

Consolidated rock – rock with strong structure that is hard to break down.

Dredging – taking sediment from river or sea bed.

ACTIVITIES

1. Use annotated diagrams to explain 'cliff retreat'.
2. Describe the effect of wave energy on the wooden sea defences and cliffs at Happisburgh.
3. A resident of Happisburgh said, 'Unless the village is protected, the community will die'. Explain this statement.
4. It was decided to build coastal defences at Sea Palling, a village 6 km south-east of Happisburgh. Why may this decision have caused conflict in the local area?

PLENARY ACTIVITY

Suggest reasons why some coastal areas are protected, whilst others are not.

How are coastal landforms created by deposition?

GET STARTED

How can you tell that sediment on a beach is moved as waves break?

As waves approach the shore, they pick up sediment. The force of the breaking waves moves sediment up the beach as swash. Movement of waves up the beach stops when the wave energy has been used. At this point, the water flows back down the beach as backwash. If the backwash does not have enough power to move the sediment back down the beach, ridges of material will form along the beach at the point of highest swash. These ridges are called berms.

Waves that deposit sediment are referred to as **constructive waves** because they help to build up beach material (Figure 1.63), often producing wide, flat beaches that are so important for the coastal tourism industry.

Where waves approach the coastline parallel to the beach, the swash and backwash move material up and down the beach, creating an even profile along the beach.

In many coastal areas in the UK, waves approach the beach at an angle and this moves material along the beach by the action of **longshore drift**. (Figure 1.64)

In some coastal areas, the movement of material along the beach is slowed down by the building of **groynes** (Figure 1.65). These trap sediment on the drift side and give the beach an uneven profile.

THINK ABOUT IT

How many ways can a beach help to protect coastal cliffs from marine erosion?

KEY TERMS

Constructive waves – waves that build up beach material to create landforms.

Groyne – wooden or concrete construction built across a beach.

Longshore drift – the movement of material along a coastline by the action of the waves.

Prevailing wind – most frequent wind direction.

Figure 1.63 Constructive waves.

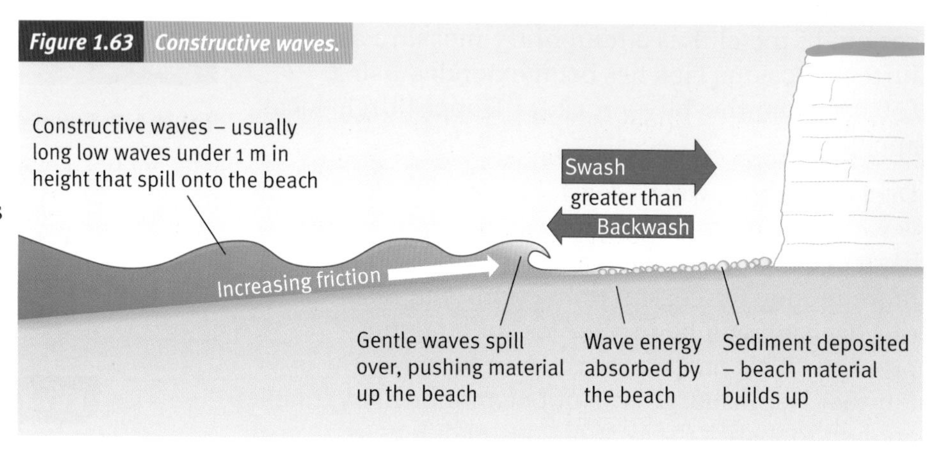
Constructive waves – usually long low waves under 1 m in height that spill onto the beach
Swash
greater than
Backwash
Increasing friction
Gentle waves spill over, pushing material up the beach
Wave energy absorbed by the beach
Sediment deposited – beach material builds up

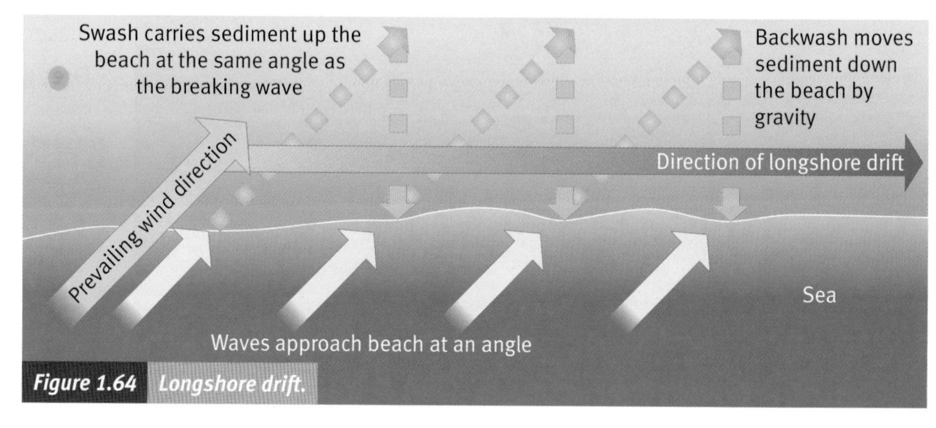
Swash carries sediment up the beach at the same angle as the breaking wave
Backwash moves sediment down the beach by gravity
Direction of longshore drift
Prevailing wind direction
Sea
Waves approach beach at an angle

Figure 1.64 Longshore drift.

Figure 1.65 Groynes.

Figure 1.66 Hurst Castle Spit.

The formation of spits and bars

Spits and bars are ridges of sand or shingle that have been formed from material deposited by breaking waves. They usually develop where:

- large amounts of sediment are being moved along the coast by longshore drift
- there is a sudden change in the direction of the coastline (in the case of spits)
- the sea is shallow enough to allow deposited material to reach the surface.

Example: Hurst Castle Spit, Hampshire

Beach material is moved along the Hampshire coast by the action of longshore drift. When it reaches Milford-on-Sea, there is a sudden change in the shape of the coastline. At this point, the material has been deposited in the same direction as the original coastline, forming Hurst Castle Spit – a shingle spit nearly 2 km long. The seaward end of the spit has been shaped by wave action and ocean currents, giving it a smooth, curved profile.

ACTIVITIES

1. Explain the term 'constructive coastline'.
2. Explain the difference in energy between swash and backwash.
3. Use an annotated diagram to explain how longshore drift moves material along the beach.
4. Describe and explain the features of Hurst Castle Spit.
5. Why might constructive coasts be difficult to manage?

Tombolos

Where deposited material joins two land areas, the resulting feature is called a tombolo. An example can be seen at Chesil Beach, which joins the mainland of Dorset with the Isle of Portland.

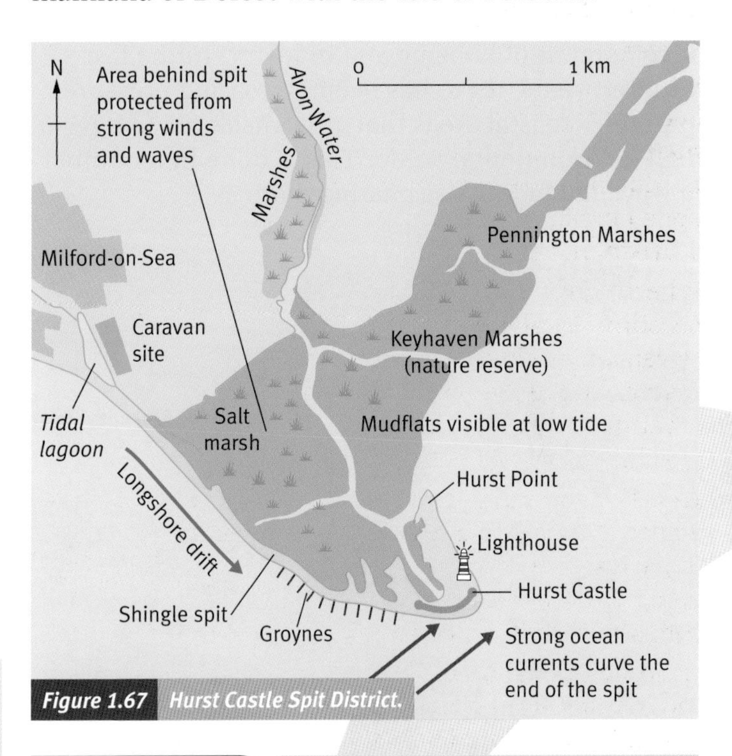

Figure 1.67 Hurst Castle Spit District.

RESEARCH LINK

Find out more about Hurst Castle Spit on the New Forest Hurst Castle Spit District Council website.

PLENARY ACTIVITY

Why are areas of constructive coastline often used for recreation and leisure activities?

CASE STUDY Blakeney Point, Norfolk

GET STARTED

Discuss with a partner why coastal areas might be valuable environments.

Why are constructive coastal areas valuable environments?

Coastlines with sand or shingle spits and bars are often distinctive landscapes. The landward side of a spit or bar is protected from strong winds and tides, and is an area where fine **silt** is deposited by rivers as they reach the sea. As levels of silt increase they become colonised by plants, creating **salt marshes**, partially submerged at high tide and visible at low tide. These environments provide important breeding grounds for migrating birds and often contain rare plant species.

In many areas, salt-marsh environments have been lost as the land has been drained for the development of housing and industry. In the last hundred years, there has been increasing pressure to protect coastal areas that are considered to have a high environmental value. This is done in a number of ways, including designating them as:

KEY TERMS

Nature reserve – area set aside to preserve plants and animals.

Salt marsh – an area of tidal mudflats, partially flooded at high tide.

Silt – very fine sediment deposited by flowing water.

- **nature reserves:** areas set aside for the purpose of protecting rare plants and animals, and preserving the environment
- **Areas of Outstanding Natural Beauty (AONB):** areas controlled by planning authorities where development is restricted
- **Sites of Special Scientific Interest (SSSIs):** areas designated by Natural England as having special features that must be protected.

Many protected coastal areas are used by the public for recreation and education activities. The following example looks at Blakeney Point National Nature Reserve in North Norfolk, which is managed by the National Trust. It is one of the most visited nature reserves in the UK.

Figure 1.68 *Blakeney Point.*

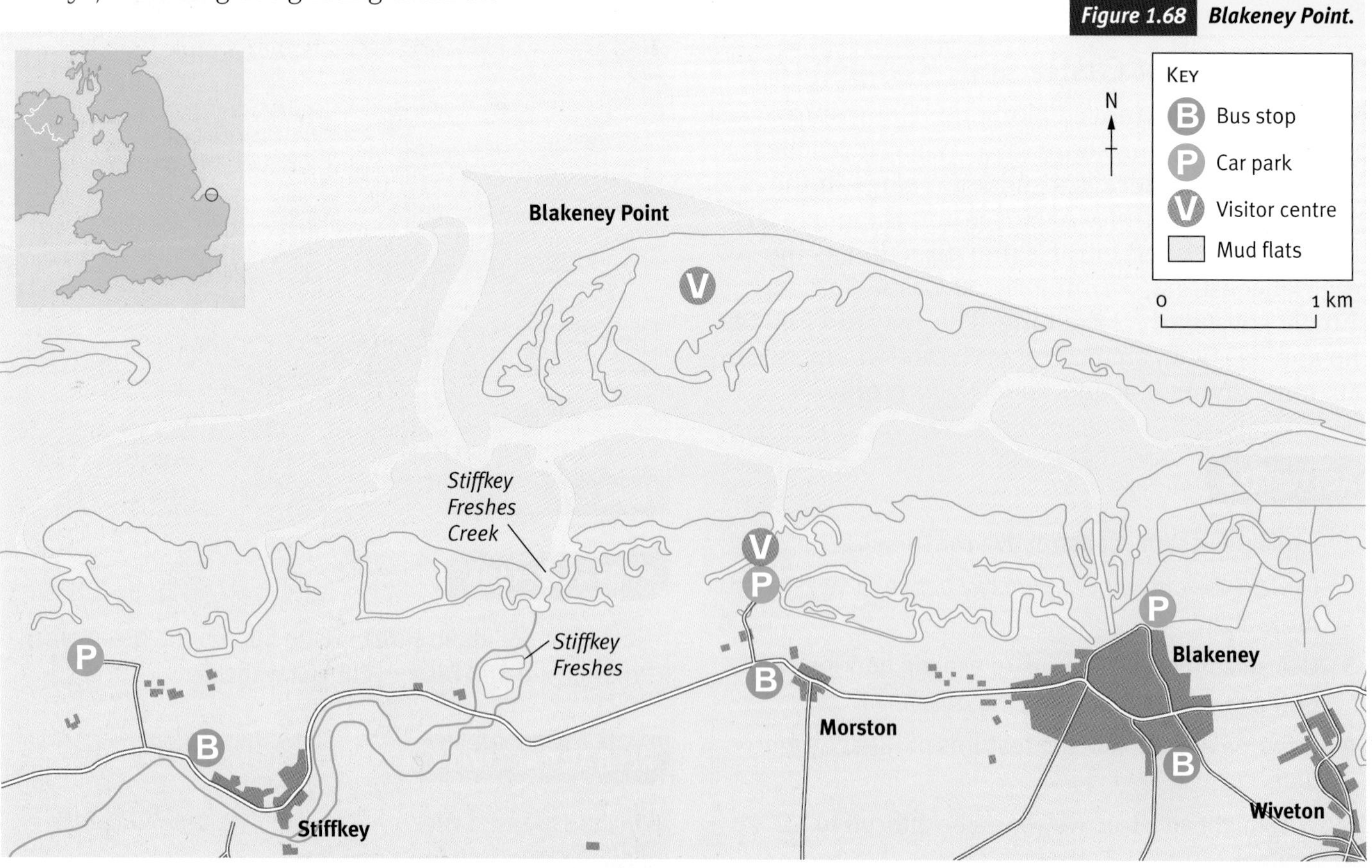

What is Blakeney Point?

Blakeney Point is a sand and shingle ridge that extends westwards from the Norfolk coast for approximately 5 km. It is separated from the coast by sheltered tidal waters, where fine sediment is deposited by rivers, helping to develop a rare salt marsh environment.

Figure 1.69 **Blakeney Point visitor centre.**

THINK ABOUT IT

The idea of preserving environments and protecting wildlife only really began in the 19th century. Why do you think this was?

Why is Blakeney Point a rare environment?

Blakeney Point is one of the largest unspoilt coastlines in Europe, and has a number of special environmental features including:

- areas of rare habitats including salt marsh, sand dunes and vegetated shingle
- an international breeding area for seabird colonies including arctic terns, oyster catchers and brent geese
- a breeding area for grey and common seals
- rare plant species including sea lavender and sea campion
- uninterrupted coastal views and long stretches of sandy beach.

ACTIVITIES

1. a What is the direction of longshore drift at Blakeney Point? Explain your answer.
   b Why is the northern side smoother than the southern side of the spit?
   c Explain why salt marshes form on the landward side of spits.
2. a Explain why salt marsh and sand dune coasts are often seen as:
   - valuable environments
   - valuable economic resources.

   b Why may the environmental and economic pressures cause conflict in these areas?

PLENARY ACTIVITY

Why are salt marsh and sand dune environments increasingly rare?

Figure 1.70 **Blakeney Point spit.**

How does the coastline provide a natural defence against flooding?

GET STARTED

Why is the coastline called 'the frontier between land and sea'?

The coast has been called 'the frontier between the land and the sea'. With increasing numbers of people living near the coast (Figure 1.72), and global warming causing changes to sea levels, the coastline will become even more important in the challenge to reduce the risk of flooding in low-lying areas.

The coastline has always provided an important buffer zone between the sea and the surrounding lowlands. Coastal salt marshes and **mangrove** forests provide a natural barrier against storms, absorbing wave energy and providing a barrier against inland flooding during high tides. With rising sea levels the sustainable management of coastal vegetation will be an increasingly important part of protecting coastal areas against flooding.

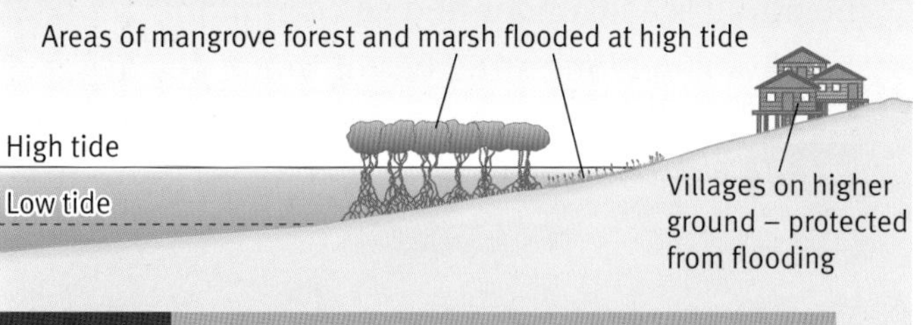

Figure 1.71 Using the natural environment as a flood defence.

Figure 1.72 Coastal development.

The pressure to develop coastal areas has meant that in many parts of the world salt marshes are being drained and coastal mangrove forests removed, often for tourism developments, as the following example of the Bimini Bay Resort in the Bahamas shows:

TOURIST RESORT DEVELOPMENT ON BIMINI ISLAND

The multimillion dollar Bimini Bay Resort and casino in the Bahamas is threatening both local communities and their environment. The project, which once built, will include a golf course, marinas and apartments has outraged local community members who have staged a protest outside the development to stop the construction. The Miami-based owner of the development has promised, amongst other things, a primary school, fire truck, and that the project would be friendly to the environment.

Some of the problems include:

- A gate has been erected denying local people access to 7 km of the 10-km-long island. This is said to leave only 3 km of land (only 500 m wide) for a population of 1600.
- Water supplies are being used for the tourism development resulting in water to local communities being frequently turned off.
- Mangroves have been bulldozed, land has been carved up, the seafloor dug and destroyed and the North Bimini lagoon has been silted with waste material.

Recent hurricanes and floods have illustrated that mangroves are the best way to protect vulnerable coastlines. The government's decision to permit a developer to destroy Bimini island's protective mangroves and replace it with a tourist resort puts the island in jeopardy. The habitats of dolphins, turtles and sharks are seriously threatened.

(Source: Tourism Concern website)

How does removing coastal wetlands increase the risk of flooding?

In a number of areas in the UK, salt marshes have been drained to provide farmland or space for building development. This drained land is then protected by a sea wall. At high tide the sea will often reach the sea wall, especially during storms, leaving no beach or marshland to absorb wave energy. In this situation the sea wall is under constant attack by waves, and will eventually fail if not regularly maintained. If the sea wall is breached by the storm waves, large areas of land will be flooded (Figure 1.73).

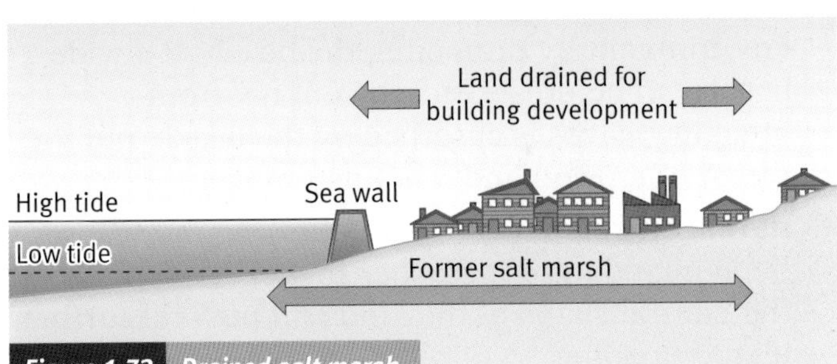

Figure 1.73 *Drained salt marsh.*

In extreme cases, it has been suggested that removing coastal vegetation can increase the effect of natural disasters. In 2004, an underwater earthquake in the Indian Ocean caused a series of tidal waves (**tsunami**) that moved towards the surrounding coastal areas. In a number of these areas, coastal mangrove forests had been removed to allow development as tourist resorts, leaving the coast open to devastating waves (Figure 1.74).

Since the tsunami, mangrove forests have been replanted in some areas as part of coastal protection programmes (Figure 1.75).

KEY TERMS

Mangroves – evergreen trees and shrubs growing in tropical coastal areas, whose roots trap sediment and aid beach development.

Tsunami – tidal wave caused by an underwater earthquake.

ACTIVITIES

1. Explain how the natural environment can help to protect areas from flooding.
2. Why is it increasingly important to manage coastal areas from the effects of storms?
3. 'The development of coastal areas is often a balance between the economy and the environment'. Explain this statement.

Figure 1.74 *The Asian tsunami, 2004.*

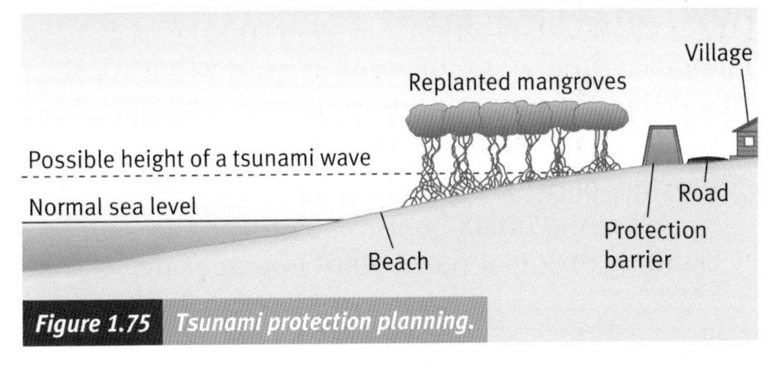

Figure 1.75 *Tsunami protection planning.*

Fact file

The 2004 Asian tsunami:

- killed over 1.3 million people
- destroyed 1 million homes
- had waves that travelled at 48.3 kph near the land and were between 3 and 9 m high.

THINK ABOUT IT

Worldwide, over 4 million people live and work in coastal areas and the number is growing rapidly. Why do people want to live on coasts?

PLENARY ACTIVITY

Suggest why managing the natural coastal environment is an important part of any sustainable coastal management strategy.

Why is there an increasing need to protect coastal areas?

GET STARTED

What types of coastal defences have you seen?

The increasing development of coastal areas has meant that a growing number of people are at risk from coastal flooding. This risk is likely to increase in the future, as climate change brings about rising sea levels and an increasing number of coastal storms. Consequently there is a need to protect coastal areas from the effects of erosion and inland flooding by putting in place 'coastal management strategies'.

How can coastal areas be protected?

There are three main methods of protecting coastal areas from the effects of erosion and flooding:

- **hard engineering:** controls the sea by building barriers between the sea and the land, or forces the waves to break before they reach the land by building artificial reefs. Hard engineering involves building large concrete structures that create a totally artificial environment and require ongoing maintenance. Figure 1.76 shows examples of hard engineering techniques
- **soft engineering:** works with the natural environment by protecting the beach. If a wide beach can be preserved, the energy from breaking waves is absorbed by the beach and the threat of erosion and flooding reduced
- **managed retreat:** involves allowing the sea to flood inland until it reaches higher land, where the natural shape of the land will prevent further inland flooding.

What is meant by 'sustainable management'?

Sustainable coastal management means that a coastline is managed in the long term without significant change to the environment or large ongoing maintenance costs.

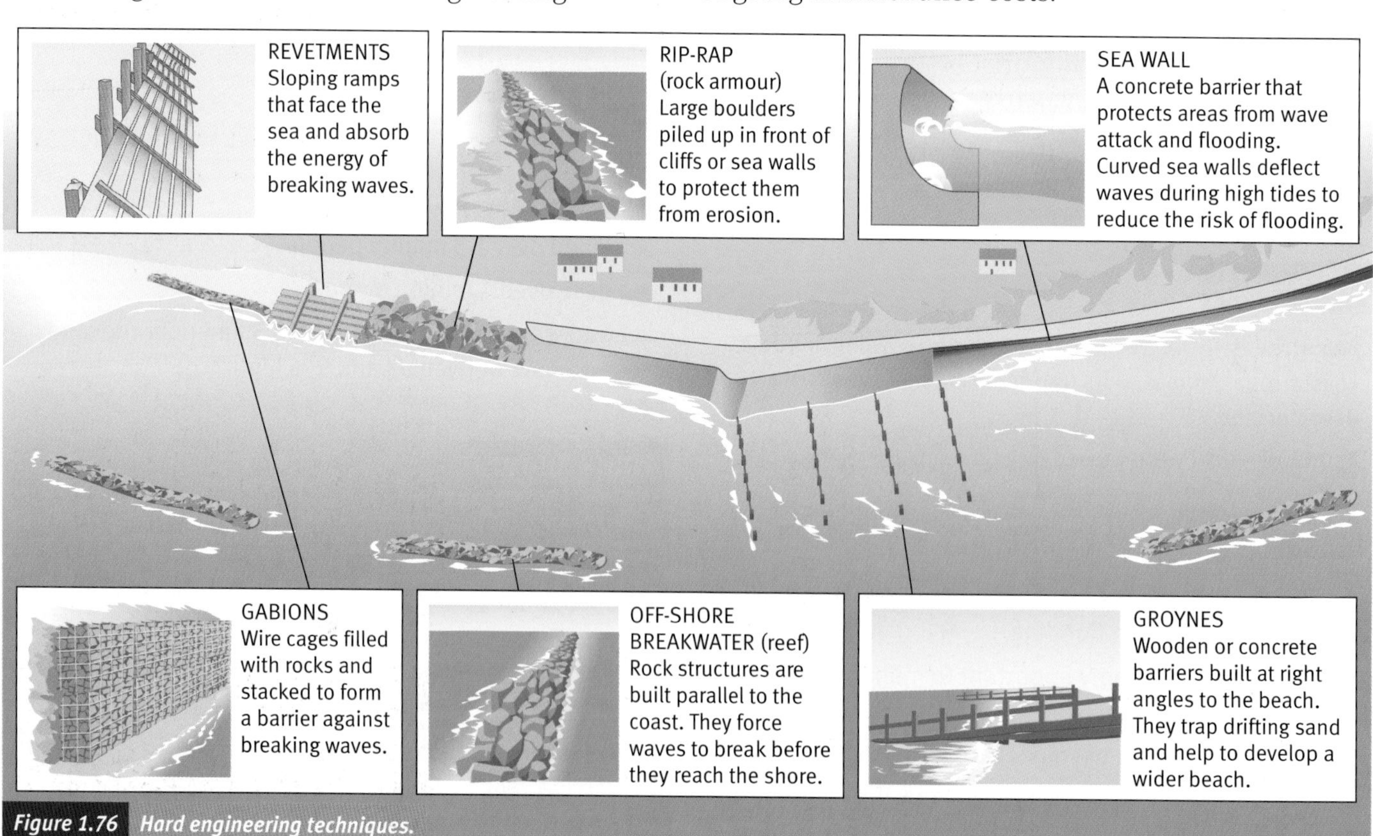

REVETMENTS
Sloping ramps that face the sea and absorb the energy of breaking waves.
RIP-RAP
(rock armour)
Large boulders piled up in front of cliffs or sea walls to protect them from erosion.
SEA WALL
A concrete barrier that protects areas from wave attack and flooding. Curved sea walls deflect waves during high tides to reduce the risk of flooding.
GABIONS
Wire cages filled with rocks and stacked to form a barrier against breaking waves.
OFF-SHORE BREAKWATER (reef)
Rock structures are built parallel to the coast. They force waves to break before they reach the shore.
GROYNES
Wooden or concrete barriers built at right angles to the beach. They trap drifting sand and help to develop a wider beach.

Figure 1.76 Hard engineering techniques.

West Bay Coastal Defence Scheme, Dorset

West Bay is a coastal settlement in Dorset, built around a small harbour. The pier and sea defences at West Bay had suffered considerable storm damage over many years, and the town had been affected by serious flooding on a number of occasions. In 2001, a study was carried out to assess the value of the properties at the risk of flooding and calculate the lifespan of the existing coastal defences. It was estimated that a serious flood would cause millions of pounds worth of damage. Engineers thought that the existing defences were in such a poor state that they had a 50 per cent chance of failure within 5 years.

Figure 1.78 *Hard engineering.*

Figure 1.77 *The West Bay Coastal Defence Scheme.* (Source: West Dorset District Council and Hilary Bosworth)

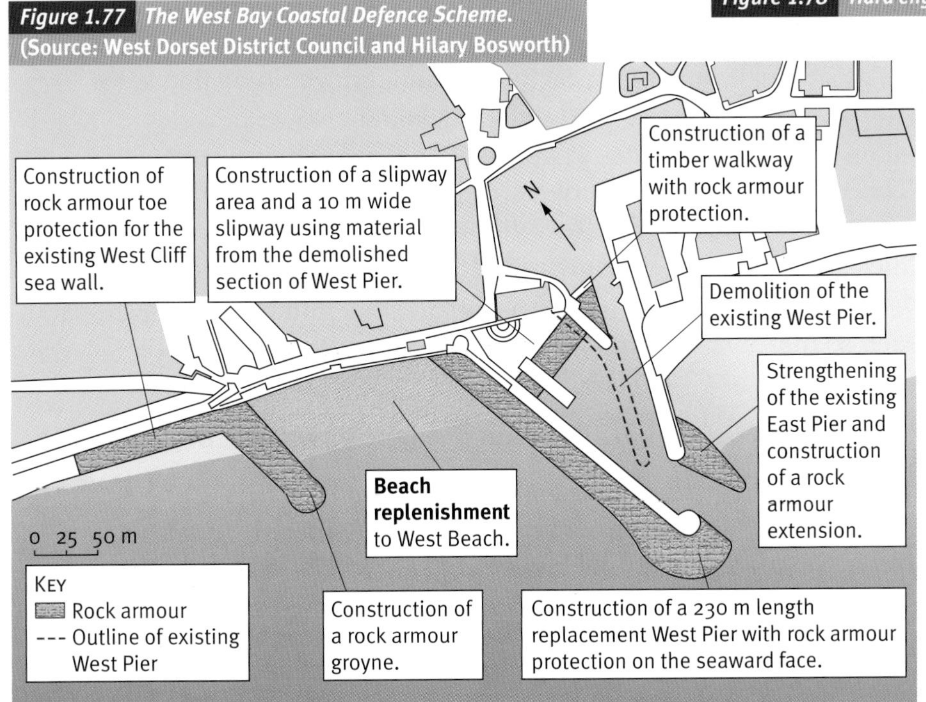

Construction of rock armour toe protection for the existing West Cliff sea wall.
Construction of a slipway area and a 10 m wide slipway using material from the demolished section of West Pier.
Construction of a timber walkway with rock armour protection.
N
Demolition of the existing West Pier.
Strengthening of the existing East Pier and construction of a rock armour extension.
Beach replenishment to West Beach.
0 25 50 m
KEY
Rock armour
--- Outline of existing West Pier
Construction of a rock armour groyne.
Construction of a 230 m length replacement West Pier with rock armour protection on the seaward face.

It was decided that a new hard engineering coastal defence scheme should be built. The aims of the new scheme were to:

- provide flood protection for homes, businesses and a caravan site
- improve the existing harbour for commercial and pleasure boats
- develop the beach to the west of the harbour
- prepare for higher sea levels.

The coastal defence scheme can be seen in Figure 1.77. It was completed in 2005 at a cost of £17 million.

ACTIVITIES

1 Explain how:
   a hard engineering tries to control coastal processes
   b soft engineering tries to work with coastal processes.
2 Why is hard engineering sometimes seen as unsustainable?
3 a Local people in West Bay were consulted about how the area should be protected and what facilities should be included in the scheme. Why is it important to get the views of local people when coastal defence schemes are being considered?
   b Describe the main 'hard engineering' features of the West Bay scheme.
   c Why was a 'hard engineering' scheme chosen for West Bay?

KEY TERMS

Beach replenishment – building up the beach by pumping sand or shingle onto it.

RESEARCH LINK

Find out more about the West Bay defence scheme at the Dorset County Council website.

PLENARY ACTIVITY

Why is hard engineering the only coastal protection option in some areas?

GET STARTED

What would make a coastal defence scheme sustainable?

Sustainable coastal management: how do soft engineering methods work with the natural environment?

Soft engineering methods try to work with the natural physical processes in an area, rather than building large concrete barriers or using massive boulders to protect the land from wave attack.

One of the most successful ways of reducing the threat from storm waves is by developing and preserving a wide, gently sloping beach that can absorb most of the energy from breaking waves. Soft engineering methods are usually based on preserving and managing the beach. This can be done in a number of ways, including:

- **beach replenishment:** replacing the sand or shingle that has been removed by longshore drift
- **beach reprofiling:** shaping the beach so that it absorbs more energy during storms
- **fencing/hedging:** building fences or planting salt-resistant bushes at the back of the beach to reduce the amount of sand being blown inland by strong winds
- **planting vegetation:** planting grasses or bushes in low-lying sandy areas to stabilise the beach material.

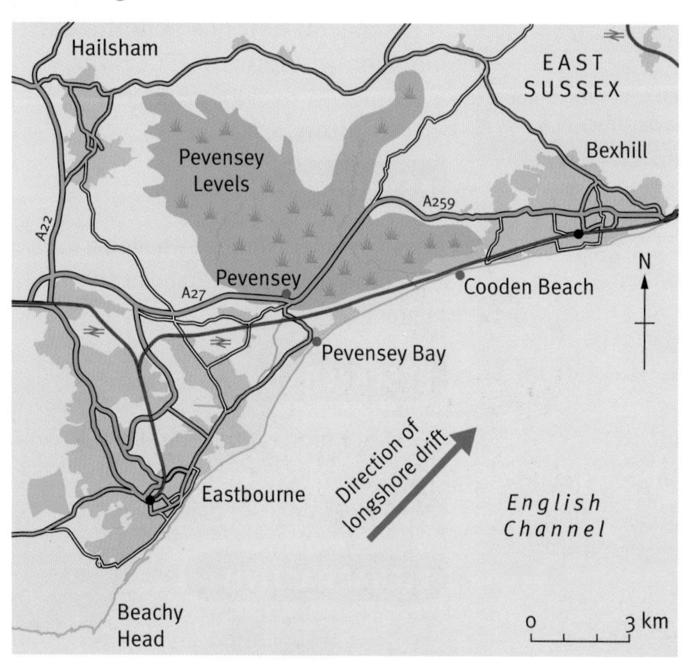

Figure 1.79 Pevensey Bay.

The following example shows how soft engineering methods are being used in East Sussex.

The Pevensey Bay Sea Defence Scheme

Pevensey Bay is a low-lying area in East Sussex that is vulnerable to coastal flooding (Figure 1.79). The use of heavy engineering methods to protect the area would not be appropriate because of the scenic and environmental value of the area.

Why does the area need protection?

- Longshore drift is removing increasing amounts of beach material.
- Over 50 km^2 of flat land would be flooded if coastal defences failed.
- The Pevensey Levels, an environmentally sensitive area (SSSI) would be flooded with salt water if coastal defences failed.
- There are over 10,000 properties in the area.
- It is a tourist area with a number of caravan parks.
- The main coast road and rail links run along the coast.

What has been done at Pevensey Bay?

Beach replenishment

Longshore drift in the area means that beach material is constantly being lost. This material is replaced by dredging it from the seabed and then spraying it back onto the beach. This is carried out by a specially adapted boat, which can come very close to the beach at high tide. This operation is carried out during the summer months when there are fewer storms.

Beach reprofiling

During the winter storms, destructive waves move the beach sediment down the beach towards the sea. This makes the upper beach levels very low and liable to wave attack. Bulldozers are used to move the material back up the beach and reshape the beach so that it has an even, gentle slope.

Figure 1.80 Beach replenishment.

Figure 1.81 Beach reprofiling.

Beach recycling

The natural processes of longshore drift at Pevensey Bay act like a conveyer belt. Beach material is moved from west to east, resulting in the western end of the beach becoming increasingly depleted. Three times a year, large trucks are used to move the beach material that has accumulated in the east back to the western end of the beach. This ensures that the beach has an even profile and no particular point is subject to wave attack.

Figure 1.82 Beach after recycling.

Why is the scheme seen as sustainable?

- It works with the natural environment.
- It does not involve massive building costs.
- It does not damage the environment.
- The resulting beach has amenity value for the local people and is an important tourist attraction.

However, the scheme does require ongoing maintenance, which is expensive.

ACTIVITIES

1. Why is there a need for coastal management at Pevensey Bay?
2. a Explain what is meant by the following:
   - beach replenishment
   - beach reprofiling
   - beach recycling.

   b 'The beach is the main protection against storms.' Explain this statement.
3. a Explain how the Pevensey Bay defence scheme is good for both the local environment and economy.

   b Is the scheme at Pevensey Bay truly sustainable? Write an extended answer to give both sides of the argument.

RESEARCH LINK

Find out more at the Pevensey Bay website.

PLENARY ACTIVITY

Why may hard engineering methods not be appropriate for some coastal areas?

Managed retreat: a sustainable way of managing the coast?

GET STARTED

Discuss with a partner how the natural landscape could be used to reduce the risk of flooding.

Managed retreat, or managed realignment, is when the sea is allowed to reach its natural position by removing existing coastal or estuary defences. The natural shape of the land then protects inland areas from flooding.

In many areas, coastal defences are reaching the end of their life and decisions have to be taken about whether they are left to fall into disrepair or reinstated. It is not financially possible to protect every coastline that is at risk of flooding, so where the natural landscape can be used to reduce flood risks, managed retreat is seen as a viable option.

How does managed retreat work?

The following diagrams show how an area previously managed through hard engineering may be realigned to use the natural landscape to protect it from flooding.

What are the advantages of managed retreat?

The method is seen as **sustainable management** because:

- it will increase the amount of salt marsh
- well-established salt marsh environments provide a natural defence against erosion and flooding
- it looks more attractive than hard engineering methods
- it can provide **amenity value**, such as nature reserves
- it does not cost much to maintain
- it is especially suitable in areas that are important for nature conservation and bird migration
- it allows farming and nature conservation to work together.

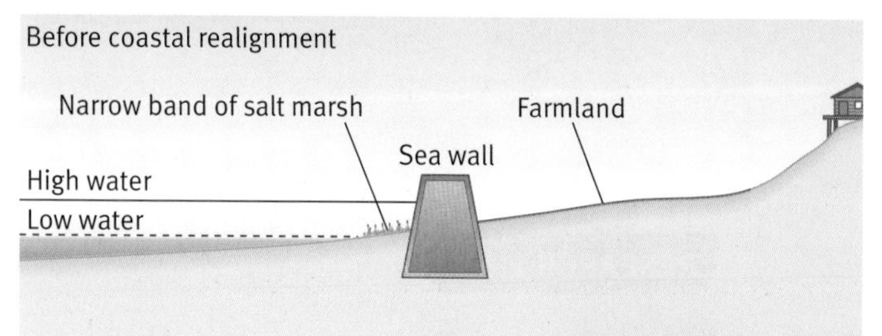

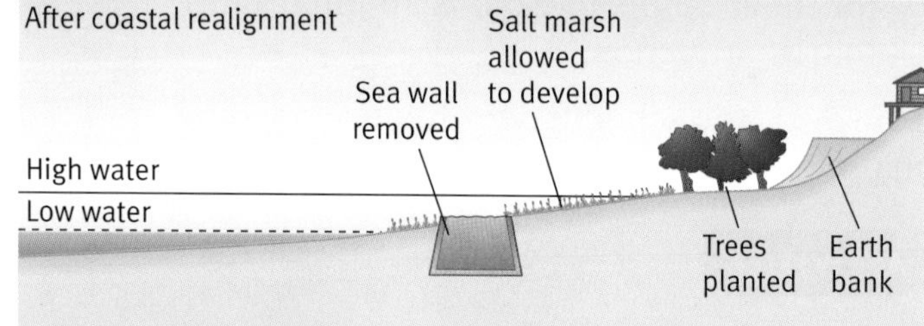

Figure 1.83 *Before and after coastal realignment.*

Fact file

Before carrying out a managed retreat scheme, the following questions have to be considered:

- What area of land will be flooded?
- Will any buildings or infrastructure be affected?
- How will the environment be affected?
- How much will it cost to complete and maintain?

RESEARCH LINK

Find out more about managed retreat at the ThamesWeb website.

KEY TERMS

Sustainable management – long-term management that does not harm people or environments.

CASE STUDY Cuckmere Estuary, East Sussex

Is managed retreat the best strategy to prevent coastal flooding?

The Environmental Agency has proposed a plan to allow the existing coastal and river defences to fall into disrepair so that the area reverts back to a salt marsh environment.

Reasons for the proposed plan

- The existing defences are coming to the end of their life.
- The cost of replacing the defences would be approximately £20 million.
- Current maintenance costs are over £50,000 a year.

How the plan will work

- Once the sea defences have been breached, the lower part of the estuary will be flooded and the natural shape of the land will prevent inland areas from flooding.
- An earth bank will be built at the northern end of the estuary to protect inland areas from flooding.
- Footpaths will be raised or moved to higher ground.
- Changes to the area will be monitored to ensure that no buildings are at risk of flooding.

KEY TERMS

Amenity value – used for recreational and leisure activities.

Managed retreat – allowing the sea to flood lowland areas.

ACTIVITIES

1. Use an annotated diagram to explain 'managed retreat'.
2. It is important to consider the four questions listed in the Fact file before carrying out a managed retreat scheme: why is this?
3. a Describe the main features of the Cuckmere Estuary Managed Retreat Scheme.
   b What are the advantages of the scheme?
   c Why may some people object to the scheme?

RESEARCH LINK

Find out more about the proposed plan at the BBC website.

Background information

- The Cuckmere Estuary and coast have been protected from flooding by a series of flood walls and artificial banks.
- The estuary attracts large numbers of visitors who come to enjoy walking the footpaths, bird watching, cycling and canoeing on the existing river meanders.

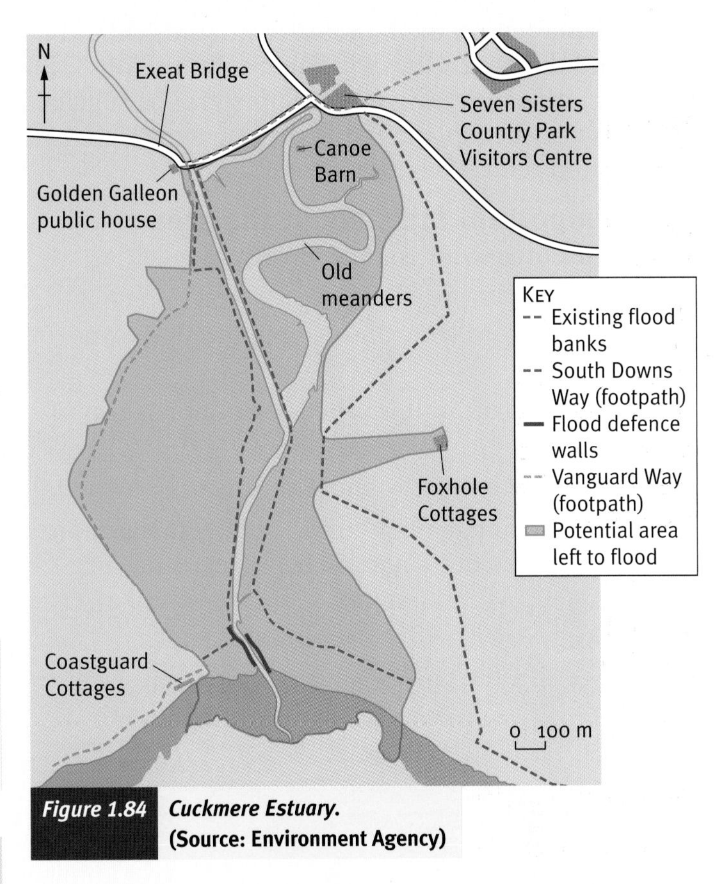

Figure 1.84 *Cuckmere Estuary.* (Source: Environment Agency)

What are the advantages of the plan?

- It will create over 200 acres of salt marsh, attracting increasing numbers of birds.
- Without artificial defences, the landscape will look more natural.
- The natural landscape will provide flood protection, especially as sea levels rise.
- It works with the environment and has low maintenance costs. This make it more sustainable.

PLENARY ACTIVITY

Discuss the view that 'Managed retreat works with the environment rather than trying to control it'.

A Decision Making Exercise – Coastal management: should Hunting Island be protected?

In this exercise, you will be asked to consider a coastal management scheme in South Carolina, USA. In recent years, the coast has suffered from severe erosion. The management scheme aims to protect an area called Hunting Island. Beach replenishment and the building of groynes should safeguard the area. However, the multi-million-dollar scheme is controversial. Some groups suggest that it is not sustainable. What do you think? Is managed retreat a better option?

Some important facts about the area

- A series of 'barrier islands' protect the mainland.
- The islands are constantly changing shape and size.
- Longshore drift in this area is from north-east to south-west.
- Many of the barrier islands are privately owned.
- Hunting Island is open to the public. It is a 'State Park' because of its natural beauty.
- In some years, 5 m of beach are lost through erosion.
- In August 2004, Hurricane Charley devastated the area: 17 m of beach disappeared within a week.

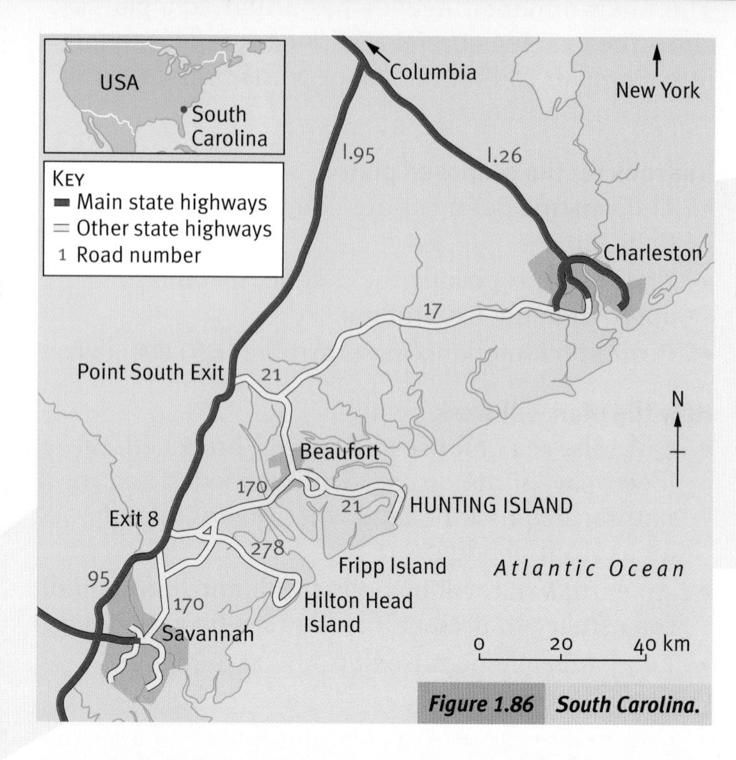

Figure 1.86 **South Carolina.**

So what is so special about Hunting Island?

The 'Friends of Hunting Island' website lists reasons why it should be protected:

- 5000 acres of outstanding beauty
- a unique maritime forest bordering the sea, but now threatened by erosion
- the beach is a breeding ground for rare turtles – recent beach loss prevents them from laying their eggs
- the forest and the sea lagoons are home to a rich variety of flora and fauna, including blue crabs, egrets and alligators
- in 2008, 1.2 million visitors came to the State Park, paying US $3 each to access it: US $3.6 million in total
- it is the only public access beach south of Beaufort. Without protection and beach replenishment, the area is under significant threat.

Figure 1.85 **Hunting island.**

Protecting the beach at Hunting Island

In 2005, a plan was devised to save Hunting Island from further erosion and to replenish the lost beach. The total cost of the scheme was estimated to be between US $8 million and US $13 million

Six 150-m stone-filled groynes would be built to capture and retain the sand along Hunting Island. They would be spaced 400 m apart and would stop the natural movement of sand through longshore drift.

Figure 1.87 **Beach protection.**

Dredged sand, sucked from the bottom of the sea 500 m offshore, would be transported to the beach. Bulldozers would be used to create a natural-looking beach profile. Regular maintenance would be required, but this would be carried out outside the turtle breeding season, to reduce the impact on the endangered species. To do nothing would mean that sand depletion will threaten the long-term survival of the turtle and the maritime forest.

YOUR TASK

Consider the evidence on these two pages and the additional information on the supporting CD. Do you think that the scheme to protect Hunting Island is sustainable? Is managed retreat a better option, despite the obvious consequences for Hunting Island?

To help in your planning, list the factors that *support* the development of the coastal protection scheme and list the factors that suggest it should *not go ahead*. Indicate if each is a Social (So), Economic (Ec) or Environmental (Env) factor. Now write a structured report outlining your views.

Figure 1.88 **Fripp Island is a private island containing luxury holiday parks and expensive apartments.**

A sustainable plan

Some experts warn that despite the probable loss of this area of outstanding beauty in the next few years, the coastline should be left to change naturally, in other words managed retreat is the only sustainable option.

The University Geologist Orrin Pilkey suggests that the area has the most unstable beach you can get anywhere. He suggests that by trapping the sand here, more erosion could be caused at Fripp Island to the south-west. It will be starved of its sand. He thinks managed retreat is the only option.

The National Oceanic and Atmospheric Administration predicts that sea levels will rise as a result of global warming. On the east coast of the USA, it is expected that the rise will be between 0.5 m and 2 m by 2100. In addition, it reports that changes in the atmosphere will cause an increase in hurricane activity.

Taxpayers for Commonsense is a pressure group in the USA that aims to reduce public spending. It campaigns on many issues, in this case on behalf of taxpayers who live far away from vulnerable coastlines.

It is particularly worried about the future cost of maintaining schemes like this.

Foundation:

Key Geographical Themes 8-mark case study question

Case study – The effects of river flooding.

i Name a river that has been affected by flooding.
ii Describe the effects of the flooding.
iii Explain the causes of the flooding.

[8 marks]

Mark scheme

Level 1: The student gives a basic description of effects and causes, but does not develop their answer at all and only shows limited relevant knowledge and information. Meaning may not be communicated very clearly because of mistakes in writing. **[1–3 marks]**

Level 2: Student describes effects and explains causes with some development of their answer, and uses some relevant knowledge based on a range of factual information and evidence. Meaning is communicated clearly. **[4–6 marks]**

Level 3: Thorough and developed description of effects and explanation of causes, with place-specific example. Demonstrates thorough knowledge, based on a full range of relevant factual information and evidence. Meaning is communicated very clearly. **[7–8 marks]**

i The River Valency in Boscastle, in 2004.

ii The river flooded suddenly and many people were trapped in their houses. They were rescued by helicopter from their roofs. The flood also wrecked many cars and damaged buildings. The repair costs came to about £300 million.

iii The main cause was heavy rainfall, over 125 mm near Boscastle. The ground became saturated and the water flowed quickly into the river. The steep valley speeded up the flow of water down into Boscastle and made the flooding worse.

Examiner's comment

A correct example has been given in (i). The place, Boscastle, and the year are not required but both make the chosen example clearer.

Three separate effects of the flooding are given: people trapped in their houses, cars wrecked and buildings damaged. People being rescued by helicopter is a good development of the 'people trapped' idea. The overall cost of damage is accurate and is a fourth piece of information about the effects of the flood.

The main cause given is the heavy rainfall. A time period would be useful, but the candidate has explained how the heavy rainfall caused the flooding. They have used their knowledge of drainage basin systems to explain how the saturated ground speeded up the surface flow of water. They have also used their place knowledge to explain how the geography of the river Valency's valley channelled the excess water downstream towards Boscastle.

Full marks have been achieved by this case study answer.

End of unit activities

Thinking about geography – classifying information

Classifying is a thinking skill that helps us to organise information and ideas. We classify in everyday life – for example, think about how people organise the music on their iPods, or how supermarkets organise and classify products on the shelves.

By being able to classify in geography not only can we make sense of a wide range of geographical information, but we are better able to show *why* we have organised information in a certain way – we are better able to *justify* our decisions.

The 'odd one out' activities below help develop information-processing and reasoning skills in geography through *classifying*.

Rivers and coasts: odd one out

In the table below, there is a list of twenty-four geographical terms that you have come across in your studies of rivers and coasts. Each term has been given a number.

1 surface runoff	9 longshore drift	17 interception
2 traction	10 attrition	18 mangrove forest
3 lagtime	11 backwash	19 watershed
4 flood plain	12 hydraulic action	20 delta
5 swash	13 beach replenishment	21 abrasion
6 infiltration	14 saltation	22 monsoon
7 groynes	15 cliff	23 meander
8 transpiration	16 spit	24 discharge

Some of the words have been placed into the following 'word sets' using their numbers:

Word Set A	*4*	*20*	*16*
Word Set B	*24*	*5*	*3*
Word Set C	*17*	*2*	*1*
Word Set D	*9*	*13*	*7*
Word Set E	*6*	*10*	*21*
Word Set F	*22*	*8*	*18*
Word Set G	*19*	*15*	*23*
Word Set H	*12*	*11*	*14*

ACTIVITIES

1 Working with a partner, find the three words that are in each of the word sets.

2 Decide which word in the set you think is the 'odd one out'.

3 Justify your decision by stating:

a why that word is the 'odd one out'

b what links the other two words together.

Developing your classifying skills

4 Now that you have started to see a pattern, add an extra word from the list to each word set, but keep the same 'odd one out'.

5 Test other partners around the classroom by putting together four word sets of your own with an 'odd one out' – see if they can work yours out.

6 Finally try to sort *all* the words from the list into four to six groups, where each group has a common feature. Make sure you can justify your decisions.

THINKING ABOUT CLASSIFYING

- From your responses to Activity 2, in which word sets was it easy to choose the 'odd one out', and in which word sets was it difficult? Explain why.
- Why is classifying important?
- In what other subjects do you classify information? Share some examples.

Chapter 2

Population and settlement

Sharing the Earth

Just over 6 billion people share the Earth's spaces and resources. This is about twice as many as only 50 years ago and likely to be about 2 billion less than 50 years in the future. Every person has a different story to tell about their past and present family, where they live and their standard of living. An aim in geography is to try to understand the trends and patterns that give these differences so that ways can be found to make life better and fairer for everyone. The use of space and resources will also need to be sustainable so that problems for the future can be avoided.

Stark contrasts

On a global scale, there are broad patterns of the wealth that people have and the amount of resources they use. The photo, however, shows an extreme case of how people with completely different life styles can live almost beside each other. The fact that this scene is in an urban area is also significant since for the first time in human history, there are now more people living in towns and cities than in country areas.

While some people live in comfortable apartments with swimming pools and other leisure facilities, others live in homes they have built for themselves in densely packed areas that lack basic amenities such as clean water and sanitation. This raises questions about how a country's population can be managed, how wealth should be shared and how the use of land can best be planned. These are the three key themes of this chapter.

QUESTIONS FOR INVESTIGATION

- How and why are there variations between the population structures of countries?
- What are the causes and consequences of natural population change over time?
- How is the pattern of land use within cities changing?
- Why does migration occur and what are its effects?
- What affects the provision of goods and retail services in rural and urban settlements?

How is the number of people changing and where do they live?

GET STARTED

Look at an online population clock to see how many extra people there are on the Earth in every minute. Work out the total for the length of your lesson.

More and more

The single most important fact about the Earth's population is the number by which it is increasing. The number can be measured in both total and density. Over the last 50 years, the total has doubled and in a few countries, it will probably double again in the next 50 years. This is called **exponential growth**.

Natural disasters, famines, local wars and outbreaks of disease will reduce the growth in some areas for a limited amount of time, but will do almost nothing to affect overall growth. The only thing that will slow down the growth in any significant way is for people to have fewer children, especially in a large number of countries where growth is fastest.

Table 2.1 The 2000–2050 increase in the world's population total and density

Year	Population	Population density per km²
2000	6,124,123	45
2005	6,514,751	48
2010	6,906,558	51
2015	7,295,135	54
2020	7,667,090	56
2025	8,010,509	59
2030	8,317,707	61
2035	8,587,050	63
2040	8,823,546	65
2045	9,025,982	66
2050	9,191,287	68

(Source: United Nations)

Table 2.2 The 2000–2050 increase in the population of the world's 50 poorest countries

Year	Population	Population density per km²
2000	679,447	33
2005	766,816	37
2010	863,394	42
2015	966,718	46
2020	1,075,104	52
2025	1,186,916	57
2030	1,300,634	63
2035	1,414,665	68
2040	1,527,425	73
2045	1,637,146	79
2050	1,741,959	84

(Source: United Nations)

Slowing down

There are signs that the rate of growth is slowing down. There are already fewer children being born in most of the richer countries, usually no more than two in each family and sometimes only one or none at all. The same trend is starting to happen in the world's poorer countries, for example in Asia, Africa and South America. This is important, since most of the world's people live in the **Less Economically Developed Countries (LEDCs)** where the population increase has been the greatest.

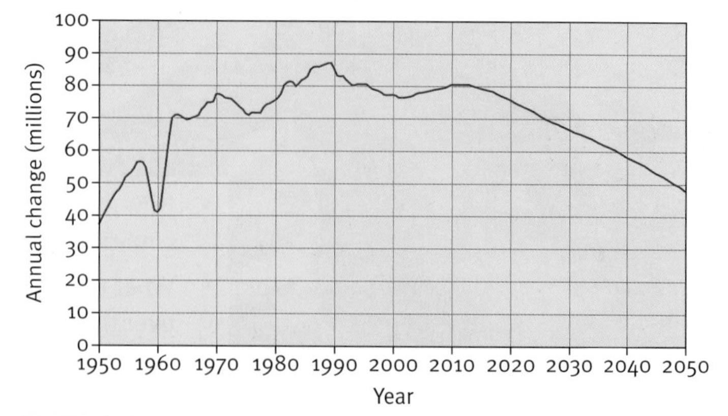

Figure 2.1 *Change in rate of growth of the world's population 1950–2050.* (Source: US Census Bureau)

Patterns of people

Some places are more densely packed with people than others. These places are often where people produce a large amount of food, such as in some river valleys and deltas. There are also densely populated places where people work in services and in manufacturing industries, such as in the conurbations of Western Europe, Japan and the north-east of the USA. There are complex historic reasons why these places became suitable as locations for these types of work.

The less densely populated places are usually where the climate or relief make it hard for people to make a living: for example, in cold environments, hot deserts and mountains. When there are too many people for them to have a good standard of living, an area is said to be **overpopulated**. It is hard to say exactly how many *more* people is 'too many' people. It all depends on the types of work they do and how they make links with people in other places such as through trade and commerce. It can be more useful to think about how people's standard of living can be sustained in the future without ruining the environment on which the standard of living depends. This is known as the **optimum population total (OPT)**.

Figure 2.2 The Mekong Delta, Vietnam is one of the world's most densely populated rural areas. It is at risk of flooding during typhoons and as the land sinks, together with the predicted rise in global sea level.

KEY TERMS

Exponential growth – a rate of increase that quickly doubles.

Less Economically Developed Countries (LEDCs) – countries with low economic output per person, often measured by Gross National Product (GNP).

Optimum population total (OPT) – the number of people that an area can support in a way that allows them to have a sustainable standard of living.

Overpopulated – the idea that there can be too many people in an area for its resources to sustain.

Fact file

Deltas at risk

- There are six major delta areas with high population densities.
- Most major deltas are sinking by up to 2 mm each year because of water and oil extraction and dams upstream that stop sediment reaching the sea.
- Most are in the track of hurricanes or cyclones.
- By 2050, there could be about 8 million people in these delta areas, many at risk of flooding.

PLENARY ACTIVITY

Why does the global increase of population matter to everyone?

ACTIVITIES

1. From Tables 2.1 and 2.2, what are likely to be the key figures for the world's population by the time you are aged about 30? Give reasons for your choices.
2. How do you think that the figure for population density might affect people's standard of living in different places?
3. What do you interpret from the graph in Figure 2.1 about changes to the rate of world population increase?
4. Why do you think that deltas such as the one in Figure 2.2 are often heavily populated? Think about the features of deltas such as flat land, rich soil, the sea, a river.
5. Technology can make it possible for people to live in difficult environments. Give some examples of how this can be done and if these methods are sustainable.
6. How do the ideas of overpopulation and optimum population relate to the area in which you live, or to another area that you know or can research?

Population change – a matter of life and death

Family trees

Tracing ancestors to make a family tree has become a popular pursuit, made easier by the Internet than it was in the past. In the UK, it is usual to find that the further back in time, the more children there were in each family. Even 100 years ago, families with six or seven children were not unusual, though not all the children survived to become adults. People also died earlier than the average today, which is at about 81 for women and 76 for men in the UK.

GET STARTED

For the area in which you live, suggest some ways in which a decrease in the local birth rate might affect different types of shops and services.

Population indices

The population change in a country is mainly because of differences between the **birth rate** and **death rate**.

A third population index is called the **fertility rate**. This indicates the number of children that the average woman in the population is likely to have. Another useful index is the **replacement rate**. This shows how a population is replacing itself over time; for example, the UK rate is 2.1, showing that for every two adults, at least two children need to be born if the population is to stay the same. The extra over two takes account of the fact that some children may die before they become adults, and that not all adults want to have children or are able to have them.

'Rapid population growth has occurred not because human beings suddenly started breeding like rabbits but because they finally stopped dropping like flies.'

Nicholas Eberstadt

Population change and economic development

The birth rate in the Less Economically Developing Countries (LEDCs) is usually much higher than in the **More Economically Developed Countries (MEDCs)**. One view is that high birth rates hold back economic development. However, there is also the opposite view that it is the lack of economic development that makes people need to have so many children. Children can stretch a family's resources to feed, clothe and educate them. On the other hand, children can also work to earn money, even though this is illegal in most countries. Some clothes that people buy cheaply in the UK, for example, are made by children in

DISCUSSION POINT

Do you think it is right that parents should expect their children to work to help the family, even if the family is in need of the money?

Table 2.3 Rates for births, deaths and natural increase: 2008 and 2020

Area	Birth rate 2008	Birth rate 2020	Death rate 2008	Death rate 2020	Rate of natural increase for 2008 (%)	Rate of natural increase for 2020 (%)
World	20.2	17.8	8.2	7.9	1.19	0.99
Less Economically Developed Countries	22.2	19.3	7.8	7.3	1.45	1.20
More Economically Developed Countries	11.0	10.3	10.4	11.0	0.06	0.07

(Source: UK Statistics Authority website, Crown copyright)

India and other countries. In rural areas, children can help with farm work. They are also needed to support their parents where there are no pensions. Where there are high child death rates, it is important to have enough children to make sure that some survive to look after parents when they are too old to work.

People also make choices as a result of their religious beliefs, for example on contraception and abortion. In some countries, government laws are based on religious beliefs. It is hard for individuals to make their own choices about the number of children to have if help with family planning is not available.

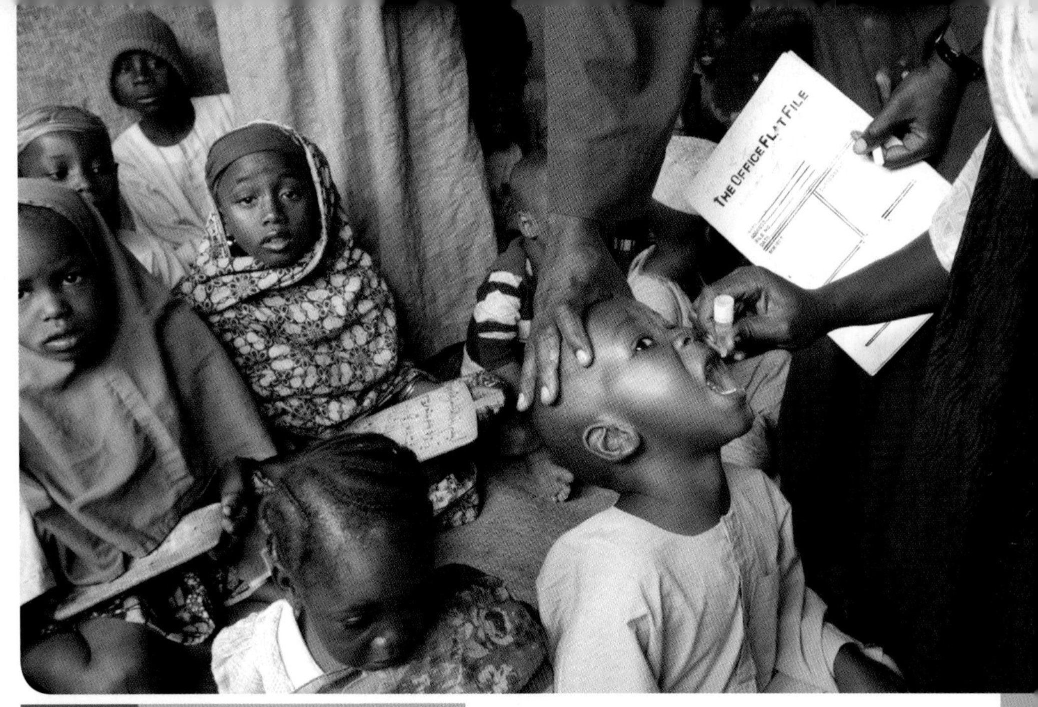

Figure 2.3 *Children in Kano, Nigeria receive better health care at a clinic, helping to reduce the death rate amongst young children.*

Lifestyle choices

In countries such as the UK, people think differently about the number of children they want to have. With better health care, there is a very low risk of a child dying before the age of five and adults cannot use their children to earn money. Houses are expensive and can become crowded with too many children. Childcare can also be expensive if both parents choose to go out to work. Paying for children can also mean fewer luxuries such as holidays, and there are both state and other pensions that adults expect to get when they stop work. These things mean that decisions about the number of children to have can be based on lifestyle choices and the kinds of family life that people desire.

KEY TERMS

Birth rate – the number of children born in a year for every 1000 people in a population.

Death rate – the number of people who die in a year for every 1000 people in a population.

Fertility rate – the average number of children to which each woman gives birth.

More Economically Developed Countries (MEDCs) – countries where there is a high level of economic activity and where there is generally a good standard of living for most people.

Replacement rate – the number of children that need to be born to replace the present population.

Fact file

Numbers and causes of child deaths

The World Health Organization (WHO) gives these figures:

- child deaths have fallen, but about 10 million still die each year before the age of five, mostly in the LEDCs
- about 4 million die within one month of being born
- a child dies of malaria every 30 seconds in Africa
- about 20 million children under five are undernourished and at risk from diseases.

ACTIVITIES

1. How do the figures for birth and death rates in Table 2.3 help explain why the world's population will still be increasing quite quickly in 2020?
2. What are the main differences you can see in Table 2.3 between the rates of natural increase in LEDCs compared with MEDCs?
3. How does the economic value of children differ between the UK and some of the world's poorer countries?
4. Use the photo in Figure 2.3 to explain why the trend is for fewer children to be needed in many of the world's LEDCs.
5. Why can the choices about the number of children to have vary so greatly between different countries, even when their general level of wealth is the same?

PLENARY ACTIVITY

Do you think that the birth and death rates in the world's LEDCs might follow the trend for the MEDCs? Explain your answer.

Is there a pattern to how a country's population changes over time?

GET STARTED

Why do governments want to know about future population changes?

Following trends

It is hard to predict what will happen to a country's population in the future based on what has happened in the past. It can, however, be useful to look at the past to see if there are any trends from other countries that can show what might happen. The study of population is called **demography**.

Although accurate predictions of numbers can be difficult, some of the future trends are fairly safe to predict. This is because trends for births and deaths have previously changed fairly slowly. The number of children in families, for example, usually falls by about two in every generation. Death rates also fall slowly in line with changes to a country's economy and with better health care.

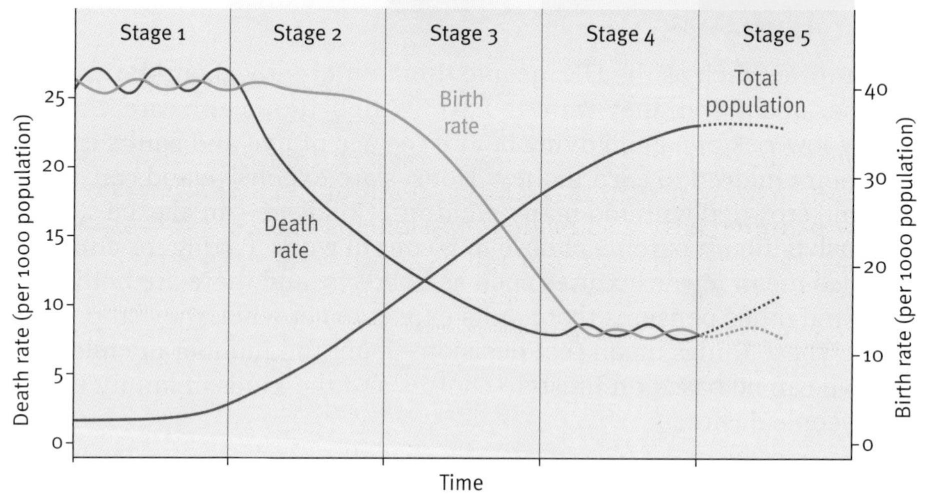

A pre-industrial society with birth and death rates about the same, so there is little population growth.
With better food supply, health care and sanitation, death rates fall but birth rates remain high, giving rapid population increase.
People have better jobs, do not need children to help with farm work and there are better living conditions so the birth and death rates continue to fall rapidly, bringing the population increase down even more.
Birth and death rates are about the same so the population total stays about the same. If death rates go below birth rates and there is no net migration, the population total can begin to fall.
A better standard of living can give increased death rates through heart disease and obesity, so a population can fall if birth rates are not increased.
Stage 1
Stage 2
Stage 3
Stage 4
Stage 5
Birth rate
Total population
Death rate
Death rate (per 1000 population)
Birth rate (per 1000 population)
0
5
10
15
20
25
30
40
Time

Figure 2.4 The demographic transition model.

A model of change

One model of population change uses experience of changes in industrialised countries over previous decades. The **demographic transition** model can be divided into stages (Figure 2.4). The model, however, only shows what happened in the past and why the changes happened. It cannot predict what might happen in the future. Governments, for example, can bring in laws to manage population numbers or there could be the widespread outbreak of a disease called a **pandemic**.

Behind the model

Some general long-term reasons for population change are easy to understand. In the past, people had more children when they depended on farming, when many children died

Fact file

HIV/AIDS infection in Zimbabwe

- About 25 per cent of the adult population in Zimbabwe is HIV-positive.
- About 40 per cent are HIV-positive in urban areas.
- Life expectancy of people infected may fall to 38 years.
- The amount of money available to spend on health care is about £5.50 per person, which is not enough to help people who are HIV-positive to overcome other infections that can kill them.

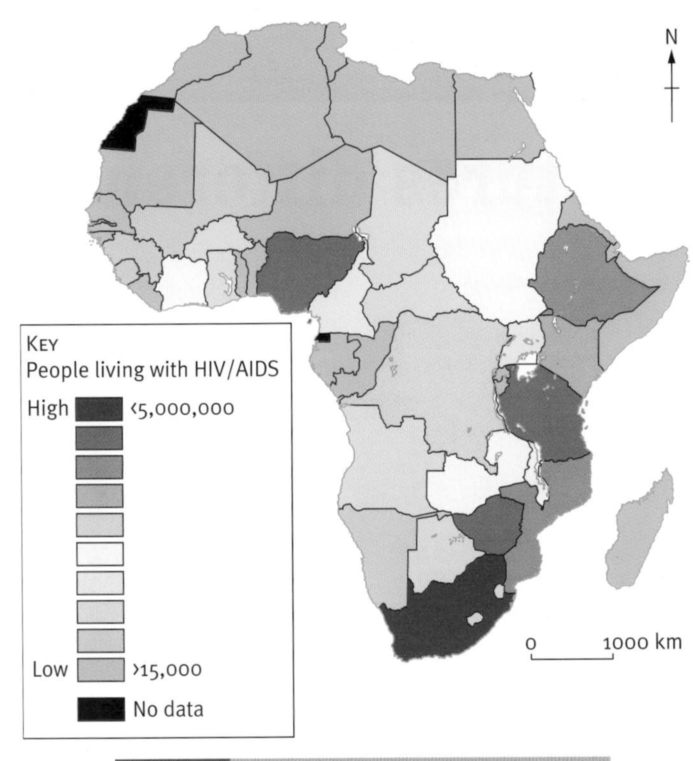

Figure 2.5 The distribution of HIV/AIDS in Africa.

young and when old people did not have pensions. As these reasons have become less important, fewer children have been needed. Better methods of family planning have also made it possible to choose family size. Death rates usually fall when there is better medical care, education and pensions for people who become too old to work.

There are also short-term reasons behind a country's demographic change. A large-scale war, for example, will increase the death rate and reduce the birth rate, though the birth rate may rise sharply when it is over. A pandemic can also increase the death rate in the short term. A new drug can reduce the death rate quickly if it can be made available quickly to everyone who needs it. A drug to help people with **HIV/AIDS** (see Figure 2.5), for example, would help millions of people who are affected by it.

Changes to the climate, such as the failure of seasonal rain, can have short-term effects on crops and animals that can cause a famine. Climatic changes can also happen over longer periods of time, as may be the case with global warming. This may change the balance of people and resources, especially in countries where many people depend on farming.

The next stage

It is also interesting to think about what the next stage in the model might be. The situation in countries where there is an increasing number of old people and a low birth rate, for example, may make governments encourage people to have more children. At the same time, growing concerns for the environment and personal wealth may act to persuade people to have fewer children!

THINK ABOUT IT

How might global warming have an effect on the birth and death rates in a country?

PLENARY ACTIVITY

Where is the UK in the model? What will happen next?

KEY TERMS

Demographic transition – model showing how the population in a country changes over time as birth and death rates fall.

Demography – the study of population.

HIV/AIDS – Acquired Immune Deficiency Syndrome (AIDS) is a set of symptoms and infections resulting from the damage to the human immune system caused by the Human Immunodeficiency Virus (HIV).

Pandemic – a disease that affects a very large area, often crossing between continents.

ACTIVITIES

1. Think of some things that have changed in the UK over the last 50 years that might have been hard to predict. How might these changes have affected the numbers of births and deaths?
2. Why did changes to birth and death rates usually happen fairly slowly over time?
3. What information that affects a country's population total is not included in the demographic transition model shown in Figure 2.4?
4. Suggest some reasons why it might not be helpful to rely on the demographic transition model to predict population change in a country in the future.
5. In 2006, Zimbabwe had a birth rate of 28 and a death rate of 19. At which stage of the demographic transition model might this put the country now?
6. Draw a consequence diagram to show the effect of HIV on Zimbabwe. Use the Fact file to start then add your own ideas and knowledge.

Why are population structures different in different countries?

GET STARTED

What are your first reactions to the photo in Figure 2.6? Why have you reacted in this way?

Figure 2.6 Child workers in North India earn money to help the family income.

Supporting each other

People in a country depend on each other in complex ways. Children depend on adults to provide for all their needs, then they will do the same for their own children. Adults depend on each other through paying taxes that are for everyone's benefit, such as for education, defence and other public services. When people become old and stop working, they can help with childcare, but they also depend on younger people who are still working to pay for pensions and the services they need. The ratio of people in work compared with those who are not in work is called the **dependency ratio**. The balance of people of different ages is called the **population structure**.

A people pyramid

The population structure of a country can be shown in a graph called a **population pyramid**. The usual pattern for an LEDC is for there to be a large number of children, making up to half the total population. The graph is like a pyramid because the numbers being born and surviving childhood has been increasing. In an MEDC, the shape of the pyramid is quite different. There are fewer children and because the child death rates are low, the shape is less like a pyramid and more like a rectangle.

The gender divide

The pyramid diagram also shows the balance of males and females in a population. The numbers of each are usually similar, though in most countries, females tend to live a few years longer than males. The difference in number between males and females can be for many reasons: for example, because of different rates of smoking, use of drugs, car accidents, war and manual labour. In some of the world's poorest countries, where females do much of the manual farm work and where there are higher birth rates, the **life expectancy** of females can be a few years less than for males.

The balance of males and females can become distorted in the younger working age group if there are high levels of **migration**. In some countries, migration from the country is mainly of young male workers. In others, it is mainly of young female workers.

Table 2.4 Life expectancy for males and females for selected countries

Country	Males	Females
Sweden	78.4	82.8
UK	76.2	81.2
China	71.1	74.9
India	66.3	71.2
Uzbekistan	61.6	68.6
Botswana	51.6	49.9
Zimbabwe	40.6	38.4

(Source: NationMaster website)

RESEARCH LINK

One insurance company in the UK has estimated that it costs about £120,000 to bring up a child to the age of 18. Work out why it might cost this amount.

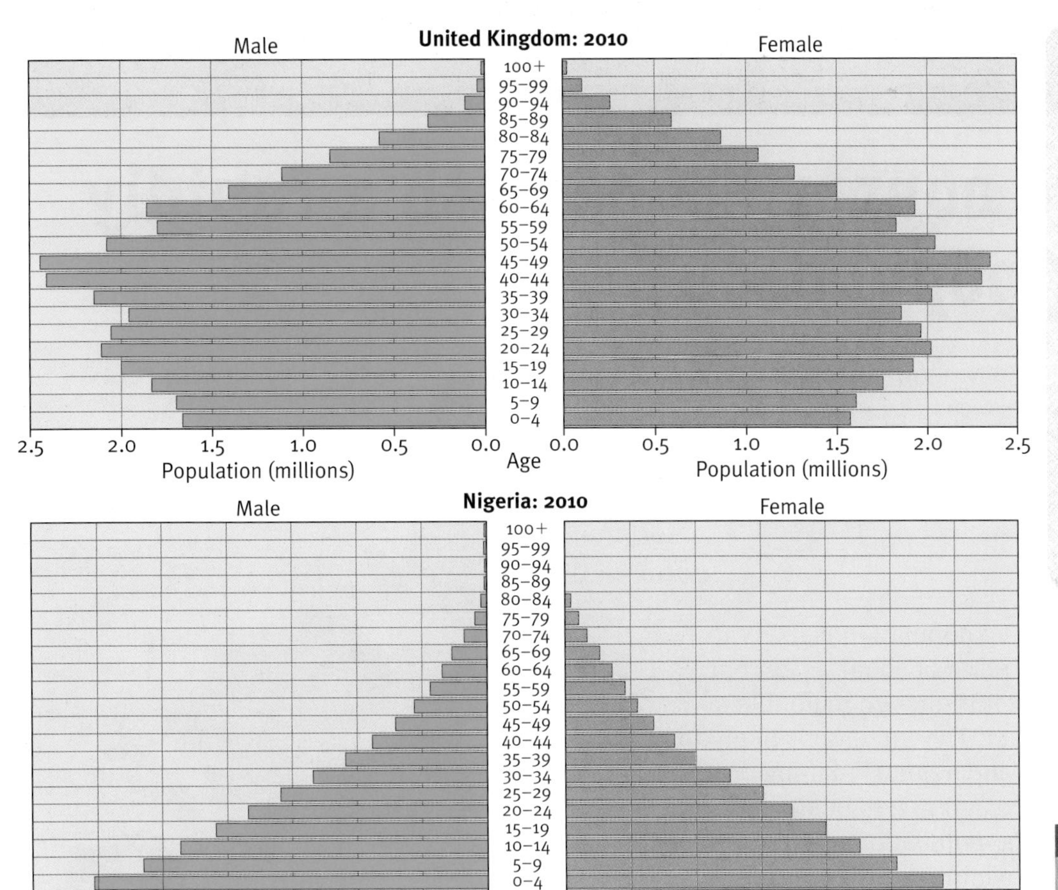

Fact file

Predictions for England's population pyramid (Audit Commission)

In 2009, around 17.7 million people (34 per cent) will be aged 50 or over and 2.4 million aged 80 or over.

By 2029, 22.9 million (about 40 per cent) will be aged 50 or over and 4.3 million aged 80 or over.

Figure 2.7 *Population pyramids of the UK and Nigeria predicted for 2010.* **(Source: US Census Bureau)**

Changing shape

The population structure of a country slowly changes shape as it becomes more economically developed. People tend to have fewer children as health care improves and they do not need as many children. The top of the pyramid also becomes higher and wider because more people live for longer.

Sometimes, the structure changes for short-term reasons, for example if there is a war or large-scale famine. There can also be exceptional times when there is a 'baby boom', such as happened in the UK after the Second World War. These 'baby boom' children are now becoming pensioners. This is causing a rise in the dependency ratio, a situation that the government has to manage so that working people are not taxed too much to support people who have retired.

KEY TERMS

Dependency ratio – the ratio between the economically active population and those who are dependent on them.

Life expectancy – the average number of years a person may expect to live when born, assuming past trends continue.

Migration – to move from one country to live in another.

Population pyramid – a graph to show a country's population structure.

Population structure – the balance of people of different ages and genders in a country's population.

ACTIVITIES

1 Compare the population structures of the UK and Nigeria as shown in their population pyramid diagrams in Figure 2.7. Explain how they show the different dependency ratios in each country.

2 Suggest some reasons why a population pyramid might be unbalanced towards one gender or another.

3 What problems might there be in a country where the numbers of each gender were not about the same?

4 Draw a diagram to show the general shape of a population pyramid for a country in which there is a dependency ratio that would make it easiest for young and old to support each other in a balanced way. Add notes to your diagram to explain it.

PLENARY ACTIVITY

Explain why a population pyramid can give such a good guide to the future population of a country.

How does a country cope with a rapidly increasing population?

GET STARTED

Look at the diagram of the decreasing amount of land per person in Africa in Figure 2.8. What are likely to be the effects of these figures?

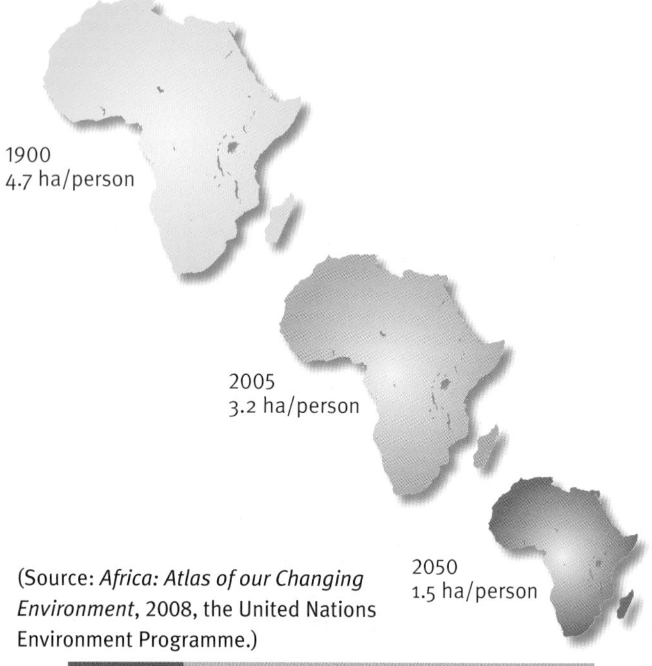

Figure 2.8 *Population pressure on the land in Africa.*

Statistics of growth

A country with a rapidly growing population is likely to have an annual percentage increase of about 2 per cent. This does not sound much, but it is enough to double a population in as little as 30 years. The increase is likely to be because of a high birth rate and a quickly falling death rate. The population structure diagram will be the shape of a pyramid, with up to half the population as children. About half the world's people live in countries with this population structure. These are countries in which people face daily challenges to improve their standard of living and, in many cases, to survive. Climatic change caused by global warming adds to these problems.

Pressures on the land

A rapidly increasing population is almost certain to cause problems in countries that are already short of money. There are bound to be problems of housing, health care, pollution and every type of resource, including land. When governments are not able to cater for the increase, people have to find their own answers to their problems. This is one reason why they live in places that are not safe: for example, where there can be flooding, landslides or where there are diseases in the water. These are the places that people avoided in the past, but are now used by people with few other choices. Geographers call them marginal sites.

THINK ABOUT IT

Divide 70 by the growth rate of a country to find the time it will take to double the population. What is the shortest time you can find?

Table 2.5 Population data for Kenya

Total population	34,700,000
Aged 0–14 (%)	42
Birth rate (per 1000)	39
Death rate (per 1000)	11
Fertility rate (per 1000)	4.8
Annual rate of increase (%)	2.3
Urban population (%)	39
Rural population (%)	61

(Source: NationMaster website)

Figure 2.9 *Degraded land in Kalacha, Kenya is often caused by population increase and overgrazing by animals – climate change is also an increasing problem.*

The effects of overpopulation

Children's needs

There are special demands on services such as health and education in a country with a large proportion of children. In some countries, the education of girls can be treated with less importance than that of boys. The result is that female **literacy rates** are often much lower for girls. Yet for both boys and girls, education can give them more choice of better jobs and where to work.

Cheap labour

An effect of the competition for jobs is that workers can be paid low wages. People in rich countries can benefit from this because it keeps prices down. When people cannot get full-time jobs, they do short-term jobs, for example in services and labouring. Some migrate to countries where even low-paid jobs are better paid than what they can expect in their own country.

Benefits from growth

While many people do have extreme problems living in countries with rapid population increase, it also means there are opportunities for businesses to grow. Some people with better jobs can afford a good standard of living, with all the consumer goods that people have in richer countries. A problem, however, is that this can widen the gap between rich and poor and give tensions in both the political life of a country, and also on the streets. While large numbers live in **shanty town** areas, people with more money may need to live in separate areas that are protected by fences and security guards.

ACTIVITIES

1. From the data about Kenya in Table 2.5, use a bubble map to explain why population change is likely to be an underlying cause of other problems there might be in the country.
2. Study the photo in Figure 2.9. For each environmental factor it shows, explain how it will affect people and their work.
3. How does the photo in Figure 2.10 show that having too many children in a country can be a problem? Think about the people for whom there are likely to be the greatest problems.
4. Study the Fact file. For one item from the list, draw a cartoon or diagram to make it meaningful to someone in the UK.
5. Who is most likely to get any benefits from a rapidly increasing population in a country? Explain your answer.

PLENARY ACTIVITY

What do you consider to be the two most immediate problems and the two longer-term problems for people who live in a country that has a high birth rate and a rapidly increasing population? Give reasons for your choices.

RESEARCH LINK

Which country has the biggest gap between male and female literacy? Find out why.

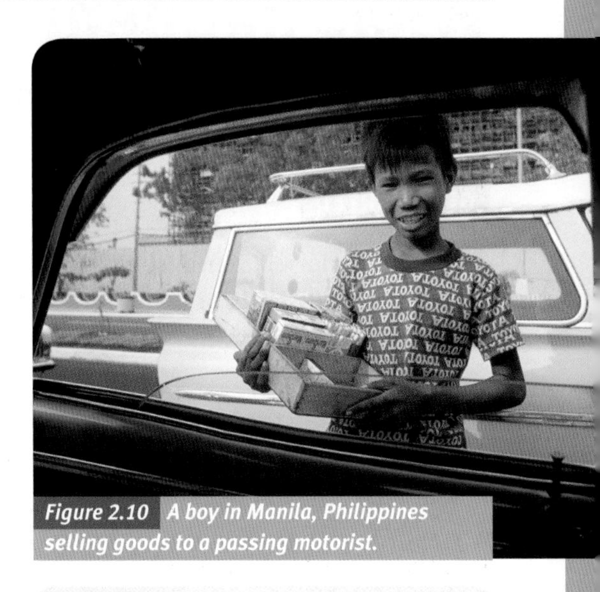

Figure 2.10 A boy in Manila, Philippines selling goods to a passing motorist.

Fact file

Issues for children

- There are 50 countries in which more than 40 per cent of the population is aged 0–14.
- The United Nations (UN) estimates that 1.2 million children are trafficked every year, often to work in another country.
- About 5000 children die every day because of diseases they get from unsafe drinking water.
- About 26,000 children under the age of 5 die every day, mostly in the world's 60 poorest countries.
- About 9.7 million children under 5 died in 2006, the first time in recent years that the figure had dropped below 10 million.

(Source: Unicef)

KEY TERMS

Literacy rate – the percentage of people who have basic reading skills.

Shanty towns – areas of housing that people build themselves, often on the edges of big cities in LEDCs, on land they do not own.

What happens when the population stops increasing?

GET STARTED

At what age do you think a person should be made to retire from their work? Or do you think that there should not be a set age that applies to everyone?

Figure 2.11 *Elderly people in Japan taking morning exercise.*

When growth stops

A stable population in a country brings a lot of advantages; for example, it is easier for a government to plan and pay for new facilities. But there can also be problems in countries with **zero growth** or, in some cases, where the population is falling, **negative growth**. These are usually the richer countries where the birth rate and death rate are either the same or where there are fewer births than deaths. This is what is now happening in countries such as Japan, Italy and the UK.

If there is no **natural increase** and little **net migration**, the population stops increasing. This causes a change in the population structure and therefore a change in the dependency ratio. When fewer children are being born and people live longer, those who work and pay taxes have to support an increasing number of people.

Table 2.6 Japan's demographics

Total population in 2008	127,000,000
Total population in 2025 (estimate)	124,140,000
Birth rate (per 1000)	8.1
Death rate (per 1000)	8.9
Rate of natural increase (%)	–0.09
Population over 65 years old (%)	20

The need for workers

A benefit of living in a country with a decreasing population total is that there are likely to be jobs for everyone. But when there are not enough people or when people have the wrong skills, essential jobs may not be done. Encouraging migrants to come is one solution to this problem.

A moving pattern

A trend in the UK has been for **counterurbanisation**, when people move out of the big cities, even while they are still of working age. People who retire can also move to places where they want to enjoy their retirement. Seaside resorts such as Bournemouth are especially popular for this. This has several effects on the town, for example in the need for more health and care workers. Planning in the town will also need to cater more for people with mobility problems and other disabilities.

Living apart

A new trend is for separate **retirement villages** that are designed to cater for older people. These create secure and supported environments for old people, though not all old people want to be separated from the rest of the community.

A different solution for some older people is to move to another country, such as Spain, where the weather is warmer and winter heating bills are less. In spite of the advantages, there are also problems such as the different laws, language and health care. Separation from their friends, children and other relatives can also cause problems.

Fact file

Countries to which people from the UK go to live when they retire

- In 2007, about 200,000 people left the UK for good, mostly retirees.
- Cyprus was the main country to which people retired because it has an income-tax rate of just 5 per cent on pensions for retired residents and low property prices and no inheritance tax.
- Panama was also popular. Pensioners get a 15 per cent reduction in the cost of hospital services in private clinics, 10 per cent reduction in medicine, 20 per cent reduction in medical consultations and surgical procedures and 15 per cent reduction in dental and optical services.

THINK ABOUT IT

Why might some parts of a country be attractive to an elderly retired person as a place to live?

RESEARCH LINK

Find out why it can be an advantage for some retired people in the UK to move to live in another country.

KEY TERMS

Counterurbanisation – movement of people away from a city or town.

Natural increase – an increase in population when there are more births than deaths in a year.

Net migration – the difference between people moving into a country and those who move out.

Retirement village – a small settlement designed for elderly people.

Zero growth – when the birth rate and death rate in a country are about the same and the population is not increasing.

PLENARY ACTIVITY

Draw a spider diagram or bubble map to show the ways in which a country's geography can be changed by an increasing percentage of older people in the population. Think about: housing, transport, employment, social and leisure facilities, migration, where people live.

ACTIVITIES

1. The photo in Figure 2.11 shows a scene in Japan. What questions might you want to ask to find out more about the people and what they are doing?
2. How do the information and figures for Japan in Table 2.6 show that the country now has zero growth and a high dependency ratio?
3. Read the news article on page 66. Is this what you might expect to happen in a country with an increasing percentage of older people, or might it be something that is unique to Japan? Explain your answer and suggest some ways that the problem could be reduced.
4. Explain why the population pyramid for a country may not help in working out the needs of people in different parts of the country.
5. What do you think of the idea of older people living in separate retirement villages? Should the government and local councils be encouraging developers to be building more of them, or fewer?
6. Older people who are no longer economically active still need an income. What do you think about where the money should come from?

Some strategies for population management are more sustainable than others

GET STARTED

Think of at least two arguments in favour and two reasons against a government taking actions to control the number of people in a country.

Only one child

Population control is being tried by governments in many countries. One of the strictest approaches has been in China where, since 1979, most couples have only been allowed to have one child. The government were concerned that the country's rapidly growing population would mean shortages of food and other problems.

The policy has been carried out by a combination of laws, taxes, better health care and persuasion. Parents who refused to have only one child, for example, have been fined and made to pay for their children's health and education. Women have been encouraged to be sterilised after having their first child.

The effect of the policy has been to reduce the growth of China's population, though even now it is still set to rise by about another 100 million between 2005 and 2025. The policy is still in place, though it is not as strictly enforced in all areas as in the past. It is easier, for example, to get permission to have two children in rural areas than in the cities.

Figure 2.13 Laws in China restrict most families to having only one child.

UN Charter of Human Rights: Article 16

Men and women of full age, without any limitation due to race, nationality or religion, have the right to marry and to found a family. They are entitled to equal rights as to marriage, during marriage and at its dissolution.

THINK ABOUT IT

Some people believe that it is a human right to have as many children as they like. What do you think?

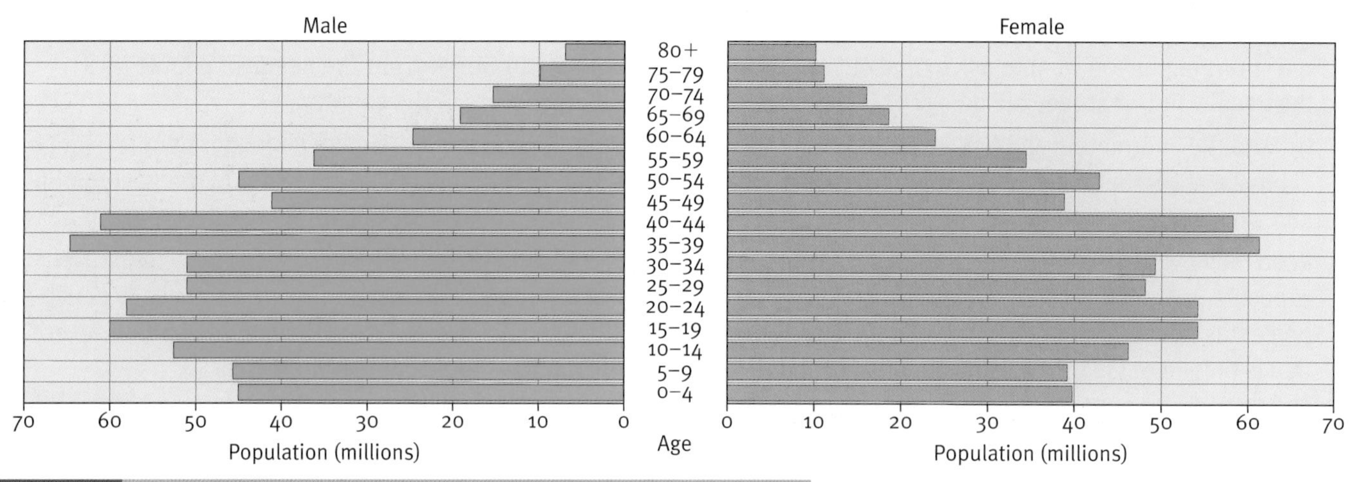

Figure 2.12 Population pyramid for China, 2008. (Source: US Census Bureau)
Note: the one-child policy was introduced when the 30-year-olds of today were born.

Future troubles

Although the population increase in China has been cut, there may be problems in years ahead when the one child becomes an adult and may have to support two parents and four grandparents. This is called the '4-2-1' problem. Another concern is that the single child 'little emperor' who has not had brothers or sisters may not be able to work so easily with others when they grow up. A third problem is that many Chinese couples have preferred to have a boy rather than a girl and have used selective termination to ensure that they do. The result is that there are now 30 million more boys than girls. They are known as 'spare branches' because they may not be able to find wives when they grow up! Sustaining the policy would face many challenges.

Beliefs about babies

In some countries, religious beliefs can affect the laws that a government makes. In many countries where the population is mainly Roman Catholic (such as Ireland or the Philippines), there are often laws that prevent abortion, or strict controls on it. The belief of Roman Catholics is that an unborn child has a right to life from the moment of conception, so abortion and most 'artificial' contraceptive methods such as condoms should not be allowed.

ACTIVITIES

1. Describe how the Chinese government's one-child policy has affected the population structure as shown in Figure 2.12.
2. Describe or sketch and label what the population pyramid for China will look like in 30 years' time if the one-child policy stays in place over that time.
3. Why are rural couples allowed more children?
4. What criteria for success do you think the Chinese government would have had for the effects of its one-child policy? Think about its effect on: jobs, food, the environment, family life, social and leisure conditions.
5. What do you think the challenges might be to the Chinese government in sustaining its one-child policy into the future?
6. What do you think are the arguments for and against a government making laws about family planning based on the views of a religion? Should this be a matter for individuals or is it best that everyone accepts the views of the majority?

The development contraceptive

The best way to manage a country's population is something about which people have widely different views. It used to be thought, for example, that population increases caused poverty and kept people poor. Now, many people believe that the best contraceptive is economic development that includes better education and health care, together with family planning. Population control based on a better standard of living is more likely to be more sustainable in the long term than policies that force people to have fewer children.

Figure 2.14 *Population increase, especially in urban areas, is causing high density living in Manila, capital city of the Philippines.*

RESEARCH LINK

Find out about the beliefs that different religions have about family planning methods and how these beliefs affect laws made by governments.

PLENARY ACTIVITY

Now that the economy in China has become more developed and as health and education have been improved, do you think that there is still a need for its one-child policy?

CASE STUDY How has the rate of population increase been controlled in Thailand?

GET STARTED

Look at some photos of tourist areas in Thailand, a country in South-East Asia. What impressions do these give you about the landscape and standard of living of people in the country?

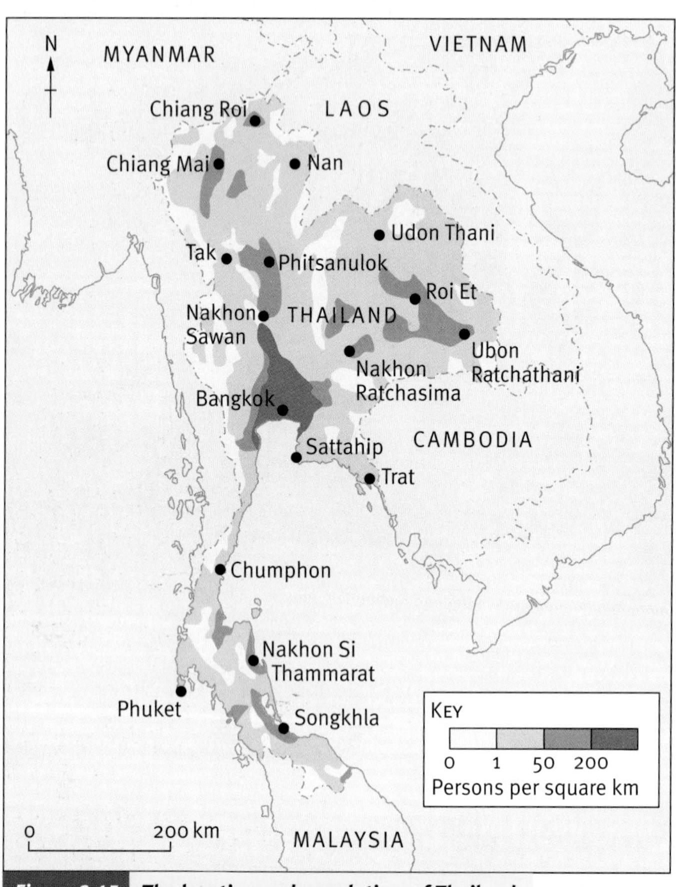

Figure 2.15 ***The location and population of Thailand.***

Thailand's economic change

Tourist brochures and holiday websites show Thailand as a country with warm, palm-fringed beaches, picturesque countryside and ancient temples. While it does have these features, it is also a country that has gone through 40 years of rapid change. It is called a **Newly Industrialised Country (NIC)** because of the rate at which its businesses have grown (for more on NICs see pages 162–163). In spite of its economic growth, about 42 per cent of people still work in farming, many as subsistence farmers. The remainder work in various types of industry and services.

Population change

Another change has been in how its population increase has been brought under control. In 1970, the population of Thailand was increasing at one of the fastest rates of any country, with a birth rate of 37 and a death rate of only 10. But within just under four decades, this had completely changed. Now, the birth rate has dropped to 15 per 1000 while the death rate has fallen slightly. These statistics show that something remarkable has happened to change people's ideas about the number of children they want to have.

Table 2.7 Population data for Thailand 1970–2006

	1970	1990	2006
Birth rate (per 1000)	37	19	15
Death rate (per 1000)	10	7	9
Fertility rate (per 1000)	5.5	2	1.8
Infant mortality (under 5) (per 1000)	151	102	31
Life expectancy (years)	59	67	70

Causes of change

Part of the change in Thailand's birth rate has been because of the country's economic growth and the fact that more people are now living in the cities, especially in Bangkok. People who live in cities do not need as many children and besides, the living conditions for many are crowded and often unhealthy. The other reason for the changing

RESEARCH LINK

Find out the names of some other NICs in South-East Asia. Check their birth and death rates to see how they compare with those in Thailand. This will help show whether Thailand is unique or if other countries are following the same pattern of population change.

KEY TERMS

Newly Industrialised Countries (NICs) – countries that have moved rapidly from having limited economic development to having many new industries producing goods for both the home and export market.

Figure 2.16 Population pyramids for Thailand: 2000 and 2020.

(Source: US Census Bureau)

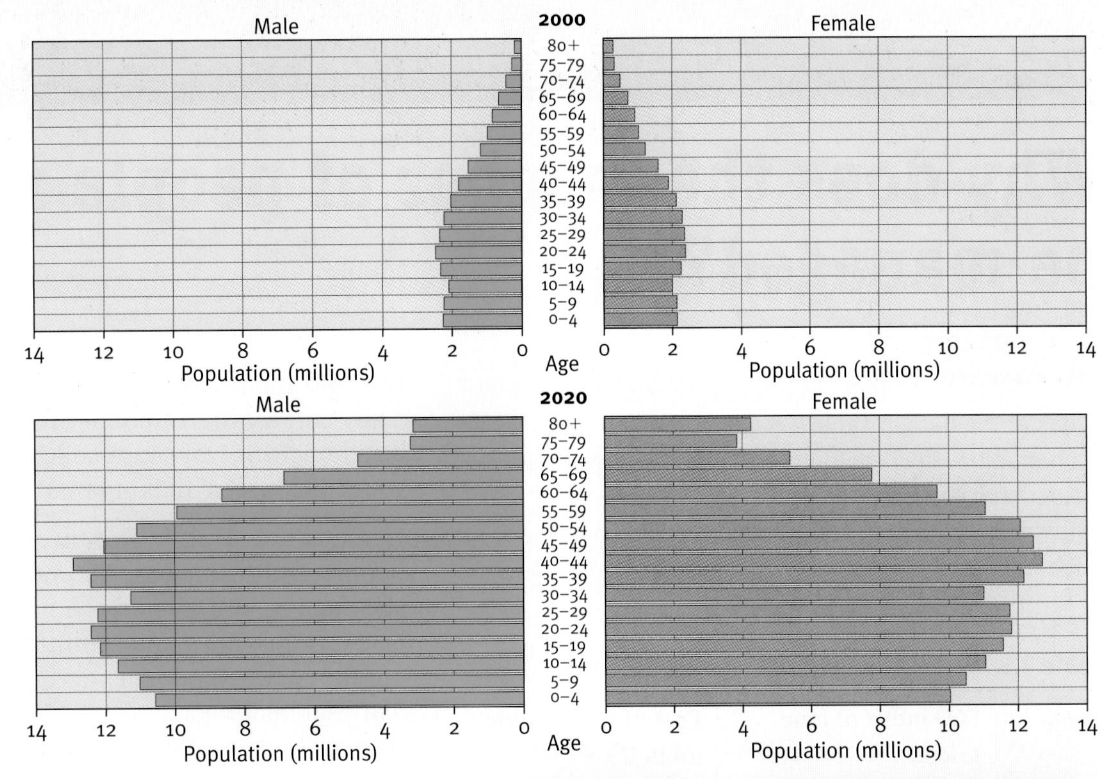

population has been the success of the government and other organisations in introducing family planning methods and in convincing the people to use them through advertising.

A particular success has been in encouraging people to use condoms. The population, however, is still increasing and is set to rise from 65 million in 2008 to 70 million in 2025, with an annual percentage increase of less than 1 per cent each year, reaching only 2 per cent by 2025. This is considerably less than in many other countries in South-East Asia and elsewhere.

Stopping HIV/AIDS infection

Using condoms, however, also has a different and equally serious purpose from controlling population increase. It is a key to fighting the spread of HIV/AIDS. This infection is a special cause for concern because of sex tourism in Thailand. The government has made it illegal for anyone working in this trade to do so without using a condom. There are estimated to be at least half a million people already infected in Thailand, though the number has been falling because of the law.

Cabbages and condoms

One Thai businessman has been running a national campaign to help make the use of condoms acceptable to everyone. To help do this, he has set up a chain of restaurants in Thailand named 'Cabbages and Condoms'. The condoms are given away free after the meal. In a country where even talking about contraception was not acceptable in the past, ideas such as this are a way to break down these kinds of barriers to population control.

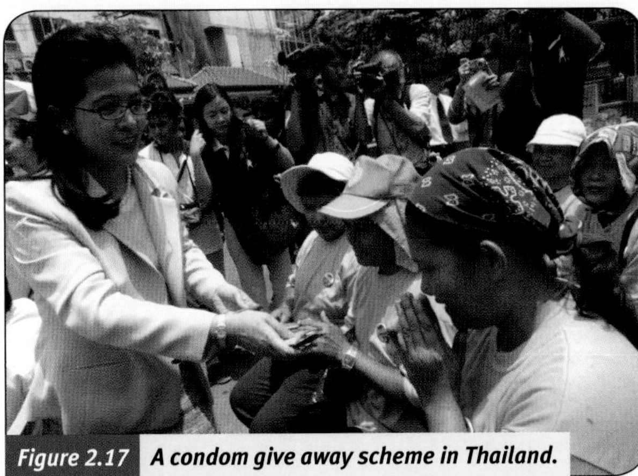

Figure 2.17 A condom give away scheme in Thailand.

PLENARY ACTIVITY

Cabbages and Condoms is a success story as part of controlling Thailand's population. How is it a sustainable story for Thailand?

ACTIVITIES

1. Look at the population data in Table 2.7. In what stages of the demographic transition model do you think that Thailand was in each of these 3 years: 1970, 1990 and 2006? Explain your answer.
2. List the approaches to population control in Thailand and explain how each of these has played a part in helping Thailand to reduce its rate of population increase.

Why does the number of people need to be managed?

GET STARTED

Look at the data in Table 2.8 about changes to prices for commodities such as foods and energy. How would you expect prices to change as the world's population increases and as people in more countries become richer?

Table 2.8 The changing price of goods on world markets

Year	FAO index of food prices	Cost of Brent crude oil in US $
2000	92	31
2001	94	26
2002	93	26
2003	102	28
2004	113	34
2005	126	57
2006	156	74
2007	196	71
2008	213	142

(Sources: Food and Agricultural Organization of the United Nations/ Thomson Reuters)

The FAO index of prices is for meat, dairy, cereals, oils and fats, sugar. Brent crude is one of the main oil prices. The prices are for July in each year. Note that all prices vary during each year.

Since 2008, the price index for food and crude oil has fallen back. The food price index had dropped to 148 by the end of the year. In January 2008, Brent crude was back down to around US $43.

Issues in town and country

The number of people in any area on any scale is important to the quality of life that they can have. This is true in the rural areas of countries where people mainly make a living by farming. In these kinds of places, there is a direct link between people and the natural environment on which they depend.

If the number of people goes up by too much, they can find it hard to grow enough food for everyone. They can try to overcome this by growing more food, but doing this can cause soil erosion and destroy water resources, making matters even worse.

In cities, there are different kinds of links between people and resources. Too many people can mean a shortage of houses and too much traffic, which causes congestion and pollution. There is not enough open space for recreation and waste disposal becomes a problem. Money is needed to resolve these problems, though solutions also need people who can make good decisions about how to spend the money.

The national scale

The number of people is also important on a national scale. Some **regions** can become more attractive for work and quality of life than others and encourage people to move there. Too many people and too much business growth in a region, however, can lead to it becoming 'overheated' with pressures of congestion, pollution and high prices for houses.

Governments can try to give everyone in a country equal access to jobs and a good standard of living by using money from taxes as **subsidies** and **grants** to businesses to support them in the poorer regions. However, governments also have to bear in mind what voters will think about how and where they spend money from taxes.

KEY TERMS

Region – a large area in a country that has some common features, for example of landscape and the type of economy.

Subsidies and **grants** – money that a government can give or loan to a business.

Tata Nano – world's cheapest new car is unveiled in India

It is 3 metres long, seats four comfortably or five at a squeeze, does 65 mph and aims to revolutionise travel for millions. The 'People's Car' is also the cheapest in the world at 100,000 rupees (£1300).

Green campaigners point to India's terrible road system and rising pollution levels. 'Even if they claim it will be fuel efficient, the sheer numbers will undermine this,' an air pollution specialist at the Centre for Science and Environment in Delhi, said. 'India's infrastructure doesn't have the capacity.' The centre estimates that the five million vehicles on Delhi roads today meet only a fifth of the capital's transport needs. Most people travel by bus but could be convinced to buy a car at such a low price. Delhi, where air pollution levels are more than twice the safe limit, is already registering 1,000 new vehicles a day.

(Source: Ashling O'Connor, *The Times*, 2008)

Figure 2.18 The low cost Tata Nano car, made in India.

Global population issues

Many people believe that the biggest environmental problem facing people everywhere is that of global climate change. One cause of this is the rise in carbon dioxide (CO_2) and other greenhouse gas emissions that are being produced by the industrial, transport and other activities of so many people.

Managing people on a global scale is beyond the power of any single government. However, the problems caused by an increase in population in one country can affect people in other countries. More people and a growing economy in China, for example, means more demand and higher prices for oil and other resources that are traded internationally. Of course, it is also true that more people in some countries can mean lower labour costs, so goods can be made and sold more cheaply.

DISCUSSION POINT

What do you think about the new cheap car that is on sale in India? In your answer, think about whether it is likely to help improve people's lives in the short term, but cause damage to the local and global environment in the long term.

ACTIVITIES

1. Why do you think that governments often want to make sure that people in different parts of the country should have the same standard of living? Think about the effects when this does not happen, for example, on housing, jobs, transport and the environment.
2. Do you think it would be a good idea for people in the UK to buy cars such as the Tato Nano? Or would it be better if people were encouraged not to buy a car at all?
3. Study the information in the Fact file about industrialisation in China. In what ways might the growing economic wealth of people in China affect the world's environment?

Fact file

Population and industrial change in China

- China has slowed its rate of population growth, but the total number will still grow by about 127 million up to 2025.
- Nearly 80 per cent of the country's electricity comes from coal and there are plans to build another 500 coal-burning power stations.
- These will push CO_2 concentrations right up to the 400 parts per million level, at which scientists expect dangerous climate change.
- Many athletes refused to compete in the 2008 Beijing Olympics because of fears about their health from severe air pollution.

PLENARY ACTIVITY

Should people in the rich industrialised countries tell people in other countries about how to develop their industries and power supplies? Explain your answer.

Push and pull factors influencing migration

GET STARTED

What questions would you want to ask about a country before deciding to move there to find a job?

UN Charter on Movement: Article 13

UN Charter on Movement: Article 13.

1 Everyone has the right to freedom of movement and residence within the borders of each state.

2 Everyone has the right to leave any country, including his own, and to return to his country.

Crossing borders

Every day around the world, tens of thousands of people migrate between countries. In some ways, it has never been easier to migrate, for example, by aircraft, trains, buses, trains and ships. It is also easier to find out about other places via the Internet, television and other media. But in other ways, migration can be incredibly difficult because of international borders and laws that stop **immigration**. Managing immigration is one of the most controversial issues in the world today.

KEY TERMS

Asylum – to claim safety in another country.

Economic migrants – people who move to another country to get a better job and improve their standard of living.

Immigration – the movement of people into a country, to live there.

Refugees – people who have fled from their homes in one country, usually against their will, to seek a more secure life elsewhere.

Pushed and pulled

People must have very good reasons (factors) to leave their home area or country. There are also usually strong reasons to attract the migrant to somewhere else. These are called 'push and pull factors'.

Reasons to leave an area can be extreme, such as a war or a natural disaster that put lives at serious risk. People who leave for these kinds of reasons are called **refugees**. Others leave because they cannot find work or because wages are too low for them to have a good quality of life. These called **economic migrants**. There are also people who move for political reasons when a government's laws put their lives at risk. They can claim **asylum** in another country where they think they will be safe.

Pulled by places

There are usually important reasons why a migrant chooses to move to one place rather than to another. Some are recruited to do jobs that need to be done in another country. Many migrants, for example, have been recruited to work in the health service or to work in temporary jobs on farms in the UK. Migrants also have to think about whether they will be allowed into a country. Laws in the European Union (EU) countries allow people to move freely for work within the member countries. This is how migrants from the EU countries in Eastern Europe have been able to move to the more wealthy EU countries where even poorly paid jobs can be worth doing.

Often, migrants with university degrees and other skills take poorly paid jobs in the hope of a better life in the future. Others earn money to send home, then return when they have earned enough to have a better life. A problem is that many migrants are misled by what they think about a place. They can find that conditions for them can be as hard as in the place they left.

Figure 2.19 About 4.3 million Iraqis have moved from their homes as refugees since Iraq was invaded by US and UK troops in 2003. About 2.2 million moved to another country such as Jordan and Syria. About 70 per cent of those who moved came from Baghdad.

Migration risks

Migrants who have the greatest problems are those who try to move from desperately poor countries to countries that are richer, but that will not give them permission to enter. This affects, for example, people from many countries in Africa and Central America who want to migrate to the richer EU countries or to the USA. Many have died when attempting sea crossings, for example across the Mediterranean or from Africa to the Canary Islands. When migrants do arrive, they can be taken to detention camps and then sent back. Even if allowed to stay, they are likely to have the lowest paid jobs that make it hard to find accommodation and earn a living.

It is worth asking why their home country is so poor and whether there is anything that richer countries can do better trade arrangements and help with health care, for example. What do you think?

Figure 2.20 Migrants cross from Africa to the Canary Isles in open boats in an attempt to enter an EU country.

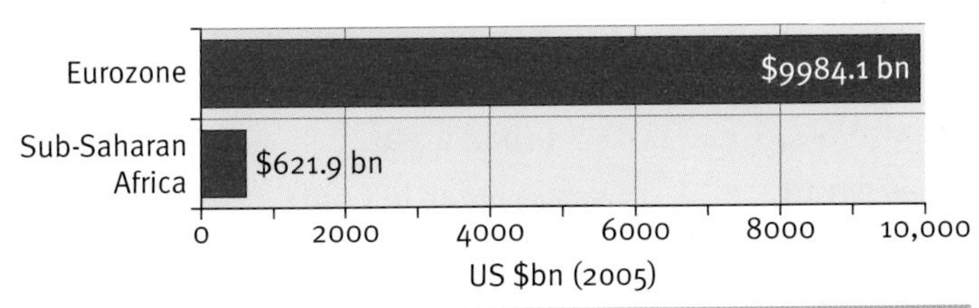

Figure 2.21 Contrasts in wealth between EU countries and countries south of the Sahara Desert as measured by GDP.

PLENARY ACTIVITY

Do you think that the number of migrants in the world will increase or decrease over the next 10 years? Give reasons for your answer.

DISCUSSION POINT

In the UK, some politicians want to set a fixed number of immigrants that would be allowed to come into the country each year. What do you think of this idea?

Figure 2.22 Routes taken by migrants from African countries to Europe. (Source: Hein de Haas, International Migration Institute, University of Oxford)

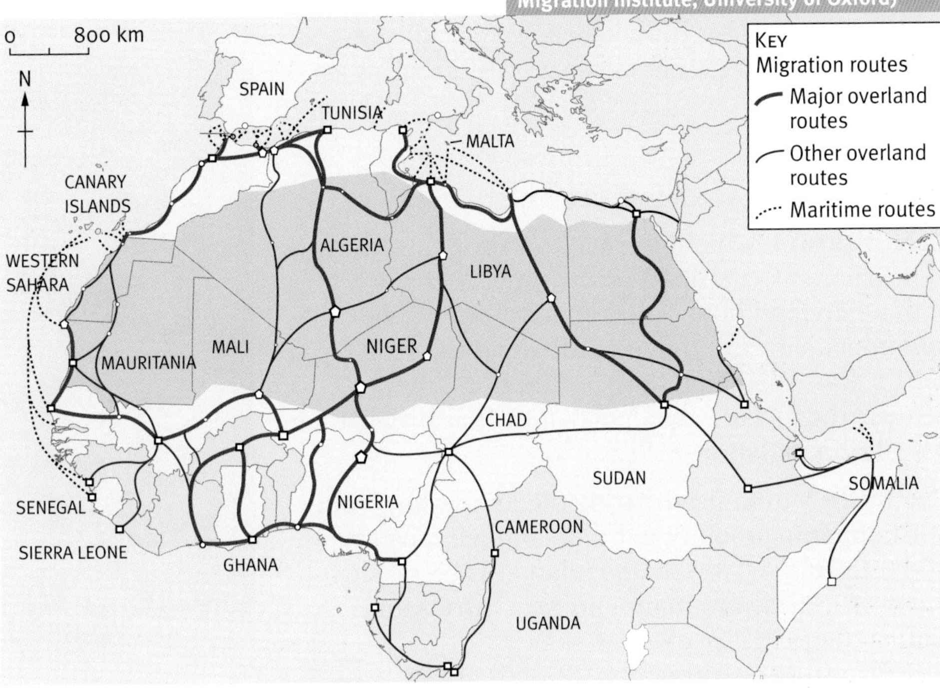

ACTIVITIES

1. Read the statement about migration in the UN Charter of Human Rights. What do you think it aims to achieve?
2. Why do you think that governments regard economic migrants differently from refugees and people who are seeking political asylum?
3. Do you think it is a good idea for workers to be able to move to work between EU member countries? Think about who might benefit and lose from this movement.
4. Use Figure 2.22 to describe the pattern of movement for migrants.
5. Why do countries in the EU attract migrants from poorer countries such as those in Africa, especially from the sub-Saharan countries?
6. How can migrants get a false impression of a country to which they want to move?

Urbanisation: why do so many people want to live in cities?

GET STARTED

Look at the photo in Figure 2.23 of the Chinese city of Shanghai. Write down five questions you would ask a resident of this area to find out what living conditions are like in the city.

Figure 2.23 *The population of Shanghai is about 18 million. Although the city's own population is not replacing itself, migration continues to add to the total. About 4 million people living in Shanghai have migrated there.*

Some of the causes of urbanisation

Urbanisation is the movement of people from rural to urban areas, a human process that is changing both where and how people live. In 2008, about 3.3 billion people were living in urban areas. On 23 May 2007 it was calculated that, for the first time in history, the global urban population was bigger than the rural population.

The MEDCs urbanised during the last century. Now it is happening rapidly in other countries, especially in the LEDCs, where the population is increasing the most. These are mainly in Asia, Africa and South America.

A lack of options

Urbanisation is usually the result of migration inside a country as people move from rural to urban areas. Although the percentage of people in the cities is increasing, the number of people living in rural areas is often also increasing. This makes it hard to sustain a living from the land, either by subsistence farming or by growing and selling produce. Farming becomes unsustainable when trees are cut down, water supplies dry up and soil is eroded. Global climate change, much of it caused by pollution and CO_2 produced by people in rich countries, is adding to these problems. Problems are sometimes made worse by short-term events such as war, drought or other disasters. With few opportunities for work and a better standard of living, staying in rural areas is usually not an option.

KEY TERMS

Informal sector – jobs that are often without regular hours and payment.

Slums – areas of poor, crowded housing in cities.

Urbanisation – the process through which an increasing percentage of a country's population lives in urban areas compared with rural areas.

Fact file

Urbanisation in China

- About 45 per cent of people in China live in cities.
- About 200 million people have moved from their home area, mainly moving from rural to urban areas.
- Incomes in China's cities are just over three times more than in rural areas.
- In 2006, the cities grew by an extra 7 million people.
- Another 300 million are predicted to leave rural areas for the cities over the next 20 years.

(Source: *Migration News*)

Some consequences of urbanisation

Moving to a city is an option that many people believe will give them some hope of a better life. Usually, it is the young, most educated and economically active who move to the cities. A few find work in offices and factories, but most work in what is called the **informal sector**. These are jobs, for example, in local services, as temporary labourers or in making goods from scrap materials. There are usually no social security or unemployment payments, so everyone has to be inventive and work hard to survive.

Fact file

Urbanisation in Africa

- About 41 per cent of people will live in cities in African countries by 2010.
- The urban population will be about 52 per cent in 2025.
- It was predicted that about 12 million people would move to live in cities in 2008.
- About 62 per cent of the urban population live in slums and shanty town areas (called bidonvilles in some African countries).
- About 300 million people in urban areas will be without clean water and sanitation by 2020.
- About 60 per cent of the GDP for countries in Africa is produced in the cities.

(Source: Global Land Tool Network (GLTN) as facilitated by UN-HABITAT)

Figure 2.24 *A poor area of Lagos in Nigeria where people build their own houses but lack sanitation, electricity and a clean water supply.*

DISCUSSION POINT

Which of these things do you think that people who live in shanty towns would most like to be improved first: housing; education; water supply; sewerage? Choose one and give your reasons.

Living on the edge

Vast areas called shanty towns have grown up on the edges of the big cities. These are where most of the poorest people have built their homes. Other smaller spaces are also used throughout the cities, sometimes next to a modern high-rise building. There are also areas of **slums** in older, crowded parts of the cities.

Shanty towns that are still growing do not have the basic amenities that people in a city really need, such as water supply, sewerage and electricity. Instead, water has to be bought from water tankers or may come from polluted streams. Human waste is put wherever can be found to dispose of it, often in the same rivers that are used for drinking and cooking water. At the moment, it is estimated that about 1 billion people in the world live in shanty towns and slums where conditions are poor.

PLENARY ACTIVITY

Imagine yourself as a young man or woman in a rural area of an LEDC. What questions would you ask yourself and others in your village so that you could make up your mind about staying or leaving? What answers would make you decide to move to a city?

Benefits from urbanisation

In spite of the problems, urbanisation is bringing many advantages to people in countries where it is growing rapidly. Cities can become centres of business and enterprise. There is a source of cheap labour for factory work that helps to build the country's economy. It also means that people in rich countries can benefit from low-cost goods. It is useful to remember that problems in one part of the world can not be separated from what people do in other places.

ACTIVITIES

1. What data from the Fact file might suggest that urbanisation in China will continue into the future, perhaps for the next 50 years?
2. Which types of area are people coming from to live in the cities and why are they migrating from these places?
3. What problems of rapid urbanisation are shown in the shanty town photo in Figure 2.24?
4. Why do you think that governments don't do more to stop people moving to cities from rural areas?
5. Can urbanisation be considered sustainable? Argue your case.

Management of urbanisation: what can be done to make life better in fast-growing cities?

GET STARTED

Look at the photo in Figure 2.25. Imagine you live there. What would be your top five concerns?

Figure 2.25 *Rocinha is a poor part of Rio de Janeiro called a favela (shanty town). Nobody knows exactly how many people live there! It has existed for several decades so there has been time for people to make their own improvements to their houses and living conditions.*

Runaway growth

One way to manage urbanisation is to stop people moving to the cities. Another is to make life better in the country areas so people will not want to move. Both of these ideas are hard for a government to do in practice.

Managing a city means building new roads, sewerage, schools, hospitals, water and energy supplies and planning how land can be used in different places. These things are all part of a city's **infrastructure**. When a city is growing by thousands of people every week, especially in a country that does not have the money to pay for the infrastructure, it is hardly surprising that there are problems.

Fact file

Improving Rocinha

- There is some local government organisation and there are community groups that help plan the area.
- About three-quarters of the people have access to electricity and most have basic sanitation and a water supply.
- There are banks, shops, bus routes and many other types of business.
- It is known for its high murder rate, drug dealing and sometimes for deaths when the steep land it is built on collapses.
- Tourists can visit it on a guided tour.

The upgrade challenge

The shanty towns and slums are the areas with the worst living conditions. The shanty towns have often been built on marginal sites: marshy land or steep and dangerous land that nobody else wanted.

No matter where they are, the people have often just moved onto the land as illegal **squatters** who have no rights of ownership. Every type of poor area needs to be **upgraded**. It might be easier to plan and manage an area if it could be done from the start, but upgrading means changing what is there already. This means changing not only buildings and roads; it can also mean changing communities and ways that people make a living. The Rocinha district in Rio de Janeiro is an example of how improvements can slowly be made, often by the local people who live there.

KEY TERMS

Infrastructure – the basic amenities that people need in a city, such as roads, sewerage, electricity and water supplies.

Squatters – people who occupy land illegally and may build houses here.

Upgrading – making conditions better.

The Dharavi solution

The Dharavi district of Mumbai in India is one of the world's biggest slums and has similar problems to the slums of Rocinha. There is a plan to redevelop the area by building new 20-storey blocks of flats with 100,000 new apartments, as well as providing the rest of the infrastructure that people need. To do this, the slum houses are being pulled down, together with the many workshops and other facilities that people have built for themselves. Anyone who can prove that they have lived in the district since before 1995 is offered a new home for free in one of the new housing blocks.

Impact of change

Not everyone agrees with how the redevelopment is being done. Private building companies have been given land in exchange for building the new homes. They will be able to build more houses on this land and sell them at a profit. The redevelopment also involves breaking up the social networks of friends and families that help people to survive. Many of the small-scale businesses will also be affected. This model of large-scale redevelopment will bring benefits to many, but it seems that not everyone will share in these benefits.

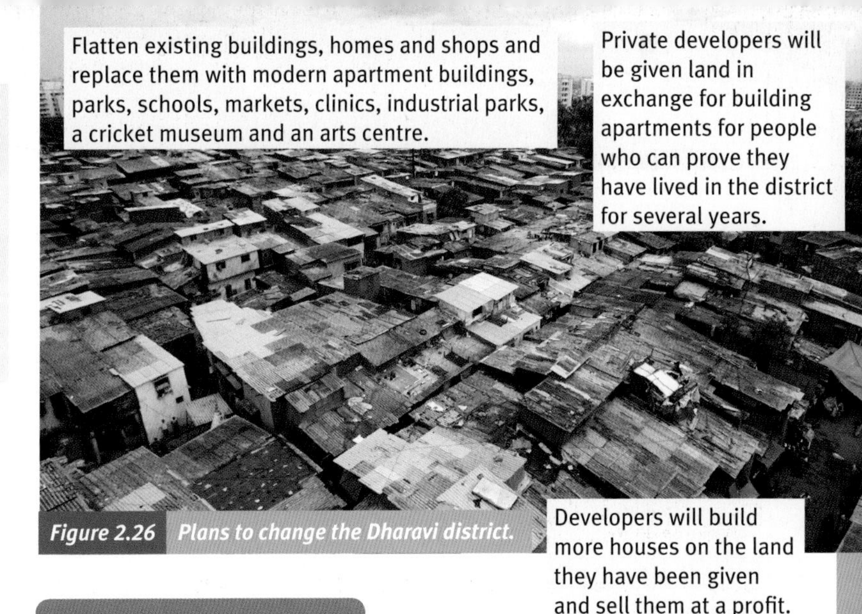

Flatten existing buildings, homes and shops and replace them with modern apartment buildings, parks, schools, markets, clinics, industrial parks, a cricket museum and an arts centre.
Private developers will be given land in exchange for building apartments for people who can prove they have lived in the district for several years.
Developers will build more houses on the land they have been given and sell them at a profit.

Figure 2.26 *Plans to change the Dharavi district.*

'India must eradicate its ubiquitous shanty towns if it is to become an economic success story': Indian architect

'We bring the entire city's dirt and create a livelihood from it': resident of Dharavi

'It is a unique, vibrant, thriving cottage industry complex, the only one of its kind in the world where all the raw materials used, the processes that make the final product are carried out at the same location and the value added is very high': critic of the redevelopment plan

Fact file

The Dharavi district

- Has a population of between 600,000 and 1 million people.
- Covers an area of 525 acres.
- There are about 15,000 small workshops and factories.
- People recycle cardboard and plastic to make a living.
- The recycling businesses make about £350 million each year.

ACTIVITIES

1. Why is the basic infrastructure usually not there in a shanty town?
2. What improvements do you think have been made to the buildings and local environment that you can see in the photo of Rocinha?
3. Suggest what else might be done to improve the area shown in the photo so that life could be made better for local people.
4. Study the information about Dharavi. Explain why it would be hard for local people to do this kind of redevelopment themselves.
5. Why do you think that developers wanted to be involved in redeveloping this area?
6. Why do you think that different people have different views about how the redevelopment is being done at Dharavi? You can find more viewpoints at the Dharavi, Mumbai website.

DISCUSSION POINT

Are the plans for Dharavi's redevelopment in the best interests of its inhabitants?

PLENARY ACTIVITY

From what you have read and seen about ways to upgrade a slum or shanty town area, make a list of ideas that you could present to people in another area that needed to be upgraded. Also, list which ideas should be avoided!

The process of counterurbanisation

GET STARTED

Name at least two television programmes that are filmed in rural areas of the UK. What impressions do they give you of what it is like to live in the countryside?

Move to the country

In most of the MEDCs, people are moving between urban and rural areas in a variety of ways. While some people still move to urban areas, others are moving out of the cities. This process is called **counterurbanisation**.

The areas that are attracting most people from cities are mostly within **commuting range** so they can travel daily to their work in the city. A railway line or motorway can give fairly quick and affordable access to places that are up to 60 miles away from a city centre. Many people travel even further. The places that people move to in rural areas, however, can be quite varied. These can be to villages, new housing estates and to country towns.

KEY TERMS

Commuter range – the distance people will travel from their homes to places of work.

Housing association – an organisation that manages the building of houses for local people.

Rurbanisation – the process of bringing features of a city to a rural area.

THINK ABOUT IT

Do you think that people who are born in a rural area should be given special help to be able to stay living there?

Figure 2.27 *The Vale of Glamorgan between Cardiff and Swansea has become a popular area for commuters. The M4 motorway that links the two cities is in the background.*

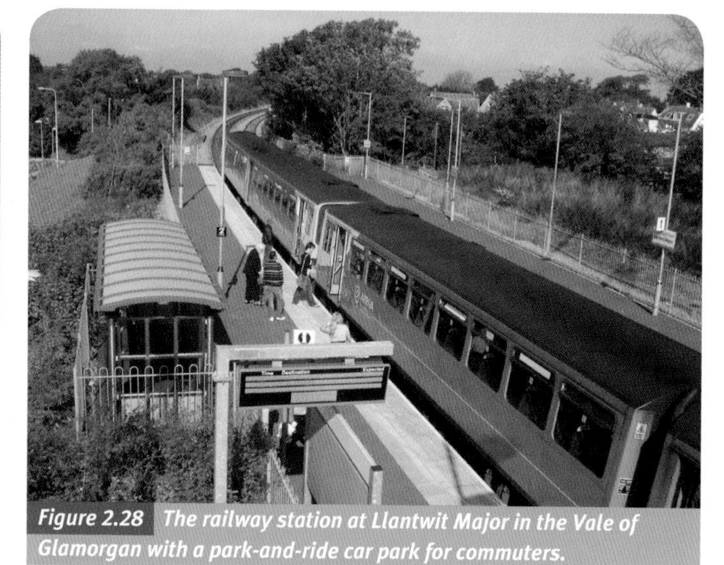

Figure 2.28 *The railway station at Llantwit Major in the Vale of Glamorgan with a park-and-ride car park for commuters.*

Table 2.9 Population statistics for Cardiff and Swansea 1971–2006

Year	Swansea	Cardiff
1971	226,406	290,227
1981	223,260	274,500
1991	233,145	272,557
2001	223,293	292,150
2006	227,100	317,500

(Source: UK Statistics Authority website, Crown copyright)

Table 2.10 Population change in the Vale of Glamorgan

Year	1971	1981	1991	2001	2006
Population	106,300	112,120	116,980	119,292	123,300
Children aged 0–4	8,833	7,451	8,165	7,086	7,045
Children aged 5–15	19,629	19,945	17,165	16,607	15,744
Pensioners (to 74)	11,559	12,838	14,304	13,396	16,959
Pensioners (75+)	4,907	6,075	7,986	9,786	10,597

(Source: UK Statistics Authority website, Crown copyright)

Understanding the data

It is also important to know that cities are defined by the boundaries of administrative areas. When a new housing area is built just outside this boundary line, the statistics will not include the new residents in the city's total. So while the built-up area continues to grow, the city's population total can show it is declining.

Reasons to move

A major attraction of living in the country used to be the lower cost of houses. Because of demand from commuters, this is no longer true in some cases. A difference, however, is that there is likely to be more garden space in a country area, the local scenery is likely to be more attractive, crime rates are generally lower and pupils in the local schools may achieve higher exam results than in some city schools. There are fewer places for some types of entertainment and shopping in rural areas, but the city is easily accessible when needed.

Pressure on property

Some new residents spend money renovating buildings and in some cases, completely change their use. Old barns, village schools and chapels can be converted into homes. But many of the new residents prefer to buy new houses. This causes extra demand that puts up their value. This makes it impossible for young people to stay in the area, unless they stay living with their parents. One solution is to build houses that only local people can buy, but it takes a special agreement with a **housing association** for this to work.

City meets country

A village can become a 'dormitory' village where people sleep, but don't work in the nearby area. But the new residents can help local shops, schools and businesses to stay open. The same areas can be attractive for businesses that need space and good access to the same communications links that the commuters use. This can bring much-needed jobs to rural areas.

Incomers can also affect the social life of the area. Some try to integrate with the local people and can bring new ideas to a community. But this is not always easy and the result can be a divide between the newcomers and the locals. The changes that people from the cities bring to rural areas are called the process of **rurbanisation**.

Fact file

Meeting local demand for housing in rural areas through a housing association

- Houses are built by a developer.
- Planning permission is given, usually within a village's existing boundaries.
- About one-third of the houses in a group are built as affordable housing.
- The housing association manages the affordable housing.
- Only people with local connections are allowed to live in the houses.
- People may pay for the houses on favourable terms, for example through Shared Ownership schemes.

Figure 2.29 ***New businesses can move to rural areas to take advantage of the space, access, attractive working environment and the local labour force. This business park is near Bridgend, in the Vale of Glamorgan.***

ACTIVITIES

1. What might attract people from a city to live in an area like Llantwit Major?
2. What style of graph do you think would be best to show the changes in population total for the Vale of Glamorgan between 1971 and 2006 in Table 2.10? Draw it then explain your choice.
3. What trends in population change do the figures show?
4. Draw a bubble map to show the ways in which new residents to a rural area can bring improvements to the area. Think about: houses and other buildings; the landscape; transport; shops and services employment; leisure and social life. Add notes to say who will benefit from the changes.

PLENARY ACTIVITY

Draw a pair of sketches or maps with symbols and notes that show how a village and surrounding land might change as a result of counterurbanisation. You can base your illustration on an area you know or research, or on one that you create from your imagination.

CASE STUDY UK internal migration

GET STARTED

The BBC is moving several of its services from London to Salford (in Greater Manchester). What effects do you think this might have on Salford and the surrounding area?

Figure 2.30 ***The old docks at Salford are being redeveloped to attract new jobs to the area that has had high levels of unemployment.***

Migration between regions

There are differences in the number and types of jobs and the amount of money that people earn in different parts of the UK. The same is true for most other countries. Migration between places is one effect of these differences. It is hardly surprising that people want to move from places where there are few jobs to places where jobs are better paid and are increasing. In the UK, there are 12 areas called Economic Planning Regions, each comprising several counties.

Inflow, outflow and balance

Recent figures for the pattern of migration between the UK's regions show big differences in the numbers of people who move to and from them. The London region has seen the greatest loss of people, while the South-West has attracted the most. Although the figures change from year to year, the pattern has been similar for at least the last 10 years.

The pattern of migration, however, is complex. In every region, people move both in and out, sometimes in almost equal numbers. So while the London region had the greatest overall loss of people, it also had the second highest inflow from other regions. People also arrive from other countries. Another complication is that the figures for migration between regions hide the fact that people also move between places within each region, especially from the big cities to rural areas and smaller towns.

THINK ABOUT IT

There is a minimum wage rate in the UK. Do you think that the rate should be different in each region?

Table 2.11 Migration between the economic planning regions in England and Wales for 2006

Region	Inflow	Outflow	Balance	Total population (thousands)
North-East	39.8	38.7	1.1	2558.3
North-West	100.9	103.5	–2.6	6846.2
Yorkshire and the Humber	93.0	92.4	0.6	5063.9
East Midlands	107.8	98.3	9.5	4306.3
West Midlands	93.2	99.9	–6.7	5365.4
East	140.5	124.4	16.0	5541.6
London	163.1	243.7	–80.6	7517.7
South-East	219.6	200.2	19.4	8164.2
South-West	134.0	107.1	26.9	5067.8
Wales	55.8	49.2	6.6	2965.9

(Source: National Statistics website)

Table 2.12 The origin of migrants to the South-East region in 2006

Origin of migrants to the South-East	Number (thousands)
North-East	3.9
North-West	11.2
Yorkshire and the Humber	9.5
East Midlands	13.9
West Midlands	13.0
East	27.6
London	90.3
South-West	32.3
Wales	7.1
Scotland	6.3
Northern Ireland	1.3

(Source: National Statistics website)

Jobs and houses

Many people move between regions to find work, sometimes when the business they work for closes down or to find a first job. Older people can choose to move to regions with attractive coastal areas when they retire. The cost of housing can also be important. Although the South-East is an area with many well paid jobs, it is also where house prices are highest. This can force some people to move away. It can also stop people moving there from other regions.

Government actions

The UK's government has a part to play in managing migration between the UK's regions. It is helpful to the country's economy if people can move to where businesses find it most profitable to locate, but too much migration to a region will mean finding enough land for houses and providing extra services such as health, education, water supply, energy and waste disposal. People who already live there may resent their area becoming more crowded and feel that politicians do not care about their needs.

Government statement on developing the regions, 2003

Regional policy is at the heart of our efforts to reach this goal, ensuring that economic prosperity reaches every part of the country and that everyone, no matter where they live, has the chance to make the most of their potential. For too long, too many nations and regions of the United Kingdom have been allowed to fall behind; for too long there have been huge differences in prosperity within regions; and for too long too many people have been left out, their talents wasted.

(Source: *A Modern Regional Policy for the United Kingdom*. © Crown copyright 2003)

Managing regions and migration

The UK government is moving some of its own workers between regions; for example, about 500 jobs in educational administration are moving to Coventry from central London. In total, about 20,000 government jobs are to be moved out of London by the year 2010. Not everyone wants to move to another region, so their jobs will be taken by people who already live there.

Figure 2.31 ***The Eden Project in Cornwall was built using money from central government and from EU funds. It has brought at least 400 jobs to this former china clay mining area and attracts about 1.2 million visitors every year.***

The government also gives grants and loans to businesses that want to locate in regions with high unemployment. This helps create jobs and it means that people are not forced to migrate. Spending money on motorways and other infrastructure can also help bring jobs to a region.

PLENARY ACTIVITY

Write a list of arguments both in favour and against the UK government spending money from taxes on trying to bring jobs to regions where there is high unemployment. What do you think would be the most important argument that would persuade you either to spend the money, or to let people migrate to find work?

ACTIVITIES

1. Look at Table 2.11. For your own region, say:
   - **a** whether there are more gains than losses
   - **b** how the inflow and outflow compares with regions that have the highest and lowest figures
   - **c** how the amount of inflow and outflow compares with the total population for the region, for example as a ratio or percentages.
2. Look at the data in Table 2.12 for the origin of migrants from other regions to the south-east region. What type of mapping technique do you think would be best to show these data? Explain your choice. Include a simple sketch map to show what it would look like.
3. What might explain that as some people are moving into a region, others are moving out? Think about:
   - different types of work
   - people of different ages
   - people with different interests.
4. Explain how the cost of houses can play a part in people's decisions about moving between regions. Remember that there are also big differences in prices within each region.
5. Look at Figure 2.31. How do you think that this project might affect migration to and from the region?

The many sides of migration

GET STARTED

Imagine you were about to go to work in another country that you have never previously visited. What do you think might be the most difficult things you would face and how might you overcome these problems?

Coming and going

Every year, millions of people **emigrate** from the country in which they were born. They become an **immigrant** in the country to which they move. Both emigration and immigration are human processes that have gone on throughout history. They are, however, processes over which people often disagree. This is why every country has laws about migration, especially immigration.

Table 2.13 Migration to and from the UK

Year	Inward migration	Outward migration	Net inward migration
Total 1998–2007	5,141,000	3,329,000	1,812,000
2007	577,000	340,000	237,000

(Source: Optimum Population Trust, an environmental think tank)

Emigration issues

Most emigrants leave their country to find better paid work elsewhere. Some are getting away from extreme poverty and sometimes, from danger. Often, it is the most qualified and skilled who choose to emigrate, such as nurses from the Philippines to the UK or scientists from the UK to the USA. This, of course, takes these skills out of countries where they are needed and that have spent money in training.

Some emigrants gain new experiences that they can use if they return to their home country. They can also send money back to support their families. Others, however, never return. When there is a high level of emigration, the population structure can be affected, leaving mainly the more elderly people behind.

Fact file

- There are about 190 million migrants between countries every year.
- One in every 35 people is a migrant.

KEY TERMS

Emigrant – a person who leaves a country to go to live in another country for longer than a year.

Immigrant – a person who moves to live in another country for longer than a year.

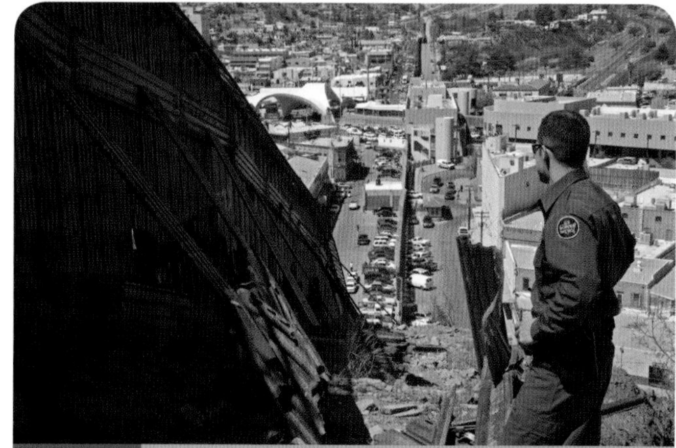

Figure 2.32 *The US government has built a security fence along its border with Mexico to try to stop illegal migration from Mexico. Between 200,000 and 400,000 Mexicans try to cross the border illegally each year. About 500 died in the attempt in 2007.*

Fact file

Migrant workers from the Philippines

- In 2007, 1.1 million people left the Philippines to work abroad.
- Filipinos account for 28 per cent of workers on ships such as cruise ships and bulk carriers.
- Almost 30 per cent of Filipino workers went to Saudi Arabia.
- About 8000 Filipino nurses went to work in other countries, with 85 per cent of newly qualified nurses going abroad.

(Source: *Migration News*)

Benefits for some

Immigrant workers often do jobs that people in the home country don't want to do. In the UK, some migrants are hired by 'gangmasters' to work at seasonal jobs on farms such as picking fruit and vegetables. Many work in hotels, other service jobs and in industries where they are paid a minimum wage. The low wages paid to immigrant workers help keep prices down for people in the UK.

In time, some immigrants settle down to a new life, paying taxes, raising families and becoming full citizens. This is a benefit, especially in a country in which there is a low or even a falling birth rate such as the UK. They also bring the richness of new ideas and new cultures to a country.

> 'Migrant workers are an asset to every country where they bring their labour. Let us give them the dignity they deserve as human beings and the respect they deserve as workers.' Jaan Somavia, Director General of the International Labour Organisation

Fact file

The effects of immigration – birth rates (1990–2008)

- Births by British-born mothers down 44,000.
- Births by all foreign-born mothers up 64,000.
- Births by mothers born in Eastern Europe up 15,000.
- Births by mothers born in Indian subcontinent up 11,000.
- Births by mothers born in Africa up 8000.

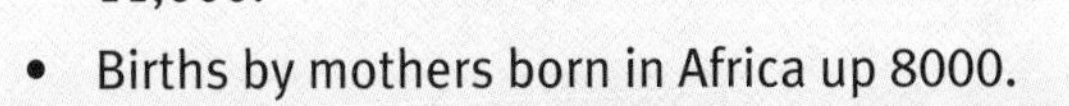

(Source: UK Statistics Authority website, Crown copyright)

Figure 2.33 *Migrant workers working on temporary farming jobs in the UK.*

Issues to address

Immigration, however, can, also bring some problems. Many of these are short-term problems that can affect some areas more than others. There can, for example, be difficulties in housing, school places and in competition for jobs. Sometimes, these difficulties are to do with what people think rather than what is true. This is when immigration can become a political as well as a practical issue that has to be addressed.

The usual way to address these issues is for politicians to make laws to control the number of immigrants and to make sure that only those with particular skills are allowed to come in. A work permit that gives permission to work for a short time is one way to do this. The number of work permits can be controlled by a government, depending on the demand for particular skills.

A problem is when people try to enter and stay in a country illegally. Those caught doing this can be sent to a detention camp, then deported. There are many different viewpoints over this and the other questions concerned with international migration.

ACTIVITIES

1. Study the figures for migration to and from the UK in Table 2.13. Are there any reasons why these figures might cause concern to anyone? Explain your answer.
2. Look at the photo in Figure 2.32. Why do you think that people in the USA are so concerned to protect their border with Mexico?
3. What problems can be caused when migrants leave a country such as the Philippines? Think about the effects on the economy of the area they leave.
4. What benefits do migrant workers bring to a country such as the UK? Who benefits from the work they do?
5. Why do you think that migrant workers to the UK often do low-paid and temporary jobs?
6. How can a country's population structure and dependency ratio be changed by immigration? What might be the effects of these changes?

DISCUSSION POINT

Do you think it is best if people who migrate to a country are kept to a small number of areas, or dispersed across the country?

PLENARY ACTIVITY

Sum up the reasons why you think that international migration is a topic that causes so much disagreement amongst so many people.

A Decision Making Exercise – Is the inward migration of workers positive or negative for the UK?

The EU encourages freedom of movement for workers amongst member states. In 2004 the EU welcomed eight new members, including Poland. The UK government placed no restrictions on workers from these states and within 2 years over 600,000 workers arrived. The workers were deemed 'economic migrants' but their arrival also had significant social effects on the UK as well.

There are very different views on the positive and negative impacts of these new arrivals. With two more countries joining the EU in 2007 (Bulgaria and Romania), the UK government decided to place severe restrictions on workers entering the country. Three more European states, including Turkey, hope to join the EU in the next 10 years, and others will follow. Consider whether the UK government should relax its policy on inward migration.

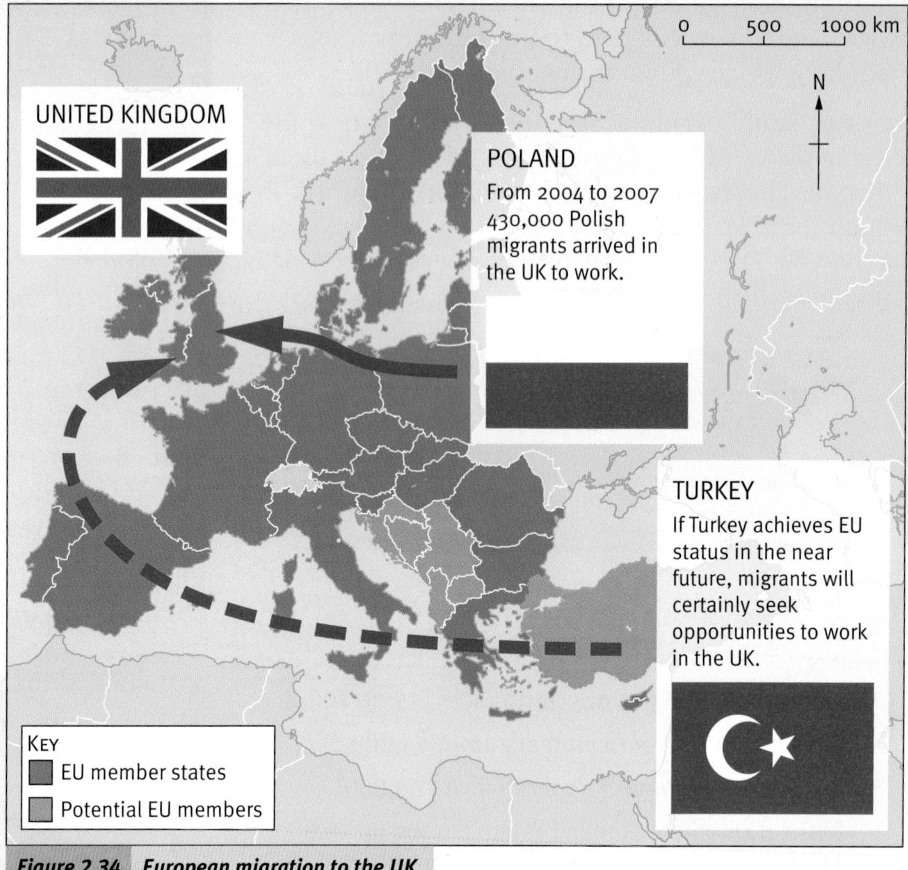

Figure 2.34 *European migration to the UK.*

Freedom of movement for workers

An important principle of the EU is to allow workers from member states to move freely between member nations.

This policy has huge economic benefits for all concerned. The UK has been able to recruit workers with a wide variety of skills. With the UK having one of the lowest unemployment rates in the EU, the economic migrants were welcomed by companies who found it difficult to recruit British workers. The migrants are expected to pay taxes in the UK from their income.

Much of the debate about the positive and negative impact of migrant workers in the UK stems from their social impact on the country. The next page shows wide-ranging views on the issue. The rights that migrant workers and their families should have to social benefits such as health, education and housing fuel the debate.

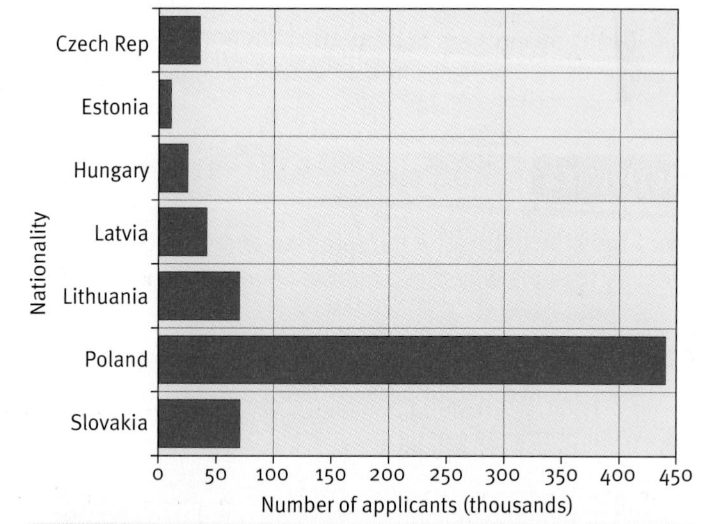

Figure 2.35 ***Nationality of Europeans applying to live and work in the UK, May 2004 to June 2007.*** **(Source: Office for National Statistics, Crown copyright, 2007)**

Read the viewpoints on this page to understand the issues and to appreciate why it is a hot topic for debate.

My life in England

'I came here because I wanted to find a new life for myself. ... Although I have a degree in politics and economics, I found it very hard to get a decent job in my home country and I ended up working as a cashier in a petrol station.

I came here and stayed with friends at first, but since my son Kamil came over two years ago I have rented a flat above a pizza takeaway shop. I consider this country to be my home country now. ...

I work here and want to stay here with my son. He can have a better life here. He wants to learn about computers and study them at university. He already speaks better English than I do. I would like to get a better job, perhaps in a hospital or a post office, when I can speak English better.

I have everything I want. There are many Polish shops and I have lots of friends, and my son. We are happy here.'

(Source: *Daily Telegraph*, 2008)

***Figure 2.36** Monica Tkaczyk arrived from Poland in 2004 and lives in west London.*

Wide-ranging views

'I live in a Suffolk village of just 235 people. When 95 East Europeans came to work in the local food processing plant our lives changed overnight. My local pub became their club! Sometimes you can't hear English being spoken anywhere in the village.'

Key industries welcomed Polish workers when there was a shortage of skilled plumbers and construction workers. Other sectors such as agriculture, health and the hotel industry would not have been able to meet the demands of customers without the Polish migrants.

***Figure 2.37** The balance of positive and negative impacts.*

Polish migrants contributed £1.9 billion to the government in tax and national insurance in 2007.

'The police are always breaking up fights between the different groups of Europeans in our neighbourhood. Most of the migrants are young men with little to do in their spare time.'

'They are bright, cheerful and polite and are prepared to work long hours to support their families.'

'The families of the migrant workers put a huge strain on our schools and hospitals. Many Polish children don't speak much English. Who pays for the additional support in our classrooms? Many of the workers decide to start a family over here. Who pays for the health visitors and the hospital care?'

'It is a fact that Polish migrants are three times more law-abiding than the average British citizen.'

YOUR TASK

Consider the evidence on these two pages and the additional information on the supporting CD.

It is 2020 and you read a headline in your local paper: 'Turkey becomes a full member of the EU'.

Write a letter to the editor of the paper giving your views on the possibility of a new influx of migrant workers seeking opportunities in the UK.

Your letter should be balanced and it should recognise both the *social* and *economic* arguments of the inward migration.

***Figure 2.38** A specialist Polish shop in Kensington, London.*

How is the pattern of land use within cities changing?

Land use in zones

The ways that land is used are broadly the same in all towns and cities. The **land use**, for example, consists of areas for houses, businesses, leisure and transport. Areas with the same types of land use are called **zones**.

Most towns and cities grow out from their historic core. An increase in population and changing technology such as in transport and energy then changed their buildings, land use and size. Often, nearby towns and villages with their own historic core are surrounded as cities sprawl and merge to become one big city, or a **conurbation** where cities merge.

Problem land use

Some zones grew at special times in history, such as during the Industrial Revolution when factories, docks and rows of workers' houses were built close to each other. Now, many of these old industries have closed down and the docks are no longer used for cargo ships. The buildings can become derelict and dangerous and the land becomes wasteland. This ruins the environment for people who live there and wastes land that may be needed for something else. Land in a city, however, is usually too valuable to be left empty for long, no matter what it costs to change its use.

GET STARTED

Are property prices likely to be higher or lower around a school? Explain why.

KEY TERMS

Conservation areas – parts of a town with historic buildings that are protected.

Conurbation – a large urban area formed when cities and towns merge as they grow towards each other.

Land use – the different ways that land is used, for example for industry or recreation.

Residential – an area with houses.

Sink estate – a housing area with a poor reputation for its living conditions.

Zone – an area with mainly one type of land use.

Figure 2.39 *Land use change in an old part of a city.*

A block of flats was built during the 1970s to redevelop the area, but not many people are living in them. Perhaps they should be demolished.
An industrial zone that provides jobs, though heavy lorry traffic travels through the area.
Rows of terraced houses from the 19th century are still suitable as homes, though there is not much open space for recreation and the area may need to be improved.
A football stadium that attracts crowds and traffic.
Trees are conserved to help make the area more attractive.
Part of the old dockland is converted and used for leisure. The historic ship is part of a museum that attracts visitors.
Old warehouses near the docks are used for activities such as offices, storage, health clubs and arts workshops.
New homes and other buildings are being built on derelict industrial land beside the docks. Decisions have to be made about the price of the homes and who can afford to live there.
Old industrial buildings can look unsightly, but are kept for manufacturing, giving people jobs.

Plans to conserve

Some towns and cities in the UK have a history that is 2000 years old. Some of the streets have survived over hundreds of years, sometimes with original buildings still there. Nowadays, historic buildings are preserved in **conservation areas** to help give variety and character to a place so they are more interesting to live in. Sometimes, old buildings and streets get in the way of new developments so decisions must be made about what to keep and what to knock down.

Land for living

Land for houses, called **residential** land, takes up the biggest amount of land in towns and cities. Some residential areas were built by councils to rent to people, often when older residential areas were knocked down in inner city areas. Housing estates are also built by building companies to sell. Though local councils do not build houses any more, they still need to make sure that there are enough houses in their area and that people can live in good quality environments with proper services and amenities.

Parts of a city can get a reputation. Some become known as what estate agents call 'desirable' areas. These are usually where houses are the most expensive and often where there is a school in which the pupils are thought to get high exam grades. There are also parts of most cities that have bad reputations, with poor quality houses, high unemployment and high crime rates. The worst areas are called '**sink estates**'. Planning how land is used is one way to give everyone a fair chance of a decent living environment and not separating people in such extreme ways.

Figure 2.40 A conservation area in Kingsbury, Buckinghamshire.

Questions about changes to land use in a city

- What type of land use is there now?
- What is the new type of land use to be?
- What effects will it have on what is near it?
- How will different people benefit and lose?
- Who owns the land and what is its cost?
- Who makes the decisions and what are their values?

ACTIVITIES

1. Think of two types of land use that can work well together. Explain your choice. Think about: landscape; transport; pollution; links between businesses; safety and sustainability.
2. Think of two types of land use that can cause conflict with each other if they are together, perhaps in an area that you know. Describe and explain the reasons for the conflict.
3. What are the advantages of land use being mixed rather than in zones?
4. What qualities would persuade you to conserve an old building in the centre of a town?
5. What are the features of a residential area that will give it a reputation as a desirable location in which to live?
6. How do you think that the reputation of an area can be changed? Who has the most influence over that change: individuals, local authorities or somebody else?
7. What do you think of the idea that different types of house should be mixed in residential areas, for example, with different sizes, costs and styles? Could it be more sustainable than an unmixed development?

PLENARY ACTIVITY

Outline the reasons why the ways in which land is used in a city need to be planned.

Write some guideline ideas that identify ways in which land used in a city can be changed so that it can be made more sustainable. Think, for example, about land use that helps give people a good quality of life, access to services and jobs, waste disposal, saving energy and transport and other ideas.

Building for a brighter tomorrow

GET STARTED

Think about the worst part of a town or city in which you live or that you know. What is wrong with it?

From decline to regeneration

Businesses come and go as technology, transport, products and markets change. This has affected places such as city docklands and land near canals. When they close, jobs are lost, people have less money to spend and the whole area can go into a spiral of decline. When this happens, an area will need to be improved. Doing this is called **urban regeneration**.

Waterfront sites in old city dockland and canal areas have become some of the most desirable places in which to live and to visit. People who live there can have a lifestyle that lets them work in the nearby centre and also enjoy the city centre nightlife.

Figure 2.42 The old docks at Gloucester Docks have been redeveloped to use for leisure and accommodation.

Demolition of old shops, replaced by the Cabot Circus centre, a new retail and leisure area of 92,900 m²
Space for 15 major stores and 120 other shop units, some with cafés
New cycle routes and bus stops
Over 2600 car park spaces
A 3000-seat cinema
Over 250 residential units and student accommodation

Figure 2.41 The Cabot Circus shopping mall in Bristol.

New life to the centre

In many cities in the UK, a lot of shoppers have stopped going to the city's **Central Business District (CBD)**. They prefer to go to out-of-town shopping malls where there is more space, the shopping area is under cover and it is easier to park a car. Besides, many of the buildings in UK city centres were built about 50 years ago and no longer meet the needs of shoppers. City councils want to keep businesses, and the money and jobs they bring, in their city centres. One approach to improving them is to knock down large areas and completely change the buildings and the whole environment.

Figure 2.43 Peckham: location map.

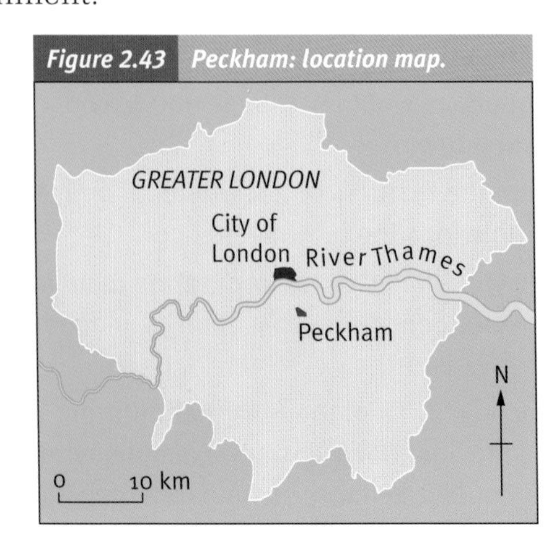

RESEARCH LINK

Choose one old dockland or canal area in a UK city. Find out about how it has been regenerated and how successful the plan has been.

Fact file

Peckham urban regeneration

The problems

- Old housing – much at least 100 years old.
- Traffic and on-street parking.
- High unemployment.
- Poor range of shops.
- High crime rate – especially burglaries.
- Low exam results in schools.
- Social problems.
- Poor reputation, so hard to attract new business.

The solutions

- About 890 new homes built and sold.
- New facilities such as a library and health centres.
- New public open spaces.
- Training for people to get qualifications and jobs.
- Spending on the transport infrastructure, such as the East London underground link.
- Helping pupils in schools to get better exam results.
- CCTV cameras helped give a 40 per cent drop in crime.
- New shops, including national chain stores.

Challenge on the estates

Improving a run-down housing area is a different kind of challenge. These can be mixed housing and industrial areas from the 19th century or peripheral housing estates that were built by local councils about 50 years ago. While buildings are fairly easy to demolish or renovate, this is not enough to improve people's lives. Getting rid of tower blocks and replacing them with low-rise individual houses is usually a good start. However, people also need jobs, local services, leisure amenities, security in their homes and to feel safe on the streets.

It is important that local people are involved in plans to redevelop their area. That is the best way to make sure that they get what they really need and value, rather than being given what someone else thinks they need.

Figure 2.44 *Urban regeneration in Peckham gives safe spaces for play in more attractive surroundings.*

ACTIVITIES

1. How do you think an estate agent might advertise a waterfront site and which groups of people would they mainly try to attract?
2. Look at the picture of the new shopping centre in Figure 2.41. What features of this design do you think that most shoppers might find to be the most attractive and will encourage them to shop there?
3. Can you think of any reasons why someone might not want to go to a new mall for shopping? Explain your answer.
4. In what ways are malls being made more than simply places for shopping? Why do you think this is being done?
5. Why is it important that local people are involved in plans to regenerate their area?
6. Study the information about the regeneration of Peckham on this page. How is the character of the area being changed? Consider its buildings; land use; scenery; facilities and amenities; security.

KEY TERMS

Central Business District (CBD) – the main shopping and business centre of a city.

Urban regeneration – rebuilding an old part of a city to improve it.

Waterfront sites – land alongside or near rivers or old docks, usually in a city.

PLENARY ACTIVITY

What do you think are the key questions to ask and the key decisions to make before an area is regenerated?

Different approaches to development: building sustainable cities

GET STARTED

How many journeys does your family make a week?

Zero carbon living

It is now a generally accepted fact that the amount of CO_2 that people are releasing into the atmosphere has to be reduced. This is so that global warming can be slowed down and eventually stopped.

About 25 per cent of the UK's carbon emissions come from homes. This is why the government has said that all new homes are to be **zero carbon** by 2016. This is a big challenge since heating, cooking and using electricity all depend on gas or electricity, much of which is made by burning fossil fuels that release CO_2.

Figure 2.45 *A thermal image showing heat loss from buildings in London. Areas with bright colours such as red and yellow show the greatest heat loss.*

Counting the carbon

Living in a way that causes global warming and that uses up finite resources is not sustainable. One meaning of sustainable living is that it can go on without causing big problems to the environment.

There are many ways in which people who live in towns and cities can live more sustainably; for example, by reducing waste, by recycling, buying goods in local shops and living near their work. Using public transport or cycling instead of going by car can also help. Some cities now have **rapid transit systems**. People need to have a lifestyle that is at least **carbon neutral**.

Low-energy homes

Inside the home, electricity can be saved by turning off standby switches and using low-energy electrical goods. Better insulation and double glazing reduce heat loss and save money on heating bills. Some houses called **eco-homes** are now being built to have zero, or near-zero, CO_2 emissions. They can do this by saving and making energy, for example, by using solar panels, insulation and other methods.

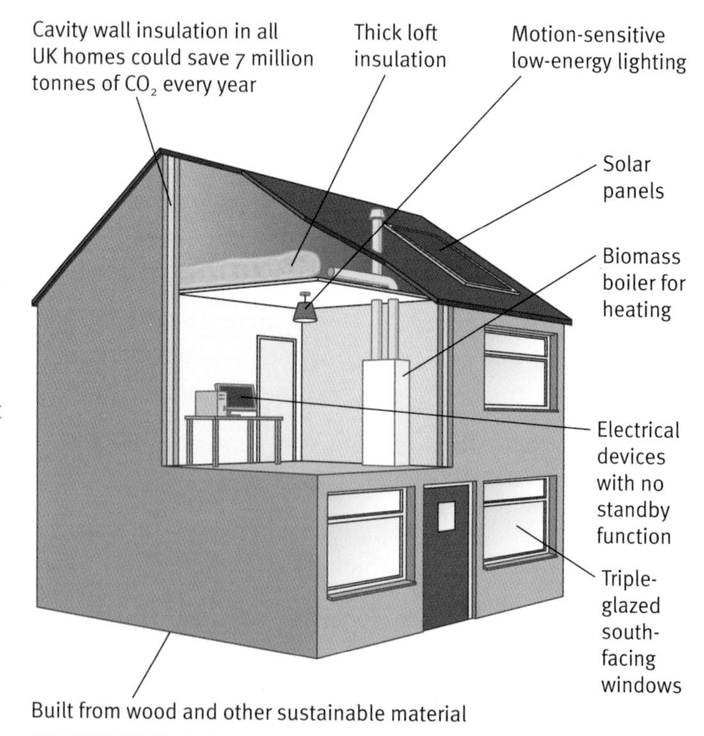

Figure 2.46 *Features of an eco-house that help conserve energy and resources.*

Eco-town plans

Building houses with zero carbon emissions is a start, but the UK government also wants to build at least 10 new **eco-towns** by the year 2020. Each is planned to have between 50,000 and 20,000 zero-carbon houses. There is certainly a need for more affordable homes and the idea of an eco-town has supporters.

However, not everyone thinks they are a good idea or would be built in the best places. One problem is that they could be too small to have the services they need, such as schools, shops and hospitals. That would mean people would have to travel out of the town, causing more CO_2 emissions. They argue that new homes should first be built on **brownfield land** in cities, so that countryside land is not destroyed.

People who live near the new towns are also likely to object as they would increase local traffic, put more pressure on local services and could lower the value of their homes. There is a suspicion that eco-towns are just a way to let house building companies build on more **greenfield land**.

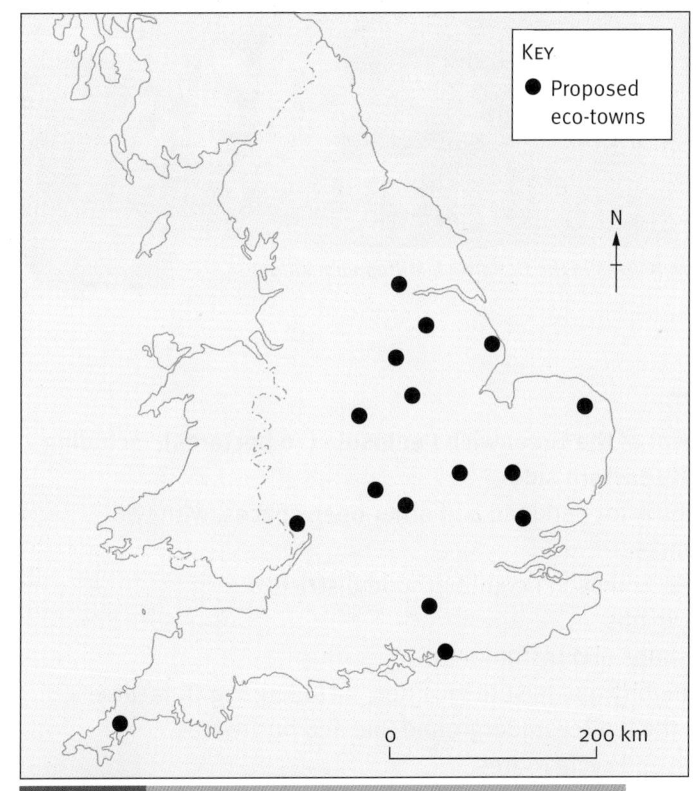

Figure 2.47 ***The proposed locations of eco-towns in England.***

> The government housing minister Caroline Flint said the new eco-towns would help to tackle climate change, as well as providing affordable new housing.
>
> The architect Lord Rogers criticised the plans for encroaching on green-belt land, adding that they were unsustainable: 'I think eco-towns are one of the biggest mistakes the Government can make … They are in no way environmentally sustainable.' Eco-towns, he argues, will damage both rural and urban environments, while increasing road congestion and carbon emission.
>
> (Source: Tom Peterkin, *Daily Telegraph*, 2008)

KEY TERMS

Brownfield land – land on which there have already been buildings.

Carbon neutral – not adding to the net amount of CO_2 in the atmosphere.

Eco-homes – houses designed in ways that conserve resources and energy.

Eco-towns – towns designed to be sustainable and that do not cause environmental problems.

Greenfield land – land on which there has not been any previous building.

Rapid Transit System – public trains or trams with short waiting times and regular stops.

Zero carbon – not releasing any CO_2 into the atmosphere.

DISCUSSION POINT

Should everyone in the UK be made by law to install double glazing and loft insulation and use low-energy light bulbs?

ACTIVITIES

1 How can a **Rapid Transit System** help reduce people's carbon footprints in a city?

2 What does the photo in Figure 2.45 tell you about the need for energy conservation in the design of buildings?

3 Thermal images of heat loss taken by aircraft or satellites could be used to tax or fine people who do not make their houses energy-efficient. What do you think of using this technology for this purpose?

4 What do you think of the government's aim to have all new homes as zero-carbon by 2016? Why might it not happen?

5 Are there any ideas being used in eco-homes that could be used in older houses to reduce CO_2?

6 What do you think of the idea of building new eco-towns? How could they be made as eco-friendly as possible?

PLENARY ACTIVITY

Can a city be sustainable?

CASE STUDY Greenwich Millennium Village – sustainable city living?

GET STARTED

Make a list of at least five things you think that people in a city can do to reduce their use of some of the Earth's resources such as oil and metals.

Figure 2.49 New houses in the Greenwich Millennium Village.

Regeneration in Greenwich

The Greenwich Millennium Village (GMV) is one of several developments in the UK where homes are being built in ways that try to let people live more sustainably. It is being built on land that was part of London's old Docklands on the south bank of the Thames. The O_2 Dome is nearby. It is an example of an urban regeneration scheme. When the scheme is finished in 2010, at least 1400 new homes will have been built on the 29-hectare site. There will also be new shops, services and other facilities.

The claims to sustain

One reason the Greenwich Millennium Village claims to be sustainable is because of its target for homes to use 80 per cent less energy and 30 per cent less water. This is being done by generating some power locally and combining this with a heating system called a combined heat and power system (CHP).

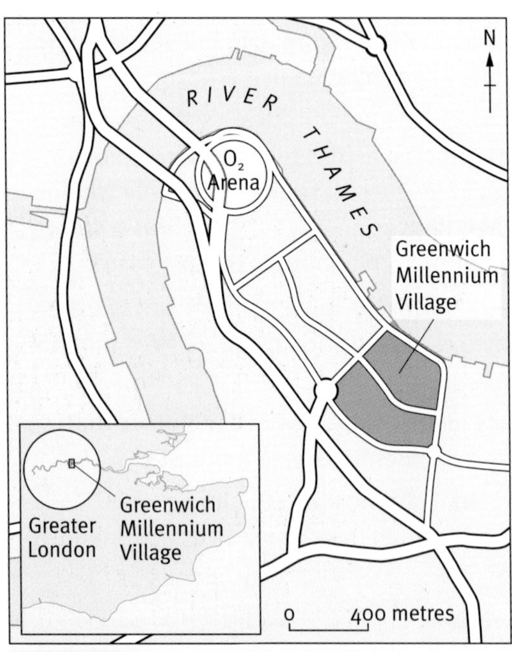

Figure 2.48 The location of the Greenwich Millennium Village development.

Fact file

Redevelopment of the Greenwich Peninsula (76 hectares), including the GMV on its eastern side

- 20 hectares is for parkland and other open spaces, with two ecology parks.
- 10,000 new homes in neighbourhood districts.
- 24,000 new jobs.
- 150 new shops and restaurants.
- New community and leisure facilities, including the O_2 facility.
- Access to the Jubilee underground line and bus routes.
- Pedestrian and cycle routes.
- An eco-friendly supermarket.
- An integrated primary school and health centre.
- 2.2 km of river walkways.

The buildings use materials that the builders say are sustainable, such as cedar wood from sustainable forests. In some buildings, aluminium is used because it lasts and can eventually be recycled. They are designed to take advantage of sunlight and to be protected from cold east winds. The site itself is on old housing and industrial land that has been cleared of pollution.

RESEARCH LINK

Is the GMV considered to be a success? Who by?

Mixing the land use

Another way to help people live sustainably is to help them to reduce their use of all forms of energy. One way to do this is to make it easy to get to places such as shops, so energy is not used in travel. This is done by having mixed land use: for example, a primary school, community centre, shops and some businesses. A Sainsbury's 'eco-store' in the GMV that uses 50 per cent less energy than other superstores has been built.

There is also an eco-park with a lake and strips of open space that connect different parts of the development. There is a new underground station nearby so people can get to other parts of London on public transport. There are cycleways and bus routes so people do not have to use cars.

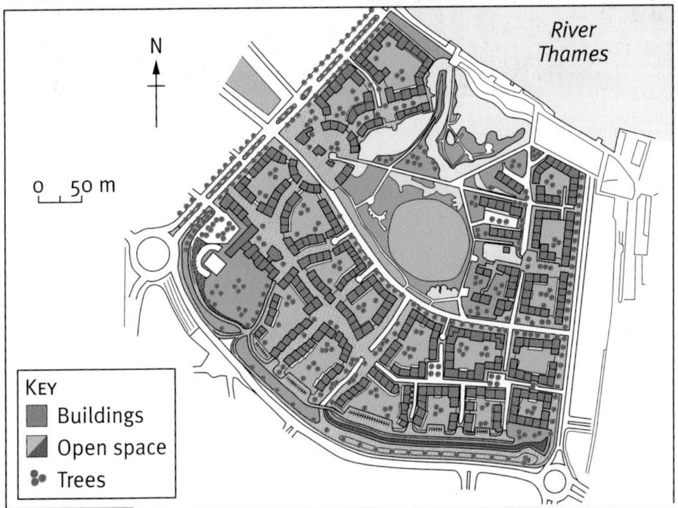

Figure 2.50 ***Land use layout at the GMV.*** **(Source: Greenwich Millennium Village)**

'I can't believe that people would pay out a third of a million to live on a toxic waste site overlooking the A2, with stunning views of Travis Perkins, The Blackwall Tunnel and the old gas works.'

'Well I certainly think that GMV is a well developed area and proves that even in the centre of a busy city there can be sustainability.'

'As far as I can see, GMV provides an excellent example of how a development can be delivered to meet really challenging environmental standards, and it looks fantastic and people appear to be very happy living there.'

'Plus the reason it is promoting walking and cycling is because by the time the buses get there, you can't get on them and you have to walk to the tube!'

(Source: Comments from members of the public to the Sustainable Development Commission)

ACTIVITIES

1. Study the photo of the Greenwich Millennium Village. What are your first impressions of the area as a place in which to live?
2. Draw up a table to show how the GMV meets all the elements of sustainability.
3. Do you think that the people who live in this area of London will feel they are part of a community? Explain your answer.
4. Study Figure 2.50. What do you think about how land has been used in the GMV and how this will encourage people to live sustainably? Consider the different types of land use; the amount of space for each type of land use; the layout of the area.
5. What does it mean for something to be eco-friendly in a city, such as a house, shop, transport or a park?
6. Read the viewpoints about the GMV plan. What are the negative points and do you think that anything could be done to overcome them?
7. What questions would you ask people who live in the area to find out if they are happy living there and the reasons for their replies?

The real test

The Greenwich Millennium Village is only one example of how ideas are changing about building not only individual houses, but also larger areas. It is likely that some mistakes will be made and not all claims will match the reality. Some people, for example, say that not even the Greenwich Millennium Village is really sustainable. People still buy goods that use finite resources that are often made elsewhere. They still use energy from fossil fuels and many still have to travel out of the area to go to work. But anything that can be done to reduce CO_2 emissions and to help people think and act more sustainably has to be worth trying.

THINK ABOUT IT

Do you think that a shop can do anything to justify calling itself an eco-store, or could it be just a name that makes good publicity?

PLENARY ACTIVITY

Is the GMV truly sustainable or just the best we can do at the moment?

Is there a pattern to shopping?

GET STARTED

Think of a way in which you can put all types of shop into no more than five groups. This can be based on what they sell, where they are, their size or any other groupings. Compare your groups with others.

A pattern of shops and services

Shops and most services have to make a profit from customers to stay in business. This means that the distribution of shops and services depends on where customers live and on how easy it is for them to get there. The area customers come from is called a shop's **catchment area**.

In the past, there was a fairly simple **hierarchy** of shopping and service centres that ranged from a city centre at the top of the hierarchy, to a large number of single corner shops or small group of shops at the bottom. Specialist goods and **consumer goods** such as furniture and electrical goods were mostly in department stores and other big shops in the city centre. Everyday **convenience goods** such as bread, newspapers and daily groceries were bought in small shops serving a very local community.

Figure 2.51 The town centre of Wells, Somerset with shops and a farmers' market, serving the needs of the customers from the surrounding area.

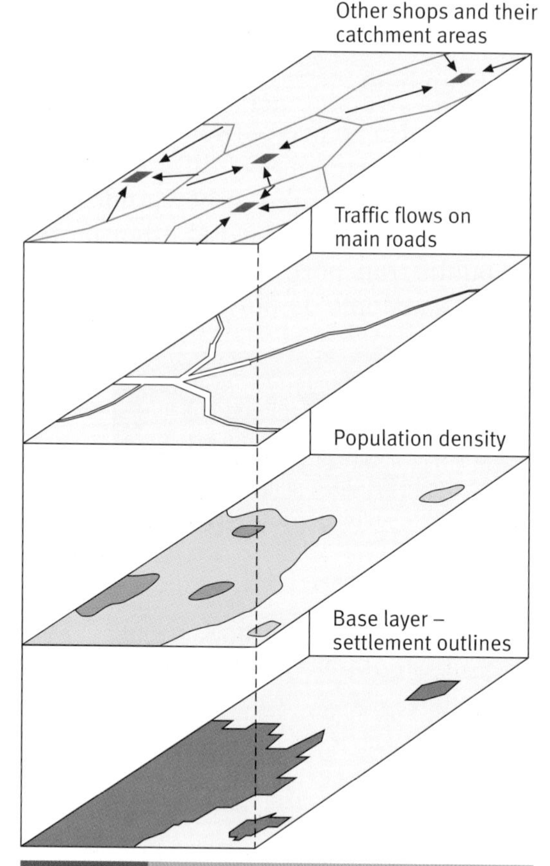

Figure 2.52 Factors that affect the location and distribution of shops.

A changing distribution

The distribution of shops is now more complex than in the past. There have never been so many different ways and places in which to buy goods and services.

There are new transport links such as motorways that can make some places more accessible. At the same time, problems of car parking in city centres

KEY TERMS

Catchment area – the area from which people come to a shop or a shopping centre.

Consumer goods – things that people buy.

Convenience goods – goods that people buy frequently for everyday needs such as groceries.

Farmers' market – a place where farmers sell produce direct to customers, usually from stalls on one day a week.

Hierarchy – an arrangement in order with one at the top and an increasing number at lower levels.

Shopping mall – an undercover area with a variety of shops.

make them less accessible. The distribution of shops also changes because of changes to how people do their shopping and how often they shop.

Complex consumer landscapes

The shops and services that customers need form a consumer landscape that has become very complicated. Petrol stations sell food, while superstores sell petrol. There are even travel agents, chemists and opticians in superstores.

New superstores and much bigger shopping areas called **shopping malls** that have as many shops as a city centre have been built on the edges of cities and further out into the countryside. These can affect the distribution of older shops, taking their customers and forcing them to close.

Markets and car boot sales

Another change has been in the growth of **farmers' markets**, often set up on a few days a week in a shopping centre. This has become popular because many people are concerned about the environmental effects of shopping and about how food has been produced. Car boot sales on a farmer's field are another way in which people can buy and sell goods.

ACTIVITIES

1. Study the photo in Figure 2.51. What are your thoughts about shopping in a place like this? Think about the range, cost and quality of goods; access; the site; the shopping environment.
2. Explain why the catchment area of a hairdresser is likely to be smaller than one for a shop that sells computer games. In your answer, use as many of the key terms as you can. The ideas in Figure 2.52 may also help.
3. What are the features of a shopping mall that make it different from a high street shopping area? Use the photo in Figure 2.53 for ideas.
4. Compare an out-of-town shopping mall to a city centre or high street shopping area. Which type of shopping area is likely to be the most sustainable? Explain your answer.
5. Would you agree that it is not necessary to have as many local Post Offices as in the past? Give reasons for your answer.
6. In what ways and for what reasons do you think that shopping has also become a leisure activity?

Figure 2.53 West Quay Shopping Centre, Southampton.

Closing Post Offices

The distribution of Post Offices is a special problem for some people. A Post Office is a special type of business that people need, but they too are being forced to close because they do not make a profit. This is mainly because more people are handling their money through the Internet and through banks.

Fact file

Closure of Post Offices

- Between 1984 and 2007, 7000 Post Offices closed, reducing their number from 22,000 to 13,500.
- In 2007, the government announced plans to close another 2500 Post Offices over the following 18 months.
- Many small shops depend on the Post Office part of their business to bring in customers.

DISCUSSION POINT

The Post Office provides a service, as well as being a business that needs to make a profit to stay open. Do you think that everyone should have equal access to a Post Office, even if this means that some need to be subsidised by the government using taxpayers' money?

PLENARY ACTIVITY

Draw a sketch map to show the locations of different types of shopping area for an area that you know. Add notes to the map to explain the locations.

Retail service provision: how and why is the shopping landscape changing?

GET STARTED

Name one shop you know that has recently opened or one that has recently closed down. Describe or draw a sketch map to show its location and write notes to say why you think it has opened or closed.

Clone town shops

Shops and services compete against each other for customers. Some become successful and grow to become national chains. There is a Tesco, Starbucks and WH Smith in almost every town and city in Britain. This brings the benefits of successful businesses to everyone.

A problem, however, is that they often replace locally run businesses so that all shopping centres can look the same. Towns that look the same are called 'clone towns' because the individual character of each place has been lost. Some businesses are multinational such as IKEA and McDonald's, bringing the same appearance to different countries as well as to different cities.

Small business battles

Small businesses have been closing down at a rate of about 2000 every year over the last 10 years. Shops owned by national chains have been taking a greater percentage of people's shopping money. Even the big chains compete against each other for the biggest market share and the best sites.

Some shop owners also say that **charity shops** are taking some of their business away. Charity shops pay less tax for their shops. This gives them an advantage over other shops when trying to make a profit and stay in business.

Figure 2.54 The same national chains, shops and cafés are now in most UK towns and cities. This is taking away the individual local character of each town and giving them the name 'clone towns'.

Fact file

UK shops in 2008

- There are about 278,600 shops in the UK.
- About half of all shops are owned and managed by individuals.
- About 103,000 shops employ fewer than five people.
- The smaller shops employ about 500,000 people between them.
- The four largest supermarket chain stores employ 800,000 people.
- One chain store has 30 per cent of all grocery sales in the UK.
- Since 2000, about 2000 small local shops have closed in the UK.

(Source: *High Street Britain* (government report) House of Commons All-Party Parliamentary Small Shops Group)

E-tailing hits the high street

The internet is changing people's shopping habits in ways that are affecting the number and location of shops and services. This is called **e-tailing** (to highlight the contrast with traditional retailing). Some shops also use the internet to sell goods, both for specialist goods and for weekly groceries. Many services such as banking and booking holidays are also done online.

In 2007, one music chain store sold off all its branches because so many people now buy music from online suppliers to play on an iPod. Superstores were also taking part of their trade. Books and many other types of goods and services are also being bought online.

Even eBay is increasingly being used for fixed-price rather than for auction sales. Shops that sell specialist goods are especially at risk from internet sales, unless they take advantage of it by selling their own goods online. With the internet, customers can from anywhere, including from other countries.

KEY TERMS

Charity shops – shops that raise money for charities, often selling goods donated by the public.

e-tailing – buying and selling goods and services online through the internet.

Food miles – the distance that food has been transported before it is sold.

RESEARCH LINK

Choose one food that you commonly eat. Find out where it comes from and how many food miles it has travelled before you buy it.

PLENARY ACTIVITY

Think about the different ways in which shopping is changing. Do you think that most people agree with these changes, or are they causing problems to others? Draw a bubble map that shows how these changes affect people of different ages and different incomes.

DISCUSSION POINT

Consider which types of shops and services have most to gain and most to lose from selling goods online.

Ethical shopping

Customers choose to buy from some shops rather than from others for a range of reasons. Some take account of where goods were produced and the conditions under which they were produced. The idea of **food miles**, for example, is now a cause for concern. A problem, however, is that some locally grown food may need more energy to grow it compared with food from another country where the climate may be warmer.

Other ethical issues that affect how and where some people shop include demand for organic products, for Fairtrade products and for products that do not use child labour. All these things have an effect on the shopping landscape.

ACTIVITIES

1. Would you agree that shopping areas and town centres deserve the name 'clone towns', or is this an exaggeration? Use examples of places you know in your answer.
2. Look at the Fact file on UK shops. Is there evidence in your local area for any of these facts?
3. Do you think it is fair that charity shops should pay less in local taxes than other shops beside them? Give reasons for your answer.
4. Would you agree that e-tailing is more sustainable? Explain your answer.
5. In what ways might the further growth of e-tailing affect the type and distribution of shops in the UK or in an area that you know?
6. Do you have any ethical concerns when you buy goods? Give reasons for your answer.
7. Are there any ways in which ethical shopping might affect the types of shops, their number and their distribution? Explain your answer.

How retail service provision has changed: superstores

GET STARTED

How would you explain the differences between the main superstores in the UK to a person from another country?

Figure 2.55 The Meadowhall, shopping mall in Sheffield.

Space-hungry superstores

Superstores and retail parks need a lot of space. This is one reason why new superstores are often on greenfield land outside the city. All types of new building have to get **planning permission** from a local council. This can be given if a superstore company can argue that there is a need for one to give more choice or to provide goods for people in new housing areas.

Often, land around a city is protected green belt land that is mostly used for farming and for different types of recreation. Permission is sometimes given to build on this land if a need can be proven. There can be a **Public Inquiry** if there is enough disagreement about a proposal. This is held like a court case in which everyone can put their case to an inspector, who reports to a government minister for a decision.

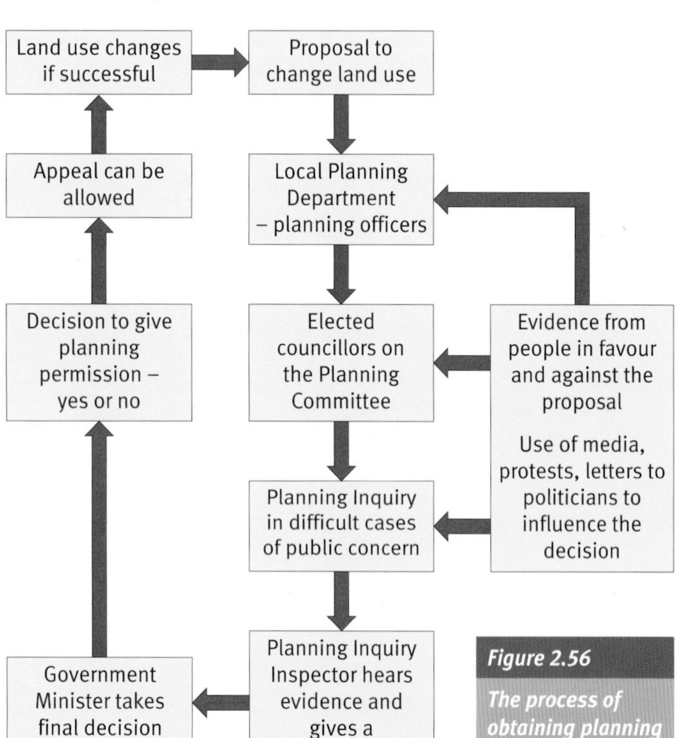

Figure 2.56 The process of obtaining planning permission.

A tough choice

People can be faced with a hard choice if a supermarket company applies for planning permission to build near their homes. A new superstore can bring the advantages of lower prices and a wider choice of some goods, together with easy and free parking. However, it also brings more cars and lorries to the area. The biggest superstores may be open 24/7, bringing noise and traffic for 24 hours a day, 7 days a week. The building may also not look right in the landscape.

DISCUSSION POINT

How important is it to have a green belt around a city, and is this more important than having easy access to shops?

KEY TERMS

Planning permission – all changes to how land is used need to be given planning permission by a local council.

Public Inquiry – a public meeting held to reach decisions about difficult planning proposals.

The result is that the quality of local people's lives can be made worse. The value of their homes can also fall if there is more traffic. In almost every case where there is an application to build a superstore, some local people form protest groups to try to stop the development. It is great to have good shops nearby, but not too near!

COMMENTS ABOUT A PROPOSAL FOR A SUPERSTORE ON THE EDGE OF MANNINGTREE IN ESSEX

In favour

'If people can't compete it's their tough luck. That's business. Who on earth would want to go back to trailing down the high street in the rain to lots of different shops for your groceries? How would you carry them or are we to go back to the wife doing a daily shop every morning? Pretty difficult, as most mums are out working to pay for their extra car and posh holidays. It's 17 miles into town for me so I think a trip there every day to the little shops would not only ruin the environment but also my bank balance! All my money would be spent on fuel!'

Against

'I will protest this, it will make it more commercial and the huge amounts of traffic will make Manningtree unbearable to live here. We live on quiet roads at the moment and it should stay that way. Why every town in the land should have a massive supermarket I have no idea. I get most of my shopping done on early Saturday mornings on the market then I visit our delightful little shops and then I come home – not even having touched a massive supermarket.'

(Source: *Daily Telegraph*, 2008)

Figure 2.57 People protest against a new Tesco store in Cambridge.

RESEARCH LINK

One way to run an effective protest campaign is to bring a problem to the attention of as many people as possible. Use websites to find out about the different ways in which people have protested about proposals for superstores in different places. Think carefully about your search terms before you start.

PLENARY ACTIVITY

Which groups of people would be in favour or against a new superstore?

ACTIVITIES

1. Look at the information in Figure 2.56 that shows how planning permission can be obtained. Do you think that this is a fair way to make decisions about how land should be used?
2. What questions should be asked to prove that people in an area need a new superstore?
3. Suggest some ways that the amount of car travel for shopping could be reduced in the UK.
4. Look at the photo in Figure 2.55. Explain why this site and location might have been chosen for a superstore. Do you think that anyone would have objected to it being built? Give reasons for your answer.
5. What evidence do you think a protest group against a superstore should try to collect to make a case for planning permission to be turned down for a new superstore near where they live?
6. Read the viewpoints that give the conflicting views of two residents about a proposal for a new superstore on the edge of their village in Essex. For each viewpoint, write a reply that tries to give a solution to the problems that are raised by both people.

A Decision Making Exercise – A sustainable future for town centres?

In this exercise you will be asked to reflect on government policy for creating sustainable retail developments in town centre locations, at the expense of 'out-of-town' locations. Read the information and then make your decision. More background information is available on the supporting CD to help you develop your ideas further.

THE NEWS TODAY

January 2008

Town Centre First – the case against new out-of-town shopping centres

In May 2006 the Department for Communities and Local Government re-emphasised their policy to promote town centres, ahead of out-of-town locations, for new retail development. Whilst the policy remains unpopular with large retail giants such as Tesco, Sainsbury's, IKEA and Comet, the government is determined to create vibrant, sustainable retail communities in the heart of our towns and cities.

The policy was developed following years of decay in town centres as the large retail giants looked for new, often greenfield sites, out of town. Government ministers were also increasingly concerned about the negative impact on small retailers such as butchers and grocers in adjacent suburbs. Legislation was passed to ensure that councils consider town centres first. Councils were instructed to refuse planning permission for out-of-town locations if land was available in the town centre. From 2006, all planning authorities were required to sequence their planning permission in favour of town centre sites.

The policy has had an impact. In February 2007, Friends of the Earth reported a number of cases where permission to build 'out of town' had been refused. In 2004 attempts by Asda and Tesco to build large stores in deprived areas outside Middlesbrough were refused on the grounds of a negative effect on the town centre. In 2006 Asda's plan for developing the site at Worthing College was denied for the same reason.

However, thousands of planning applications for out-of-town sites are still being made. This puts pressure on planning departments. Not only do the large retail giants want to move out of town, it seems that shoppers prefer to shop there. The Federation of Small Businesses recently carried out an extensive survey of public opinion. Members of their organisation should be worried, as 75 per cent of all shoppers who were asked said they preferred shopping in large stores out of town.

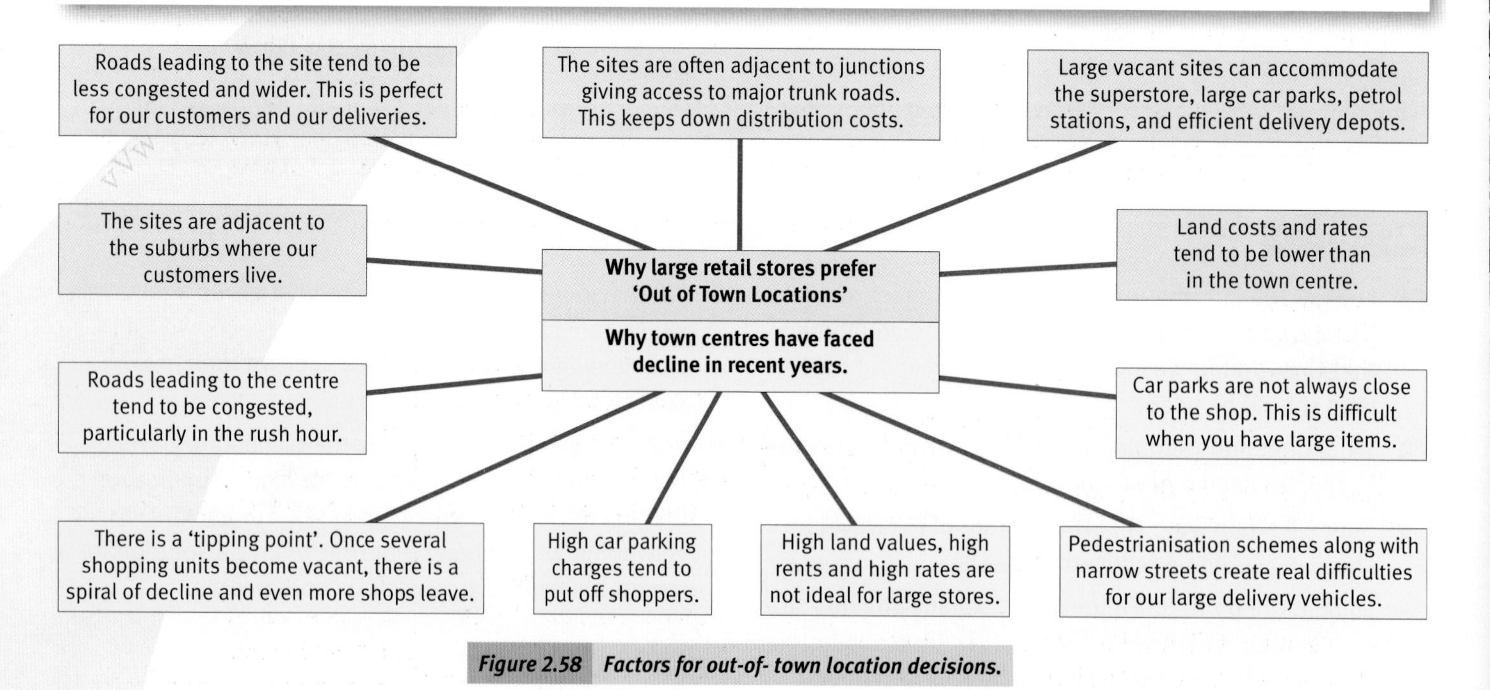

Roads leading to the site tend to be less congested and wider. This is perfect for our customers and our deliveries.
The sites are often adjacent to junctions giving access to major trunk roads. This keeps down distribution costs.
Large vacant sites can accommodate the superstore, large car parks, petrol stations, and efficient delivery depots.
The sites are adjacent to the suburbs where our customers live.
Why large retail stores prefer 'Out of Town Locations'
Land costs and rates tend to be lower than in the town centre.
Why town centres have faced decline in recent years.
Roads leading to the centre tend to be congested, particularly in the rush hour.
Car parks are not always close to the shop. This is difficult when you have large items.
There is a 'tipping point'. Once several shopping units become vacant, there is a spiral of decline and even more shops leave.
High car parking charges tend to put off shoppers.
High land values, high rents and high rates are not ideal for large stores.
Pedestrianisation schemes along with narrow streets create real difficulties for our large delivery vehicles.

Figure 2.58 *Factors for out-of- town location decisions.*

Ipswich Borough Council is sympathetic to the idea of town centre regeneration. When the former Civic Centre site became available (left), it was always going to be selected for development ahead of the redundant Crane works on the south-east edge of Ipswich (right).

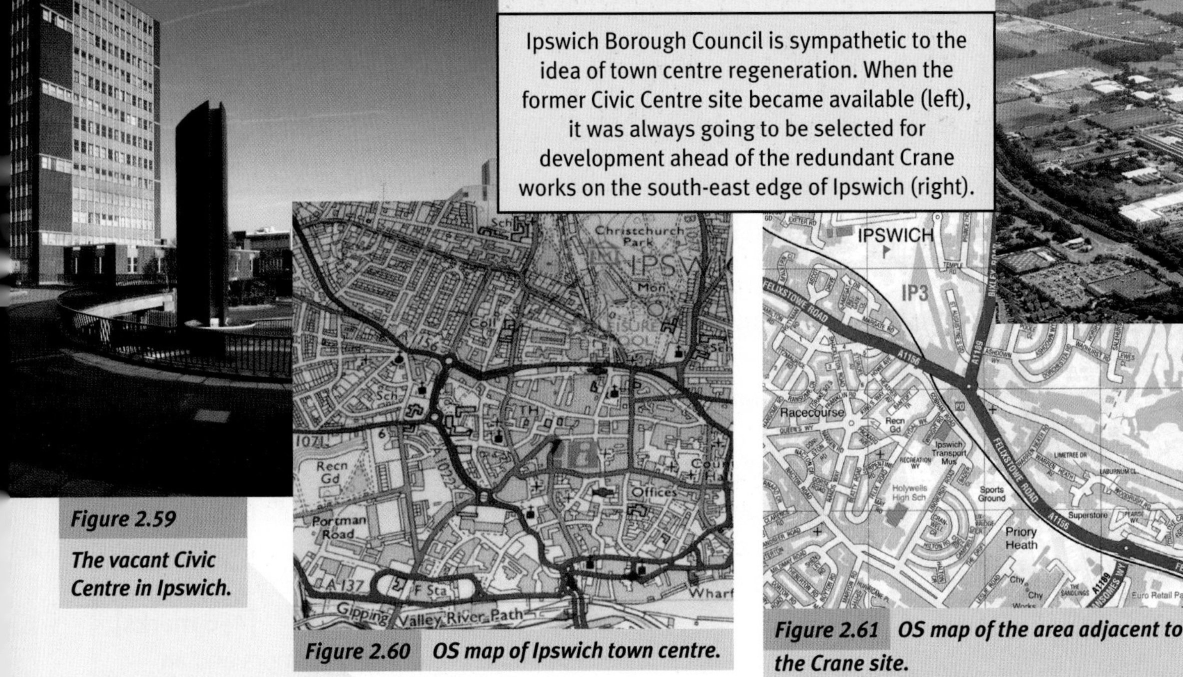

Figure 2.59 **The vacant Civic Centre in Ipswich.**

Figure 2.60 **OS map of Ipswich town centre.**

Figure 2.61 **OS map of the area adjacent to the Crane site.**

Figure 2.62 **The vacant Crane site on the south-east edge of Ipswich.**

THE NEWS TODAY

August 2008

New retail plans for redundant site in Ipswich town centre

In February 2008 Ipswich Borough Council (IBC) produced a final version of its Local Development Framework (LDF). The aim was to promote sustainable growth in the town. Developed over a number of years, the LDF outlines the need to utilise town centre sites ahead of other locations. This would ensure a vibrant town centre for future generations and bring a halt to the decline of the CBD. Retailers have closed or moved out to other favoured locations.

As early as 2005, IBC carried out a Retail Study to ensure that their framework would reflect Government policy. DTZ, experts in retail development, reported that the council should develop town centre sites, such as the Westgate Quarter, ahead of vacant plots on the edge of town, such as the former Crane industrial site. This would comply with national policies.

When the former council offices at the Civic Centre were vacated in 2007, it was no surprise that the council granted permission for Turnstone Property Developments to promote the site for retail development. This would be at the expense of developing retail outlets at sites such as the former Crane works. The decision might fit the sustainable credentials of IBC, but it would not prove popular in the eyes of retail giants who seek to develop their businesses out of town.

YOUR TASK

Consider the evidence on these pages and the information on the supporting CD. Write two letters to the editor of the *Ipswich Evening Star* regarding the plans to develop the town centre of Ipswich ahead of sites such as the Crane site.

In class, debate which of the two views you think is the strongest. Try to relate the Ipswich case study to a dispute about retail developments in your own area. The key issue is the debate about sustainable options.

Your first letter should state the views of a person who fully supports IBC's policy to regenerate the town centre. Emphasise why this is the sustainable option.

Your second letter should state the views of someone who is opposed to the policy of prioritising town centre development as it is simply not sustainable.

Higher:

Key Geographical Themes 8-mark case study question

Case study – Population control.

i Name a country in which attempts have been made to influence the birth rate.

ii Describe the attempts to influence the birth rate. What are the outcomes of these attempts?

[8 marks]

Mark scheme

Level 1: The student gives a basic description of attempts to influence the birth rate, but with only limited relevant knowledge and information. Meaning may not be communicated very clearly because of mistakes in writing. **[1–3 marks]**

Level 2: Student describes attempts to influence the birth rate with some evaluation and limited development, and uses some relevant knowledge based on a range of factual information and evidence. Meaning is communicated clearly. **[4–6 marks]**

Level 3: Thorough and developed description of attempts and an evaluation of their success – specific to a particular country. Demonstrates thorough knowledge, based on a full range of relevant factual information and evidence. Meaning is communicated very clearly. **[7–8 marks]**

i *China.*

ii *The one-child policy has worked well in big cities such as Beijing, but many families in rural areas ignored the policy because they wanted more children to work on the farm. The government introduced the one-child policy because the population was growing too quickly. The policy made it illegal to have more than one child and families were criticised and fined if they had more children. For families who obeyed the policy, the child received free education and the family received a family allowance, a guaranteed home and pension. If they had a second child all the benefits were lost. It was made law that a woman had to be at least 20 and men 22 before they could marry, and they had to get permission from the local government to marry and have a child. Also forced sterilisation and abortion were carried out in some areas. Often, when a family had a girl they tried to have a boy afterwards, because boys are thought to be more important than girls to carry on the family name. There is also concern that there may be a shortage of women in the future. The birth rate has fallen and population growth has slowed down. It is now allowed for some couples to have another child without automatically losing their privileges.*

Examiner's comment

China and India are the countries most often referred to; you could also tackle this for an MEDC that encourages higher birth rates, but for 8 marks you need to show and evaluate quite a lot of relevant information.

The answer focuses in detail on the 'one-child policy', describing benefits and penalties that the policy has brought. This answer is specific to the named country.
The second requirement of the question is a consideration of the outcomes of the policy. The answer recognises that the policy was not as successful in rural areas and explains why. It also suggests a possible future problem, that there may be a shortage of women in China. In conclusion, the answer suggests that another outcome is that the policy has been relaxed in response to its success.
Full marks have been achieved by this case study answer.

End of unit activities

Thinking about geography: reading images

Reading images is a visual strategy that helps us to interpret what we see in photographs, maps, diagrams. Examples in everyday life are when we use a road map in the car to plan a journey, or follow instructions to put together self-assembly furniture.

By reading images, not only can we develop our literacy skills by *describing* and *explaining* features and processes in geography, we can also *enquire* through asking questions and suggesting hypotheses, and apply our imagination so that we can think creatively.

Population and settlement: reading images

This activity uses a simple but effective enquiry technique called the '5Ws', which can be used elsewhere in geography or your other subjects when you need to find out information.

ACTIVITIES

1. In your group of four draw a large table with five columns, each column headed with one of the 5Ws: What? When? Where? Who? Why?
2. Study the photograph of slum dwellings in an LEDC in Figure 2.63 and, as a group, think of as many 5W questions as you can that you would want to ask about the image – put each question under the appropriate 'W' heading in the table. Feed back one question from each column to the rest of the class.

Developing your image reading and enquiring skills

3. Look again at your questions and put them into a logical asking order if possible. Get rid of any questions that you think are no longer worthwhile, but keep a minimum of 10 if you can.
4. Of your remaining questions, explain which ones would be suitable to ask of:
   - **a** the people who live in the dwellings
   - **b** the local council
   - **c** an **NGO** aid worker
   - **d** a resident from a more wealthy area near to the slum dwellings.
5. Can you imagine what these groups might say in response to your questions? Write the answers with detail.
6. Finally, do you need to change any of your questions – or bring back discarded ones – to get more useful responses from these groups?

Figure 2.63 *Slum dwelling in an LEDC.*

THINKING ABOUT READING IMAGES

- How did you create your 5W questions and how did the features in the image influence you?
- How did you decide which questions to get rid of, keep or change?
- From your responses to Activity 5 was it easy to imagine what the different groups might say?
- Why is being able to 'read' an image important?
- In what other subjects do you read images and extract information? Share some examples.
- In what other subjects could you use the 5Ws technique?

KEY TERMS

Non-Governmental Organisations (NGOs) – private organisations that work on big issues affecting humanity. They may get money from governments and work with governments, but they are private organisations, not government-controlled.

Chapter 3
Natural hazards

In a moment, lives can be changed by the devastating effects of natural hazards. People in this photograph might have been cooking, texting or working at their desks, when suddenly a cataclysmic earthquake destroyed their lives and homes and buried their loved ones in the rubble. Sometimes, natural hazards can be predicted and people can prepare for them; sometimes, as with this earthquake in Sichuan that killed almost 90,000 people, they occur with no warning, destroying ways of life and landscapes forever.

Consider this

As communications around the world have improved, we are increasingly aware of natural hazards as they happen. The news is full of catastrophic events that cause billions of pounds worth of damage and leave thousands of people dead or injured. Why is it that some places appear in these reports again and again, but other areas seem never to be affected?

People remembering their relatives killed in the 2008 Sichuan earthquake in China.

QUESTIONS FOR INVESTIGATION

a What is the global distribution of different types of natural hazards?

b What natural processes cause different types of natural hazards?

c How do natural hazards affect people and places in parts of the world with different levels of development?

d How can human activities affect the impact of natural hazards?

e How can people and places be protected from the impact of natural hazards?

What is a natural hazard?

Natural hazards

A natural hazard is an extreme natural event or process that causes loss of life and/or extreme damage to property and creates severe disruption to human activities. They include earthquakes, tsunamis, volcanic eruptions, tropical storms, floods and drought.

Some natural hazards occur all over the world; for example, floods caused by heavy rain. Others are very limited in area, such as landslides or tornadoes. Yet others are related to certain geological or climatic conditions and are concentrated in certain areas; for example, earthquakes, avalanches in mountainous areas or tropical storms.

How often hazards occur can depend on the location of the hazard; for example, Bangladesh regularly suffers from river flooding as a result of the annual heavy rains caused by the monsoon. Hurricanes and typhoons only occur during a certain period of the year when weather conditions are suitable for their formation in the tropical oceans. A major earthquake can produce hundreds of **aftershocks**, some of which can be almost as powerful as the initial ground shaking. The Sichuan area of China recorded over 400 aftershocks in the two weeks after the earthquake in May 2008.

How do hazards affect people?

On average, about 150,000 people are killed worldwide by natural hazards each year. The economic cost of natural hazards is increasing dramatically. For example, in the 1990s the average cost of all natural hazards in the USA doubled from US $25 billion to US $50 billion per year. In many developed countries, loss of life is low but economic costs are extremely high. In developing countries, loss of life tends to be very high while economic costs are usually lower.

The risk of natural disasters is increasing as a result of population growth, urbanisation and alteration of the natural environment. In many areas substandard houses and public buildings result in many deaths. Besides the direct impacts of natural hazards, such as earthquakes causing buildings to collapse, there are usually many indirect impacts, such as the tsunami (or tidal wave) caused by the earthquake off Indonesia in 2004. These indirect impacts are often more costly and can add years on to the recovery time after a disaster.

Why study hazards?

It is important to be aware of natural hazards, because human activities can increase how often a natural hazard occurs and how severe a natural hazard can become. For example, building on top of an unstable slope adds weight to the slope and will increase the probability of the slope collapsing. Understanding when, where, why, and how natural hazards occur is the first step in minimising their impacts on our lives.

GET STARTED

How many natural hazards have you heard about in the last few weeks? Whereabouts in the world did they occur?

KEY TERMS

Aftershocks – smaller earthquakes that occur in the same general area during the days to years following a larger event.

Figure 3.1 *Devastation caused by tsunami in Indonesia.*

Figure 3.2 *Buildings destroyed by landslide in Hong Kong.*

THINK ABOUT IT

Are natural hazards increasing in the world? If not, why are the impacts getting bigger and bigger?

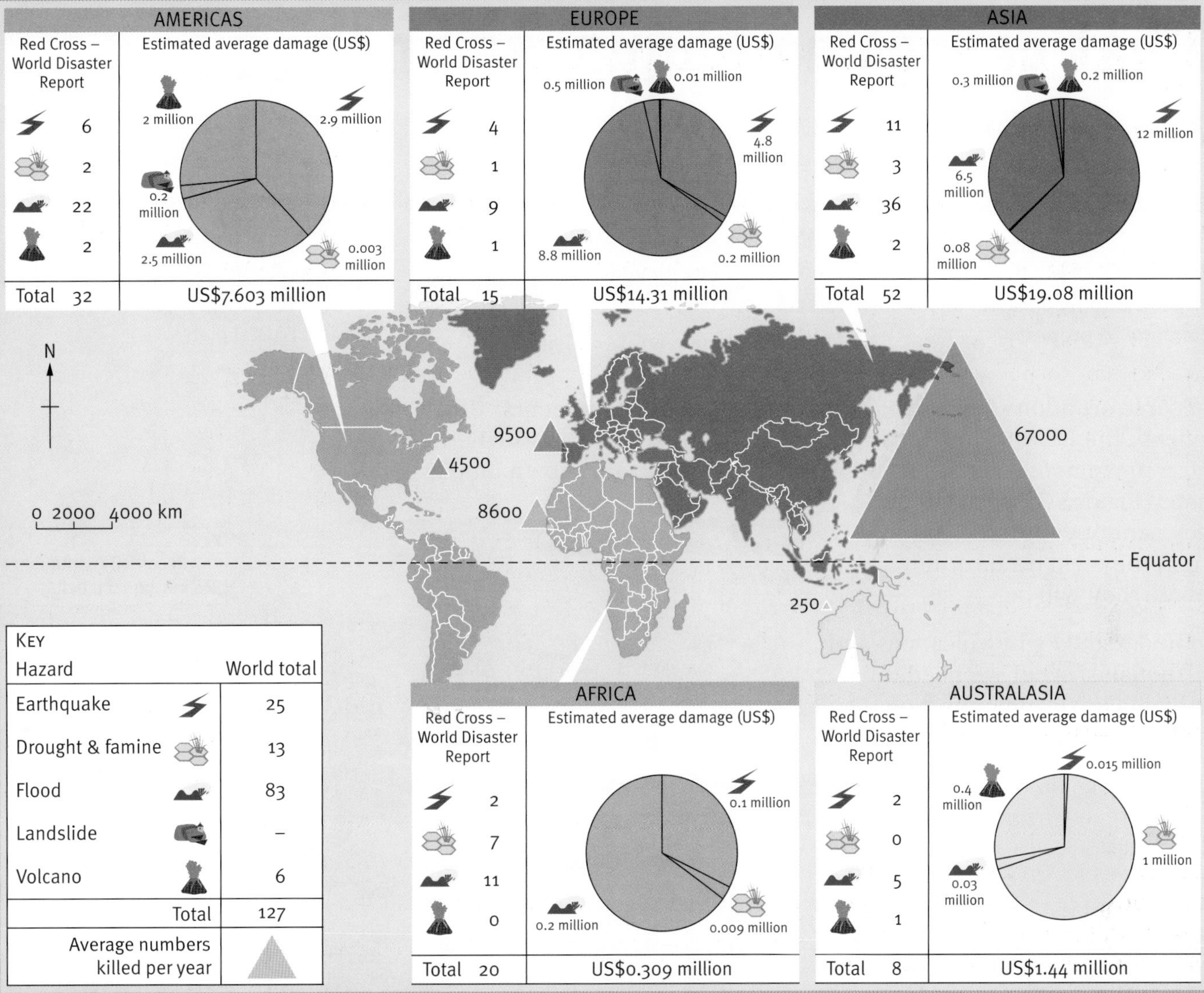

AMERICAS

Red Cross – World Disaster Report	
Earthquake	6
Drought & famine	2
Flood	22
Volcano	2
Total	32

Estimated average damage (US$): volcano 2 million; earthquake 2.9 million; landslide 0.2 million; flood 2.5 million; drought & famine 0.003 million. US$7.603 million

EUROPE

Red Cross – World Disaster Report	
Earthquake	4
Drought & famine	1
Flood	9
Volcano	1
Total	15

Estimated average damage (US$): landslide 0.5 million; volcano 0.01 million; earthquake 4.8 million; flood 8.8 million; drought & famine 0.2 million. US$14.31 million

ASIA

Red Cross – World Disaster Report	
Earthquake	11
Drought & famine	3
Flood	36
Volcano	2
Total	52

Estimated average damage (US$): landslide 0.3 million; volcano 0.2 million; earthquake 12 million; flood 6.5 million; drought & famine 0.08 million. US$19.08 million

KEY Hazard		World total
Earthquake		25
Drought & famine		13
Flood		83
Landslide		–
Volcano		6
	Total	127
Average numbers killed per year		

AFRICA

Red Cross – World Disaster Report	
Earthquake	2
Drought & famine	7
Flood	11
Volcano	0
Total	20

Estimated average damage (US$): earthquake 0.1 million; flood 0.2 million; drought & famine 0.009 million. US$0.309 million

AUSTRALASIA

Red Cross – World Disaster Report	
Earthquake	2
Drought & famine	0
Flood	5
Volcano	1
Total	8

Estimated average damage (US$): earthquake 0.015 million; volcano 0.4 million; drought & famine 1 million; flood 0.03 million. US$1.44 million

Annual average numbers of hazards by source and type per year.

ACTIVITIES

1 Look at the information shown on the world map above. Which region suffers the most natural hazards? Which hazard occurs most frequently across the world?

2 Which region of the world suffers most in terms of cost of damage due to natural hazards and where are most people killed by natural hazards?

3 a Can you suggest any reasons why there is this variation across the world? (Think about the frequency of the hazard; the place in the world where it occurs; the area that it covers; the causes of the hazard; the wealth of the people that it affects.)

b Write two paragraphs, one that describes the distribution of hazards around the world and one that gives your suggestions about why different hazards cause different amounts of damage in different parts of the world.

RESEARCH LINK

Make a list of all the natural hazards mentioned on these two pages. Find examples of these hazards occurring in the past year. Include their locations, and dates and numbers of casualties (dead or injured).

PLENARY ACTIVITY

Think of three questions you would like to ask about hazards. Write them down and see whether you can answer them at the end of this chapter.

What causes earthquakes and volcanoes?

GET STARTED

If there are many thousands of earth movements every year, why do you think that only occasionally one is violent enough to cause loss of life or damage to property?

KEY TERMS

Plate tectonics – the study of the distribution and movement of the Earth's crustal plates.

Volcanic eruptions and earthquakes are caused by movements beneath the Earth's surface. Scientists know broadly the areas where these movements will occur in the world and these are shown on the maps (Figures 3.3 and 3.4) but they cannot yet predict accurately when they will occur or how severe they will be.

Figure 3.3 *World distribution of earthquakes.*

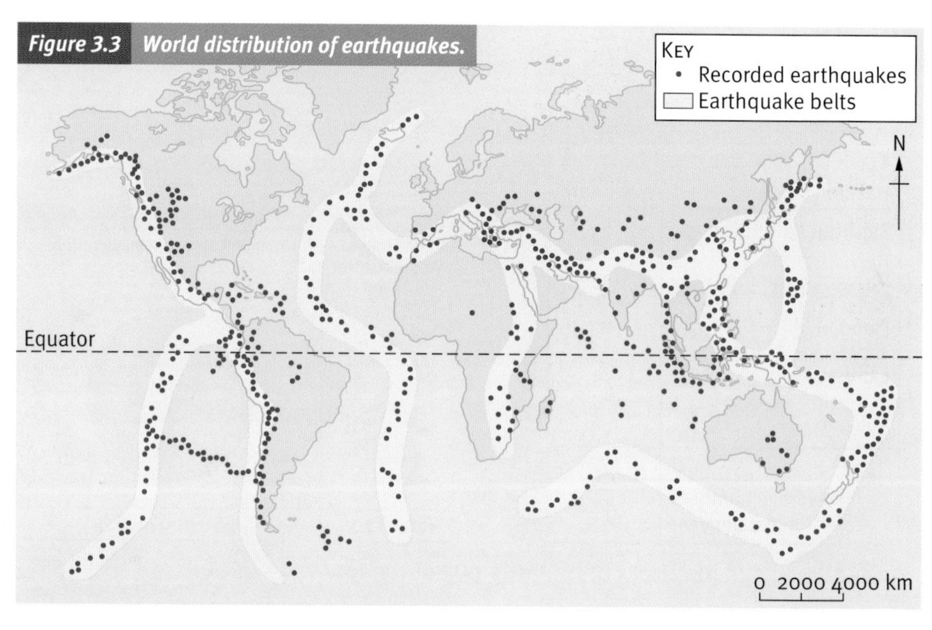

To understand why volcanoes and earthquakes occur we need to study what is happening beneath the Earth's surface.

The Earth is made up of several concentric layers of rocks of different densities and temperatures. If you were to slice the Earth in half, you would reveal three major zones. The core is the central part and it is made of two sections – the inner core, which is solid, and the semi-liquid outer core. Around the core is the mantle, where temperatures are so high that rocks exist in a semi-molten state. It is the thickest layer and makes up more than 80 per cent of the Earth's volume. The outermost layer is the crust, the thinnest layer that varies between 5 km and 90 km thick.

The crust is broken into 7 large and 12 smaller sections known as plates, which float like rafts on the less dense material of the mantle below. These plates fit together like pieces of a jigsaw and move relative to one another

Figure 3.4 *World distribution of volcanoes.*

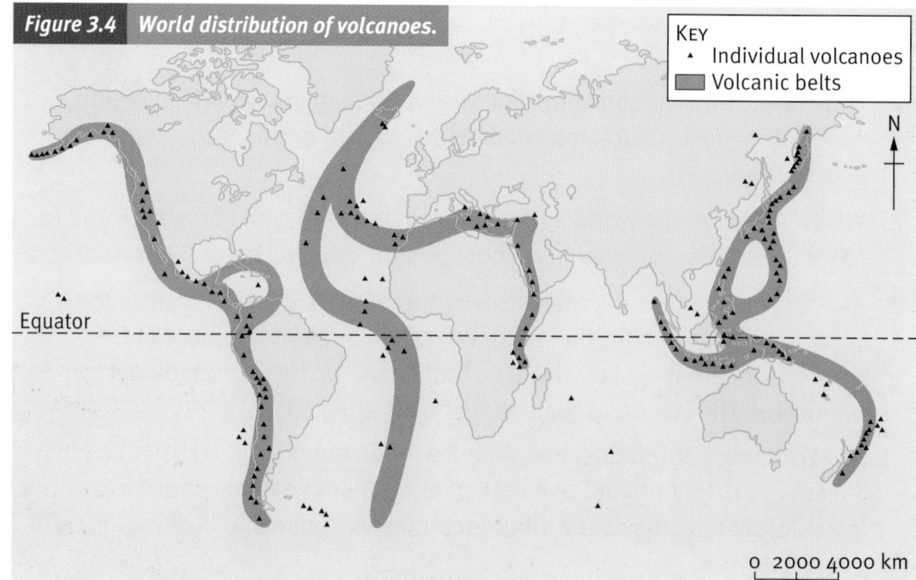

as a result of convection currents set up in the Earth's mantle (Figure 3.6). This movement may be up to several centimetres in a year and plates can move away from, towards or sideways past neighbouring plates. The movement of these crustal plates is known as **plate tectonics**.

Plates are made up of two types of crust – continental and oceanic. Older continental crust is mainly composed of granite and is less dense than oceanic crust. It does not sink and is permanent. Oceanic crust is mainly made up of an igneous rock known as basalt. It is constantly being destroyed and replaced and can sink beneath the continental crust.

Plates meet at plate boundaries (or margins) and it is in narrow belts along these boundaries where the major centres of volcanic and earthquake activity occur and where high mountain ranges are located.

Figure 3.5 *Satellite photo of San Andreas Fault plate margin.*

THINK ABOUT IT

How could you prove that plate tectonic theory actually happens? How do scientists know what the interior of the Earth is like?

RESEARCH LINK

Use the internet to research igneous rocks and how they form. What is the difference between granite and basalt rock types?

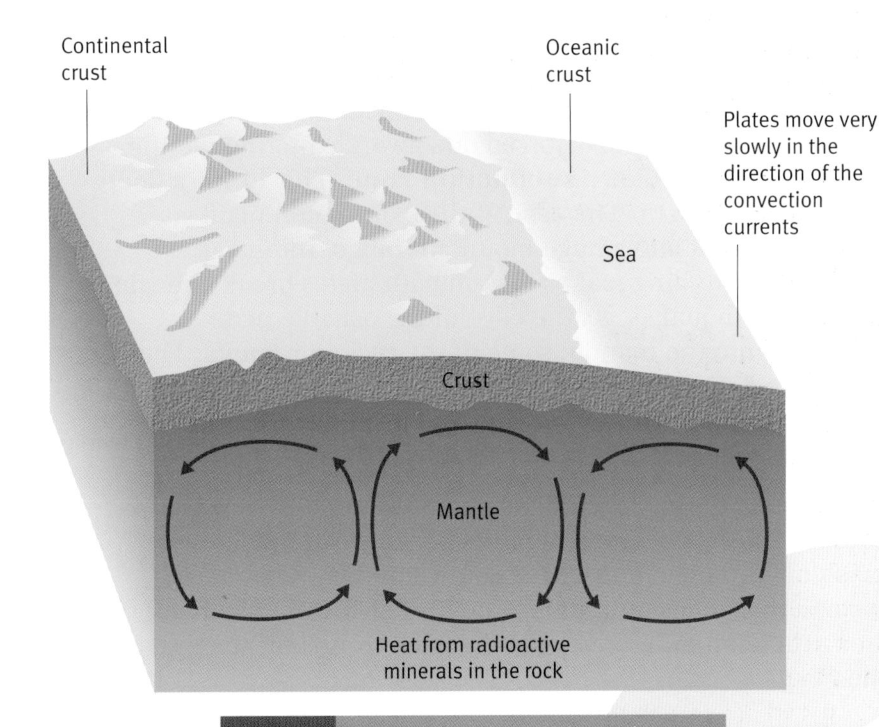

Figure 3.6 *Cross-section through the Earth's crust.*

ACTIVITIES

1 a What are plates?
  b What causes them to move? (Use a diagram to illustrate your answer.)
2 Give three differences between oceanic and continental crust.
3 Using an atlas to help you, describe five features of the distribution of earthquakes in the world and five features of the distribution of volcanoes. For example, earthquakes occur under the sea and on land; volcanoes occur in narrow belts. Use the names of continents and oceans in your descriptions.

PLENARY ACTIVITY

Look carefully at a map of the world. Are there any clues that tell you that the continents were once in different places? You can research 'continental drift' to find out more.

What happens where the plates meet?

GET STARTED

With a partner, make a list of all the people who would want to know about plate tectonics as part of their jobs and why.

The Earth's tectonic plates are all moving in different directions and at different speeds in relation to each other. There are three types of plate boundary: convergent (Figure 3.7), divergent (Figure 3.8) and transform boundaries (Figure 3.10).

Convergent boundaries

Places where plates collide with one another are called convergent boundaries. Plates only move a few centimetres each year, so collisions are very slow and last millions of years. Where an oceanic plate collides with a continental plate the oceanic plate is forced downwards into the mantle forming what is called a **subduction zone**. This forms a deep trench in the ocean. The edge of the continental plate is folded upwards into a huge mountain range. Intensive folding and bending makes rock in both plates break and slip, causing earthquakes. As the edge of the oceanic plate descends into the mantle, some of the rock in it melts. The melted rock rises up through the continental plate, causing more earthquakes on its way up, and forming volcanic eruptions where it finally reaches the surface.

An example of this type of collision is found on the west coast of South America where the oceanic Nazca Plate is crashing into the continent of South America. The crash formed the Andes Mountains, the long string of volcanoes along the mountain crest, and the deep trench off the coast in the Pacific Ocean.

Figure 3.7 Convergent plate boundaries.

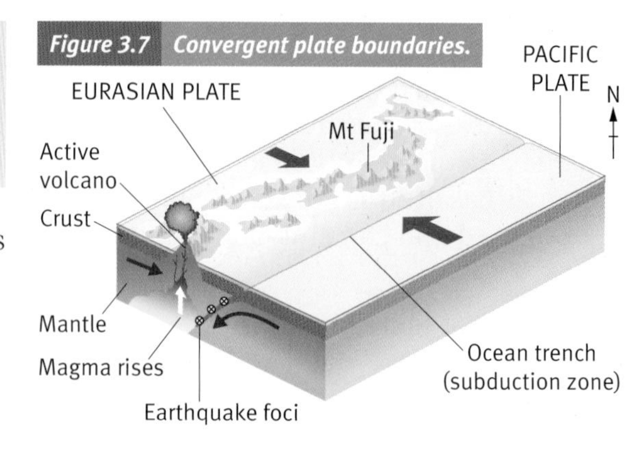

Figure 3.8 Divergent plate boundaries.

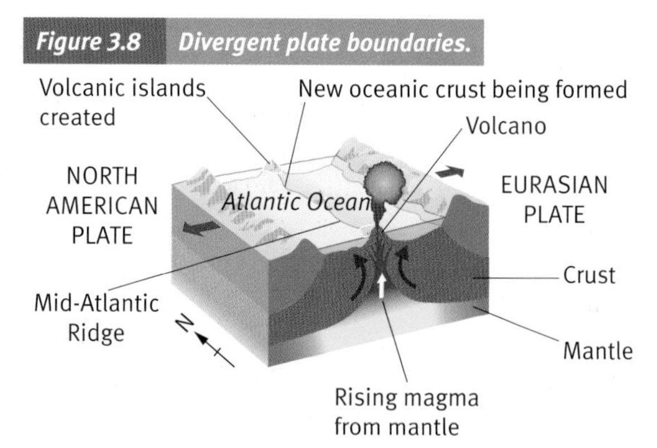

KEY TERMS

Fault – a fracture in the Earth's crust that shows signs of movement.

Subduction zone – the area of a destructive plate boundary where one plate descends beneath another.

Figure 3.9 Global plates and plate margins.

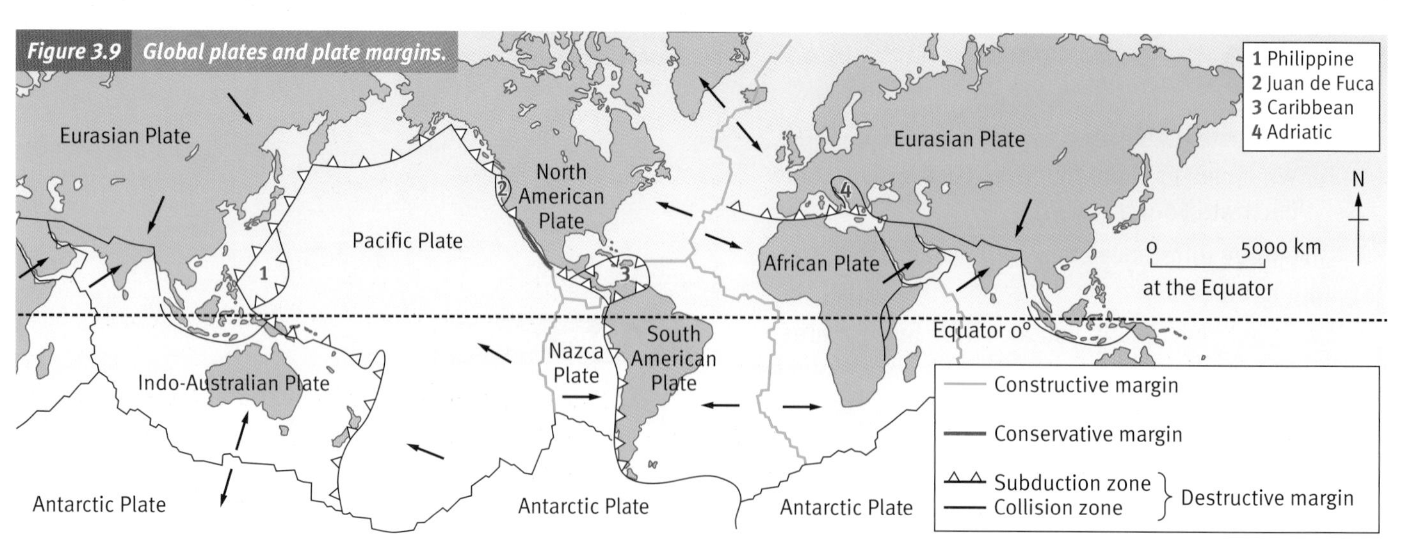

Where two plates made up of continental crust collide, both are pushed upwards and the rocks are folded and **faulted** to form mountain ranges. This occurred in Europe when the Eurasian and the African plates collided to form the Alps and in Asia where the Indo-Australian Plate collided with the Eurasian Plate to form the Himalayas.

Divergent boundaries

Places where plates are moving apart are called divergent boundaries. When the crust is pulled apart, it typically breaks along parallel faults that tilt slightly outward from each other. As the plates separate along the boundary, the block between the faults cracks and drops down into the soft, molten interior (the mantle). The sinking of the block forms a central valley called a rift. Magma seeps upward to fill the cracks. In this way, new crust is formed along the boundary. Earthquakes occur along the faults, and volcanoes form where the magma reaches the surface.

Transform boundaries

Places where plates slide past each other are called transform boundaries. Although transform boundaries are not marked by spectacular surface features, their sliding motion causes a lot of earthquakes. Instead of slipping smoothly past one another they tend to get stuck. Pressure builds up until suddenly the plates jump forward, sending out shock waves that cause an earthquake. An example of a transform boundary is the San Andreas Fault, shown in the diagram in Figure 3.10. The area of California to the west of the fault is slowly moving north relative to the rest of California.

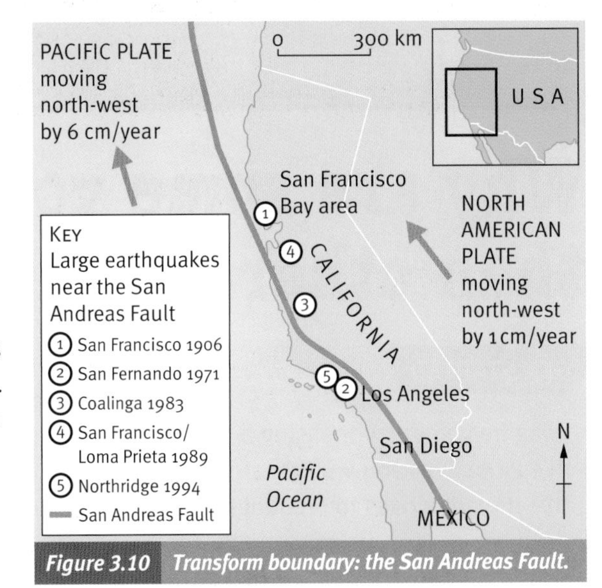

Figure 3.10 *Transform boundary: the San Andreas Fault.*

PLENARY ACTIVITY

If you were asked to explain the theory of plate tectonics to geography students in a Year 8 class, choose four of the most important points that you would want to tell them. Can you think of any resources that you might use to help you?

ACTIVITIES

1 Make a copy of Table 3.1 and use the information from the last four pages to:

a describe the changes that take place along the plate boundary

b describe the types of volcanic and earthquake activity that occur there

c give examples of where this activity occurs.

Table 3.1

Type of plate margin	Description of changes	Earthquake/volcanic activity	Examples
Convergent (oceanic and continental)			
Convergent (two continental)			
Divergent on land			
Divergent under the ocean			
Transform			

2 **a** On a copy of Figure 3.9, mark and label the three types of plate boundary in different colours. Name the plates on your map and show their direction of movement.

b What type of plate boundary (convergent, divergent or transform) exists between the following plates?

- Eurasian and Indo-Australian
- Pacific and North American
- Nazca and South American
- African and Indo-Australian
- Eurasian and African

Why are some volcanoes more violent than others?

GET STARTED

Imagine you are witnessing a volcanic eruption. Choose five adjectives you would use to describe what you see. Use the adjectives to recount that eruption to a partner.

Volcanoes cluster along the boundaries of the Earth's plates. Although the deeper levels of the Earth are much hotter than the surface, the rocks are not usually molten because the pressure is so high. But along plate boundaries there is molten rock – magma – to supply volcanoes.

Constructive margin volcanoes

At constructive margins, the sheet of ocean floor on each side of the ridge moves away from the centre as if on a conveyor belt. Hot fluid rock from deep in the mantle flows up to fill the central gap, warming shallower parts of the mantle, which begins to melt. Basalt lava then erupts, to make new ocean crust. Here the emerging magma is more fluid and the resulting volcano shape will be low and wide. This is called a **shield volcano**. Eruptions are generally much more gentle than subduction-zone volcanoes and can last for several months (Figure 3.14: 3).

Subduction-zone volcanoes

At destructive margins, a different process triggers the melting of the rock. As the slab of ocean floor – basalt – slides down the subduction zone, it warms up slowly, melting and mixing with water and CO_2 released from the rock. The hot magma which is more 'sticky' (or viscous) in this type of volcano rises in violent eruptions to form steeper-sided volcanoes. An explosive eruption results in rock, both solid and molten, exploding into tiny fragments of rock and glass. This mixes with gas and steam to form huge clouds known as **nuées ardentes**. Layers of ash and lava build up around the vent to form a **strato volcano**. Subduction-zone volcanoes form a line parallel to the plate boundary. Most of the volcanoes of the 'Ring of Fire', which forms around the Pacific Ocean, are related to destructive plate margins (Figure 3.14: 1 and 4).

Table 3.2 Volcanic eruptions of the world

Volcano	Location
Mount St Helens	USA
Cotopaxi	Ecuador
Mont Pelée	Martinique, Caribbean
Surtsey	Iceland
Mount Vesuvius	Italy
Krakatoa	Indonesia
Mount Pinatubo	Philippines
Unzen	Japan

KEY TERMS

Nuées ardentes – highly destructive, fast moving clouds of dust and ash.

Shield volcano – volcano that covers a large area with very gently sloping sides.

Strato volcano – tall, conical volcano composed of many layers.

Figure 3.11 *Strato volcano.*

Figure 3.12 *Nuée ardente from Pinatubo eruption.*

Hot-spot volcanoes

There are some exceptions to the rule that volcanoes cluster at plate boundaries and these form isolated chains of volcanic islands in the middle of the oceans. Hawaii, famous for the spectacular eruptions of its active volcano, Kilauea, is just one in a line of volcanic islands stretching north-west across the Pacific Ocean. These volcanoes result when a plate moves over an especially hot part of the fluid mantle. A plume of hot material, rising from deep in the mantle, is responsible for such a hot spot. If you imagine a piece of paper sliding above a candle, the volcanoes are the equivalent of scorch marks on the paper. The further away from the hot spot the volcano is, the older it is. These island chains provide important clues about the speed and direction of movement of the plates (Figure 3.14: 2).

Volcanoes can also be classified according to how recently they have erupted. Volcanoes that have erupted in recent times are known as active volcanoes. Volcanoes that have not erupted for a long time are known as dormant. Vesuvius, overlooking Naples in Italy, is dormant. Volcanoes that have not erupted in historical time are called extinct. An example is Mount Snowdon in Wales.

Figure 3.13 **Kilauea (Hawaii) erupting.**

Figure 3.14 **Different types of volcano formation.**

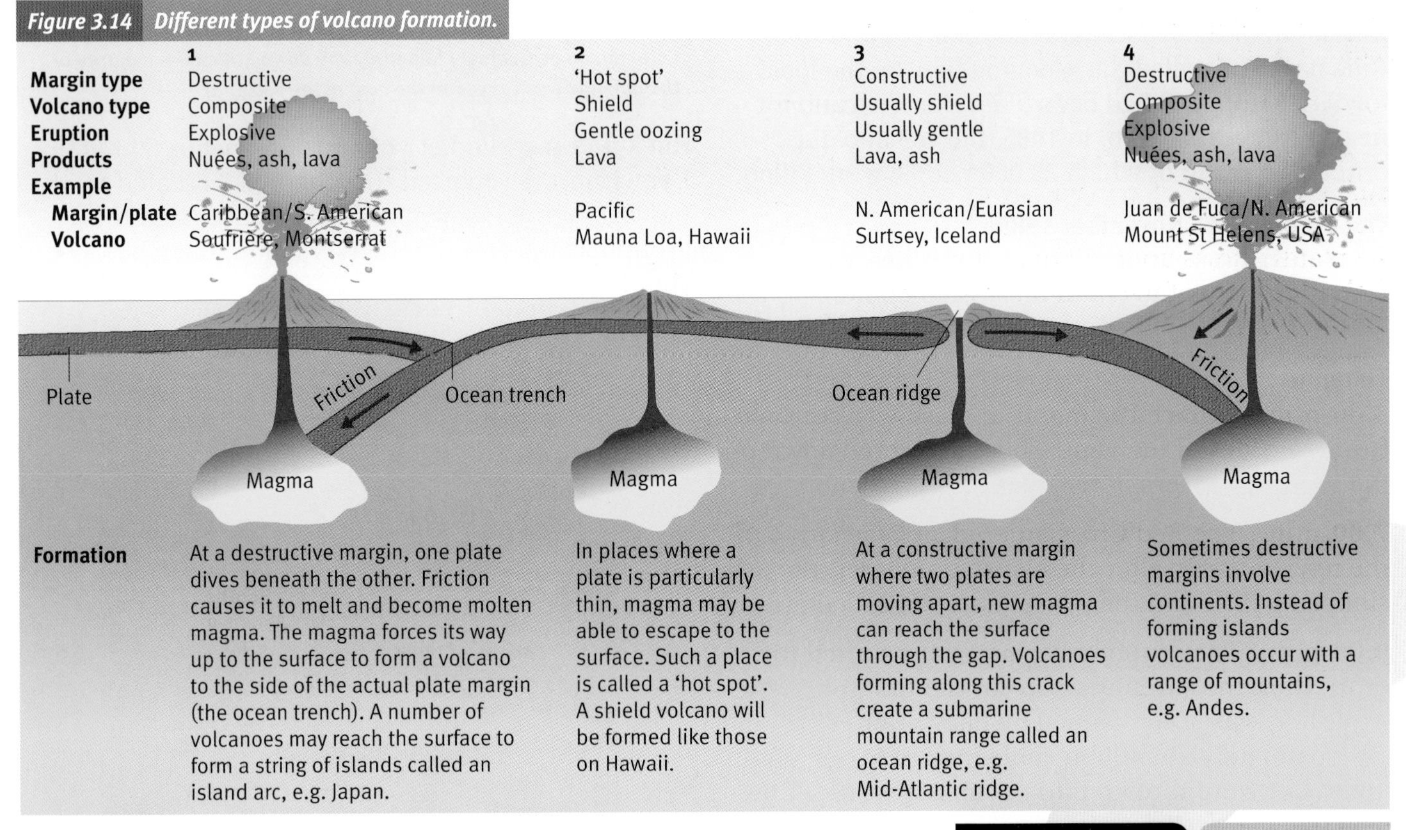

	1	2	3	4
Margin type	Destructive	'Hot spot'	Constructive	Destructive
Volcano type	Composite	Shield	Usually shield	Composite
Eruption	Explosive	Gentle oozing	Usually gentle	Explosive
Products	Nuées, ash, lava	Lava	Lava, ash	Nuées, ash, lava
Example				
Margin/plate	Caribbean/S. American	Pacific	N. American/Eurasian	Juan de Fuca/N. American
Volcano	Soufrière, Montserrat	Mauna Loa, Hawaii	Surtsey, Iceland	Mount St Helens, USA
Formation	At a destructive margin, one plate dives beneath the other. Friction causes it to melt and become molten magma. The magma forces its way up to the surface to form a volcano to the side of the actual plate margin (the ocean trench). A number of volcanoes may reach the surface to form a string of islands called an island arc, e.g. Japan.	In places where a plate is particularly thin, magma may be able to escape to the surface. Such a place is called a 'hot spot'. A shield volcano will be formed like those on Hawaii.	At a constructive margin where two plates are moving apart, new magma can reach the surface through the gap. Volcanoes forming along this crack create a submarine mountain range called an ocean ridge, e.g. Mid-Atlantic ridge.	Sometimes destructive margins involve continents. Instead of forming islands volcanoes occur with a range of mountains, e.g. Andes.

ACTIVITIES

1. On a copy of Figure 3.9 (p.112) and mark the volcanoes in Table 3.2 on the map. With which types of plate boundaries are they associated?
2. Why are Japan, Hawaii and Iceland prone to volcanic activity? With the aid of a diagram for each area, explain which type of volcano you would expect to find in each country and why.
3. Using the 'Active Volcanoes of the World' website, find a set of satellite images for a volcano that has recently erupted. Cut and paste them into your work and label the features that you are able to see on the image.

THINK ABOUT IT

What can volcanoes that occur above hot spots tell us about plate movement?

PLENARY ACTIVITY

Explain how earthquakes and volcanic eruptions are linked to each other.

GET STARTED

Find a map of Colombia that shows its physical features in an atlas or on the internet. List five facts about the relief and drainage of the country: for example, the height of the mountains; the names of some of the rivers.

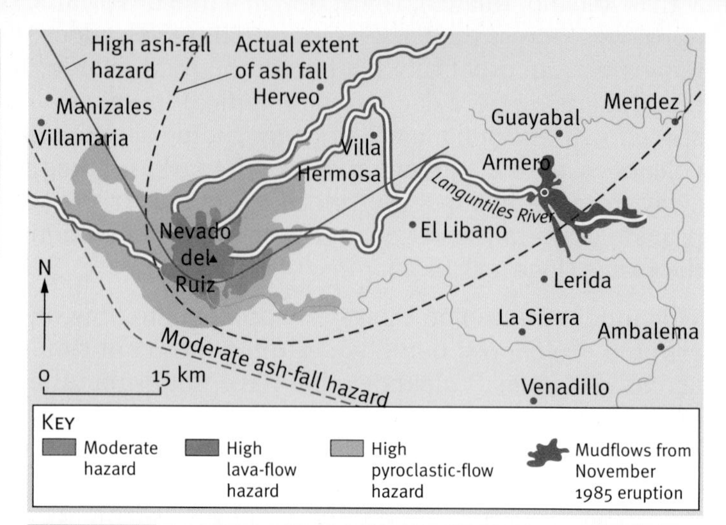

Figure 3.15 ***Map of hazard zones.*** **(Source: Ingeominas, Instituto Columbiano de Geologíy Minería)** ***Why do you think the majority of the damage was caused to the east of the volcano?***

Nevado del Ruiz is a strato-volcano in the mountain range of the Andes. It is the highest Colombian volcano with a history of violent activity. The Nevado del Ruiz was produced by subduction of the oceanic Nazca Plate beneath the South American Plate. Through the centuries a series of movements have occurred along this plate boundary, leading to the formation of the Andes fold mountains. Nevado del Ruiz has been called the 'sleeping lion' by the local towns around it. It had been a dormant volcano for nearly 150 years when, in 1985, the volcano did erupt, as a result of which 23,000 people were killed.

The eruption: 13 November 1985

After nearly a year of minor earthquakes and steam explosions from the crater of Nevado del Ruiz, the volcano exploded violently.

Timeline

3.06 p.m.: Pumice fragments and ash were thrown from the sides of the vent. These showered down on the town of Armero at the foot of the volcano.

7.00 p.m.: The Red Cross ordered an evacuation of the town. Shortly after the evacuation order, the ash stopped falling and the evacuation was called off.

9:08 p.m.: The eruption began as lava started to erupt from the summit crater for the first time. The explosion was accompanied by heavy rainfall. Approximately 20 million cubic metres of hot ash and rocks were thrown into the air across the snow-covered glacier. These materials were transported across the snow pack by **pyroclastic flows** and fast-moving, hot, turbulent clouds of gas and ash. The hot flows caused rapid melting of the snow and ice, and created large volumes of water that swept down canyons leading away from the summit. As these floods of water descended, they picked up loose debris and soil from the valley floors and walls, growing both in volume and density, to form hot **lahars**. In the river valleys further down the volcano's slopes, the lahars were as much as 40 m thick and travelled at velocities as fast as 50 km/h. Eyewitnesses reported that the noise created by the passage of the lahar made their houses and the ground shake and that conversation, even by shouting, was impossible.

Figure 3.16 ***Map of Colombia showing plate boundaries.***

KEY TERMS

Lahar – a type of mudflow composed of pyroclastic material and water that flows down from a volcano, typically along a river valley.

Pyroclastic flows – avalanches of hot volcanic debris.

Fact file

Although not the most violent explosion, Nevado del Ruiz was the second most destructive eruption of the 20th century, following Mont Pelée in 1906.

11.28 p.m.: One of the lahars reached Armero. In a few short minutes most of the town was swept away or buried in a torrent of mud and boulders. One lahar flowed down the River Cauca, submerging the village of Chinchina and killing 1927 people.

Figure 3.17 ***Nevado del Ruiz erupting.***

What were the effects of the eruption on people and the landscape?

The effects of the eruption can be divided into primary and secondary types. Primary effects are those that are a direct result of the eruption occurring in that location. Secondary effects include those impacts that result indirectly from the event.

Primary effects

- Hot clouds of ash and gas (nuées ardentes).
- Magma, thrown into the air from the vent.
- Avalanches of hot ash, pumice and rock fragments (pyroclastic flow).

Secondary effects

- Mud up to 40 m deep was deposited in the valleys in the path of the lahars.
- 5000 homes were destroyed when they were swept away or buried by the mud and ash.
- 23,000 people were killed, mainly from being buried in their homes.
- 5000 people were injured.
- The temperature of the lahars served as a fertile breeding ground for all kinds of fungi and bacteria. Somc survivors who had minor cuts were killed by the infections, which could not be treated with known antibiotics.
- Roads were blocked and rescue workers found it difficult to rescue survivors from the deep mud
- The eruption cost Colombia US $7.7 billion (20% of the country's GDP at the time).

ACTIVITIES

1. Use a diagram to explain the causes of the 1985 eruption of Nevado del Ruiz.
2. Make a timeline to show the events that took place from the moment of the eruption until the lahar reached Armero. Indicate on your timeline which are primary and which are secondary effects.
3. Describe how the effects of the eruption affected the landscape of the area.
4. Why did the earthquake cause so many deaths?
5. What do you think was the impact of the eruption on:
   a the lives of people who lived in the area of Armero
   b the lives of people of the rest of Colombia?
   Explain your answer.

RESEARCH LINK

Use the internet to find out what has happened since the eruption to the people and the environment around Armero.

PLENARY ACTIVITY

'If the slopes of Nevado del Ruiz had not been covered by glaciers the effects of the eruption in 1985 would have been far less severe for the people and the landscape.' Discuss this statement with a partner.

GET STARTED

What makes every volcanic eruption different?

What caused Etna to form?

Etna is a strato-volcano, which has formed as a result of repeated volcanic eruptions that have built up layers of lava and ash to create a cone. Figure 3.18 shows the causes of the formation of Mount Etna.

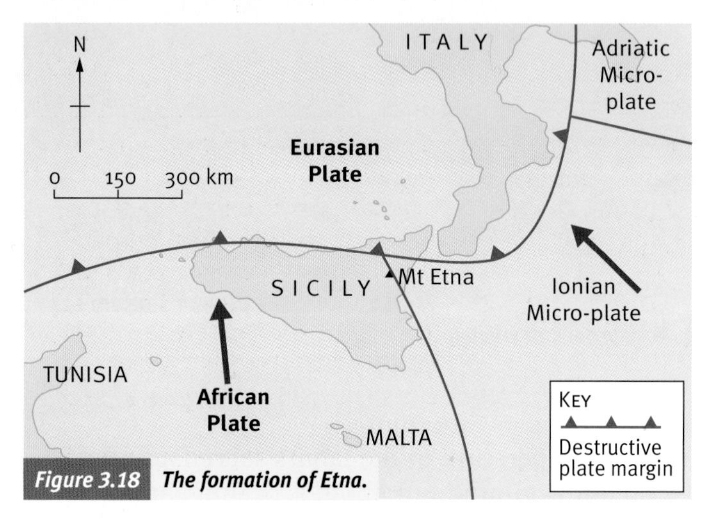

Figure 3.18 *The formation of Etna.*

Sicily is located just south-west of the Italian mainland. It is the largest island in the Mediterranean basin. It is one of the poorest areas of Italy, but the fertile soil and a long, hot growing season mean that agriculture is an important part of the economy here. A newer form of income for the local people is the more recent development of a tourist industry, which includes skiing on the upper slopes of the volcano.

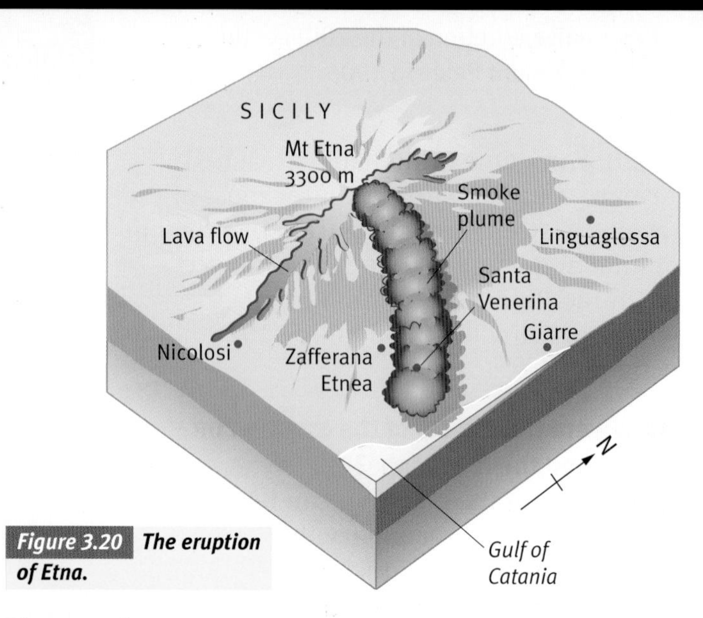

Figure 3.20 *The eruption of Etna.*

The eruption

Mount Etna erupted throughout November 2002. A series of earthquakes, measuring up to 4.3 on the Richter scale, accompanied several explosions that blackened the sky above the mountain. Clouds of gas and ash were forced from two vents, one a new one, in the side of the volcano. This was followed by magma, which was thrown more than 100 m into the air in a spectacular display. The lava then began to run quickly down the mountainside, forming two separate flows. Ash fell continuously onto the city of Catonia and drifted as far south as Libya.

Figure 3.19 *Erupting Etna.*

Fact file

Largest volcano in Europe. Height: 3329 m.

Mount Etna is one of the most active volcanoes in the world and is in an almost constant state of eruption.

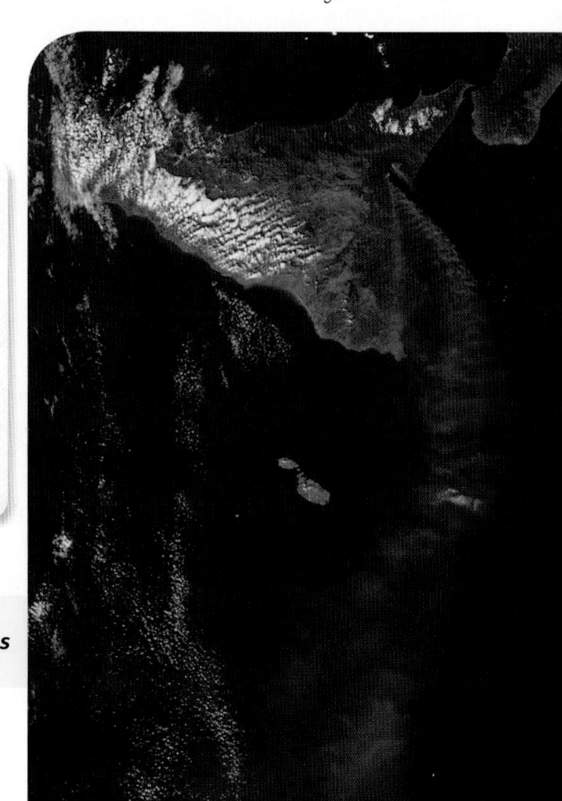

Figure 3.21 *Ash cloud from Etna blowing across the Mediterranean.*

The impact of the eruption

The earthquake damaged more than 100 homes in Santa Venerina and holiday homes were taken over by the local authorities to house the displaced people. Residential areas such as the town of Linguaglossa were evacuated because of the threat of the lava flow. One thousand people had to leave their homes. Schools in the town were shut down, although the church remained open for people to pray. Villagers also continued their tradition of parading their patron saint through the streets to the railway station, to try to ward off the lava flow.

The airport at Catania was closed for 4 days because of the ash that covered the runway and threatened to clog up the aircraft engines. The skiing season was about to start, but the area was swamped by the flowing lava. It engulfed a restaurant and pushed over three ski-lift pylons. It also destroyed hundreds of hectares of forest on the slopes of the volcano.

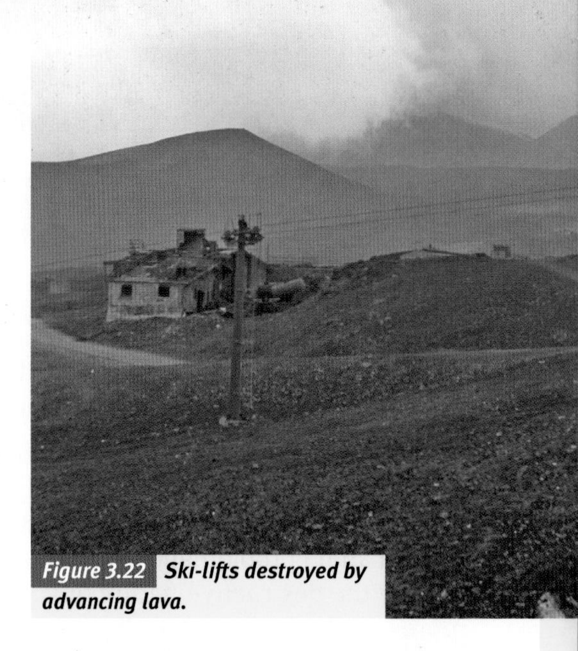

Figure 3.22 ***Ski-lifts destroyed by advancing lava.***

The response to the eruption

The Italian government declared a state of emergency in parts of Sicily during the eruption. Rescue workers battled to divert the lava that threatened to engulf a scientific monitoring centre at the foot of the mountain. The army used bulldozers to crack the tarmac and build barriers in a car park near the centre, in an attempt to create a channel that would redirect the lava away from populated areas. Emergency workers dug channels in the earth in an attempt to divert the northern flow away from the town of Linguaglossa. A ship equipped with a medical clinic was positioned off Catania to be ready in case of emergency.

The businesses of 300 families were affected by the eruptions and the Italian government gave tax breaks for villagers to help them get through the crisis and more than US $8m (£5.6m) in immediate financial assistance.

'Everyone has another job in order to survive winter. My ski school on Etna has been closed for three years and I have to work in northern Italy, but I come back here as often as I can. Etna will always be my mountain. They say it will take two years before skiers are able to use the site again.'

Figure 3.23 ***Gianicolo's ski-school has been shut for 3 years.***

RESEARCH LINK

Has Etna erupted recently? Is there still a tourist industry in the area?

ACTIVITIES

1. Why do you think people continue to live on the slopes of Etna, even though it continuously erupts for months at a time?
2. Make a list of all the people whose jobs were affected by the eruption. How have some of them tried to compensate for their loss of income?
3. Think of five questions you would need to ask about volcanoes to make a comparison between the eruption of Nevado del Ruiz (LEDC) and the eruption of Etna (MEDC). This should include the effects of the volcano on the landscape and on people.
4. With a partner, compare your questions and agree on the six best questions.
5. Use your questions to compare the eruptions of the two volcanoes by drawing up a table to show the similarities and differences.

Is it safe to live near a volcano?

GET STARTED

Volcanoes can cause violent eruptions. We know where they are, so why do some people choose to live in volcanic areas?

Volcanoes have a wide range of effects on humans. These can be problematic or beneficial, but it is usually the destructive nature of volcanoes that is more widely documented. Increasing numbers of people have died from the effects of volcanic eruptions since the beginning of the 20th century; this is mainly due to more and more people living on the slopes of active volcanoes or close to them. Many people rely on volcanoes for their everyday survival. The variety of advantages of living in areas of volcanic activity are increasingly seen as outweighing the risk of coping with the effects of an eruption.

Reasons to live near a volcano

- **Geothermal** energy can be harnessed by using underground steam resources that have been heated by the Earth's magma. This steam is used to drive turbines in geothermal power stations to produce electricity for domestic and industrial use. Countries such as Iceland and New Zealand use this method of generating electricity.
- Volcanic areas often contain some of the most mineral-rich soils in the world. This is ideal for farming. Lava and material from pyroclastic flows are weathered to form nutrient-rich soil that can be cultivated to produce healthy crops and rich harvests.
- Volcanoes attract millions of visitors around the world every year. Apart from the volcano itself, hot springs and geysers can also attract tourists. This creates many jobs for people in the tourism industry. This includes work in hotels, restaurants and gift shops. Often locals are also employed as tour guides.
- Lava from deep within the Earth contains minerals that can be mined once the lava has cooled. These include gold, silver, diamonds, copper and zinc, depending on their mineral composition. Often, mining towns develop around volcanoes. Volcanic rocks also make good building stone.

KEY TERMS

Geothermal – energy generated by heat stored deep in the Earth.

GPS (Global Positioning System) – a group of satellites that allow people to find out their exact location on the Earth's surface.

Figure 3.24 ***Despite the danger, people still live close to volcanoes (Vulcano, Italy shown here).***

How can you predict volcanoes?

- As magma rises towards the surface it can cause the earth around it to vibrate, which can be detected on an instrument called a seismograph. Any sudden change in earthquake activity around an active volcano will, hopefully, give scientists enough time to sound warnings.
- The magma also causes ground deformation. As it rises in a volcano it has to make space for itself. This can cause the ground to bulge, which can be detected by sensitive **GPS** instruments via satellite remote sensing. Any change of position – of as little as 10 cm – could mean the onset of an eruption.
- Any changes in the composition of gas escaping from the volcano can indicate that an eruption is about to take place.
- The amount of water flowing off the volcano can also be measured (hydrology). If the discharge increases or the density changes, it may be a signal of imminent eruptions.

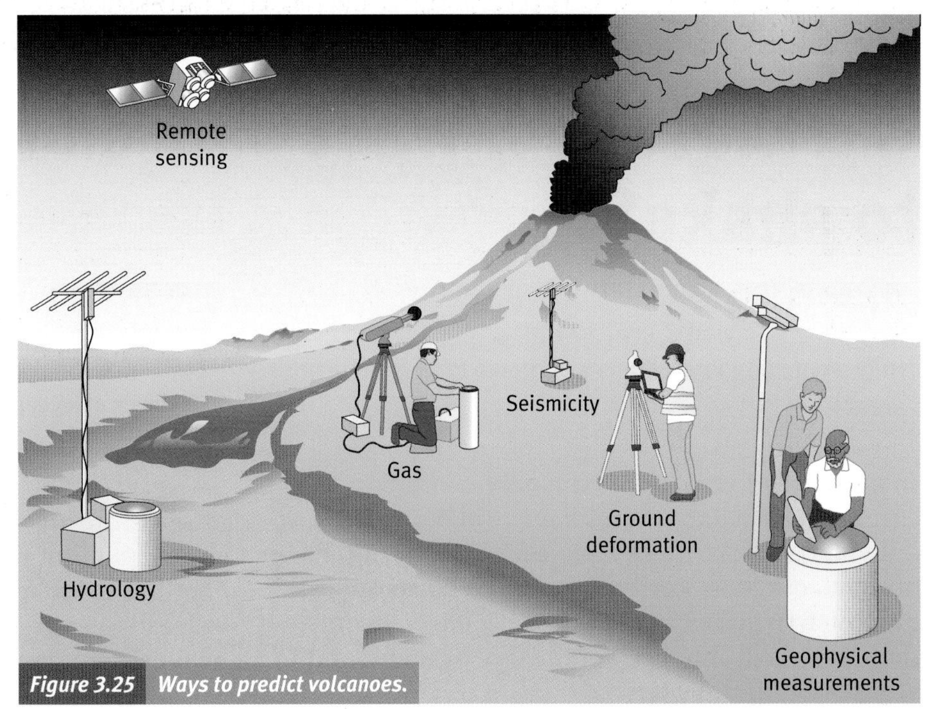

Figure 3.25 *Ways to predict volcanoes.*

Figure 3.26 *Using instruments to measure the temperature of lava.*

PLENARY ACTIVITY

Imagine that Nevado del Ruiz had begun showing signs of erupting, experiencing earthquakes similar to those that occurred in 1985. In order to save lives, what kind of evacuation plan would you implement?

Consider the following issues when outlining your evacuation plan:

- When should people leave the area? At the first earthquake, or not until the volcano starts smoking?
- What should people be allowed to take with them when an evacuation order is issued? Pets? Livestock? Household goods? What would happen if everyone tried to take everything they owned?
- What should be done about people who don't want to leave?
- What if the area was successfully evacuated and then the volcano did not erupt? When would you allow people to return to their homes?
- Where will everyone go? How will people survive without the land from which they make their living?

ACTIVITIES

1. Give three reasons why people might continue to live on the slopes of a volcano. Which factors are likely to have been the most important for people living on Nevado del Ruiz and Etna? Will these reasons have changed over time?
2. Is it possible to predict volcanoes? Give two examples of how this might be achieved.

Earthquakes: what causes the ground to shake?

GET STARTED

How would you know if an earthquake happened in your area?

KEY TERMS

Epicentre – the point on the Earth's surface directly above the focus of an earthquake.

Seismic waves – vibration generated by an earthquake.

In an average year, approximately 12 million earthquakes occur around the world. Fortunately, only about 100 of these have a significant impact on human activities. Earthquakes have no seasonal time of arrival. Because they are geologic events, they can occur any time of the day or night, and in any season of the year.

Causes

Most earthquakes are caused by sudden movement within the Earth. This could be caused along a fault plane or as a result of the upwelling of magma during a volcanic eruption. The movement is the result of the release of stress in the rocks that builds up over a period of time. This sudden release of energy causes shocks or **seismic waves**. The point at which the pressure is released is called the focus. This is the centre of the earthquake. The point on the Earth's surface immediately above the focus is called the **epicentre**. The seismic waves radiate outwards from the focus and decrease in strength as they spread.

Table 3.3 Deadliest and biggest earthquakes 1990–2008

Year	Largest earthquakes				Year	Deadliest earthquakes			
	Magnitude (Richter)	Intensity (Mercalli)	Deaths	Location		Magnitude (Richter)	Intensity (Mercalli)	Deaths	Location
2004	9.1	12	227,989	Off coast of Sumatra	2004	9.1	12	227,989	Off coast of Sumatra
2005	8.5–8.6	12	1313	Northern Sumatra	2008	7.9	11	87,652	Sichuan, China
2007	8.3–8.5	12	25	Northern Sumatra	2005	7.6	11	80,361	Pakistan
2001	8.4	12	138	Off coast of Peru	1990	7.4	11	50,000	Iran
2003	8.3	12	0	Hokkaido, Japan	2003	6.6	8	31,000	SE Iran
1996	8.2	12	166	Indonesia	2001	7.7	11	20,023	India
1995	8.0	11	3	Off coast of Chile	1999	7.6	11	17,118	Turkey
2003	7.9	11	0	Central Alaska	1993	6.2	8	9748	India
2008	7.9	11	87,652	Sichuan, China	2006	6.3	9	5749	Java, Indonesia
1997	7.8	11	0	South of Fiji	1995	6.9	9	5530	Kobe, Japan

Table 3.4 Effects of an earthquake

Ground shaking	This is the main hazard created by earthquakes. Its severity at any one point depends upon a range of factors, including the strength of the earthquake, the distance from the epicentre and local geological conditions. The longer the shaking, the more damage will be caused. This is what led to the collapse of many buildings in the Turkey earthquake of 1999.
Landslides, rock and snow avalanches	Occur in mountainous areas when slopes are weakened by strong shaking and eventually give way. Landslides blocked the roads and filled the reservoirs in the Sichuan earthquake in China in 2008.
Tsunamis	Seismic sea waves that result when earthquakes occur beneath the sea. They can travel with enormous speed and cause devastation along the coast. The tsunami caused by the Indian Ocean earthquake in December 2005 led to a tidal wave that killed more than 200,000 people.
Liquefaction	The shock waves cause groundwater to rise to the surface, turning soft ground to mud. Any buildings built on clay, for example, sink into the mud and collapse. During the Loma Prieta earthquake in California (1989), the only buildings that collapsed were built on reclaimed land in San Francisco's Marina District.

Measuring earthquakes

There are two main ways of measuring the size of earthquakes – magnitude and intensity. Magnitude is measured on the Richter scale and given a value of between 1 and 10. The scale is logarithmic, so each point is 10 times greater than the previous one. Earthquake intensity measures the degree of shaking of the ground and records the level of damage and other consequences of the earthquake. Mercalli developed a 12-point scale for measuring the intensity of earthquakes, which uses the observations of the people who experienced the earthquake.

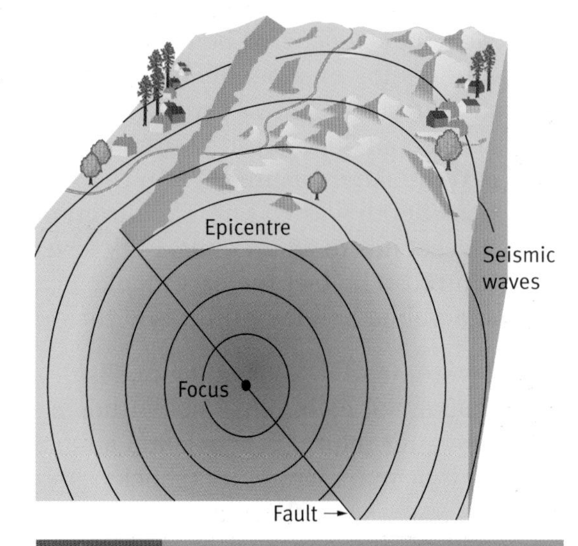

Figure 3.27 *The focus and epicentre of an earthquake.*

Damage

Several things affect the amount of damage that occurs as a result of an earthquake, including:

- the distance from the epicentre – generally the further from an earthquake the less damage there is
- the type of material (rock or sediment) that a building rests on – whether a building is built on solid rock or sand makes a big difference to how much damage it sustains. Solid rock usually shakes less than sand, so a building built on top of solid rock should not be as damaged as it might if it was sitting on sandy soil
- the design of buildings – rigid structures are very 'ungiving' when shaken. It is possible to build them more flexibly and leave space for movement.

Fact file

The largest recorded earthquake in the world measured 9.5 in Chile on 22 May 1960.

ACTIVITIES

1. Plot the earthquakes listed in Table 3.3 onto a world map. Name the location and note the date and scale of the earthquake.
2. Compare this map with the map of plate boundaries in Figure 3.9 (p.112). With which types of plate margin are the earthquakes associated? Describe their distribution.
3. Explain how an earthquake happens and the effects that it can have on the landscape. Use the following words in your answer: seismic waves, focus, epicentre, shaking, tsunami, liquefaction. Illustrate your answer with a diagram.
4. Why would knowing the intensity in a particular area be of more use than knowing the magnitude to those planning rescue work after an earthquake?

PLENARY ACTIVITY

Why do earthquakes in economically developing countries cause many more deaths than those that occur in more economically developed countries?

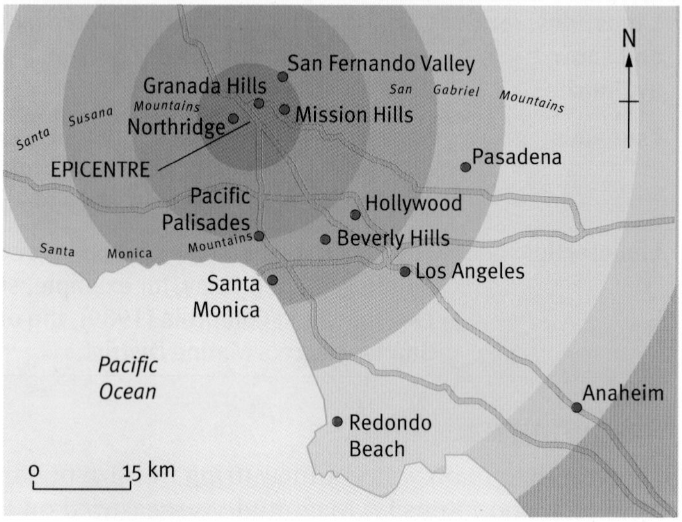

earthquake in an MEDC

GET STARTED

What reasons would make you stay in an area that suffers from many earthquakes, such as California?

The Northridge earthquake of 17 January 1994 struck a modern urban environment, generally designed to withstand strong earthquakes. There were relatively few casualties, but the economic cost was high, with losses estimated at US $20 billion.

The earthquake occurred beneath the San Fernando Valley on a deeply buried **blind fault** at 4.30 a.m. local time on 17 January 1994. Northridge is located about 30 km north-west of Los Angeles. This earthquake measured 6.7 on the Richter scale. The focus depth was 19 km. The duration ranged from 10 seconds to 20 seconds. Fifty-seven people were killed and 9000 injured. The fact the earthquake occurred early in the morning and on a public holiday minimised the death toll.

Figure 3.28 *Northridge, California: extent of damage.*

KEY TERMS

Blind fault – one where the fault line does not reach the surface.

There were nearly 15,000 aftershocks that occurred after the main earthquake. For example, a magnitude 5.9 aftershock occurred about 1 minute after the main shock and a magnitude 5.6 earthquake occurred 11 hours later. Aftershocks can trigger the collapse of structures weakened by the main shock.

Structural damage

- Most casualties and damage occurred in multi-storey wood frame buildings (for example, the three-storey Northridge Meadows apartment building). In particular, buildings with a weak first floor (such as those with parking areas on the bottom) performed poorly.
- The Northridge earthquake caused extensive damage to parking structures and freeway overpasses.
- The Northridge earthquake triggered landslides in surrounding mountain areas. These landslides blocked roads and damaged water lines and also damaged homes, particularly in the Pacific Palisades area.
- Numerous fires were also caused by broken gas pipes damaged by houses shifting off foundations or by unsecured water heaters falling over. A fire devastated an area of trailer homes in the San Fernando Valley. Several undergound gas and water mains were severed.

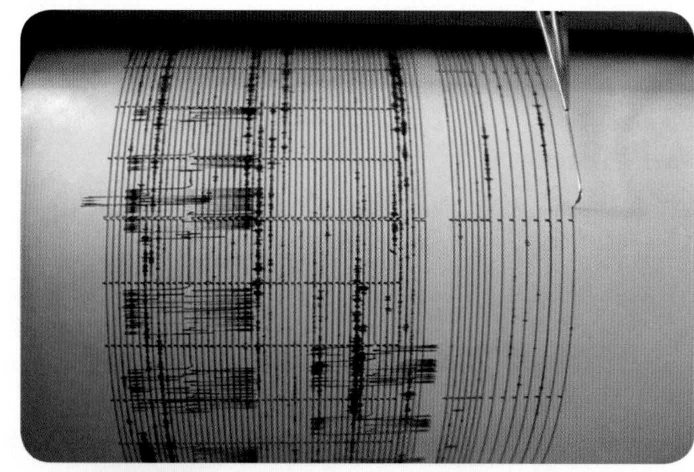

Figure 3.29 *Seismograph of Northridge earthquake.*

Fact file

The San Andreas Fault is not a single, continuous fault but a zone made up of many segments. Movement may occur along any of the many fault segments along the zone at any time. The San Andreas Fault system is more than 1300 km long, and in some spots is as much as 16 km deep.

Figure 3.30 *Multi-storey car park collapse.*

- The scoreboard at Anaheim Stadium collapsed onto several hundred seats. Fortunately, the stadium was empty at the time, due to the time of day the earthquake occurred.
- Liquefaction occurred amongst some alluvium in the upper San Fernando Valley.

The area was declared a federal disaster by President Clinton and hundreds of workers from the Federal Emergency Management Agency (FEMA) were deployed to Southern California to help the communities there recover. More than 600,000 individuals applied for state and federal disaster assistance, and FEMA spent millions of federal money helping the area recover.

Predicting an earthquake

Researchers in 2008 predicted that Southern California was much more likely to suffer an earthquake the size of the Northridge one than Northern California by 2028. The problem is that they cannot predict when or where the earthquake will happen.

The United States Geological Survey (USGS) have been investigating whether it is possible to predict earthquakes for over 30 years. Scientists analyse data to work out the likelihood of fault movements in specific areas using new information about prehistoric earthquakes, locations of hidden faults, and an increasing database of satellite-based GPS data of the Earth's crust movement.

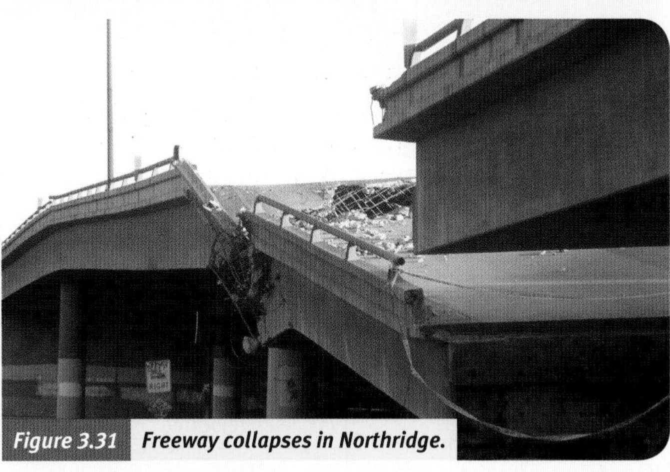

Figure 3.31 *Freeway collapses in Northridge.*

The Parkfield experiment

The town of Parkfield is located right on the San Andreas fault. Scientists have set up instruments to record all ground shaking in the area and will use the data to attempt to understand what actually happens on the fault and in the surrounding region before, during and after an earthquake. Eventually scientists hope to better understand the earthquake process and, if possible, to provide a scientific basis for earthquake prediction.

PLENARY ACTIVITY

Why is it so difficult to predict the time and location of large earthquakes?

ACTIVITIES

1. Draw a sketch map showing the location of the Northridge earthquake and label it with examples of the effects of the earthquake. Indicate whether they are primary or secondary effects.
2. Imagine an earthquake of 7.0 on the Richter scale occurred with its epicentre in your home area. List the effects that the earthquake would have on your surrounding area. Divide them into primary and secondary effects and try to give actual examples of structures that would be affected.
3. Services and cars are important to most of us in our lives. When disaster strikes many people have to go without both. How would the following disruptions affect you?
   - No power
   - No natural gas
   - No water
   - Inaccessible roads and no public transport.
4. What groups of people would experience special difficulty in confronting these disruptions?

GET STARTED

If you were to experience an earthquake whilst inside your house, what three items would you grab on your race to get outside? Think about what you would need to survive if your whole area was devastated.

On Monday, 12 May 2008 at 2.28 p.m. local time, an earthquake measuring 7.9 on the Richter scale struck the province of Sichuan in southern China. People were at work, in school and out shopping when an event that lasted only 2 minutes killed 69,180 people and changed millions of lives for ever.

Fact file

China

- Population: 1,323,350,000.
- Fastest-growing population in the world for last 25 years.
- Gross National Income per capita US $6500.

Figure 3.32 ***Sichuan area showing damage.*** **(Source: Map provided courtesy of the Relief Web Map Centre, UN Office for the Coordination of Humanitarian Affairs)**

Agricultural damage

- Irrigation systems for 100,000 hectares of paddy fields wrecked
- >50,000 greenhouses destroyed
- 7.3 million m² of livestock barns collapsed

Infrastructure damage

- 15,000,000 buildings collapsed
- 5000 km of pipes damaged
- 839 water tanks collapsed
- 1300 water treatment works destroyed

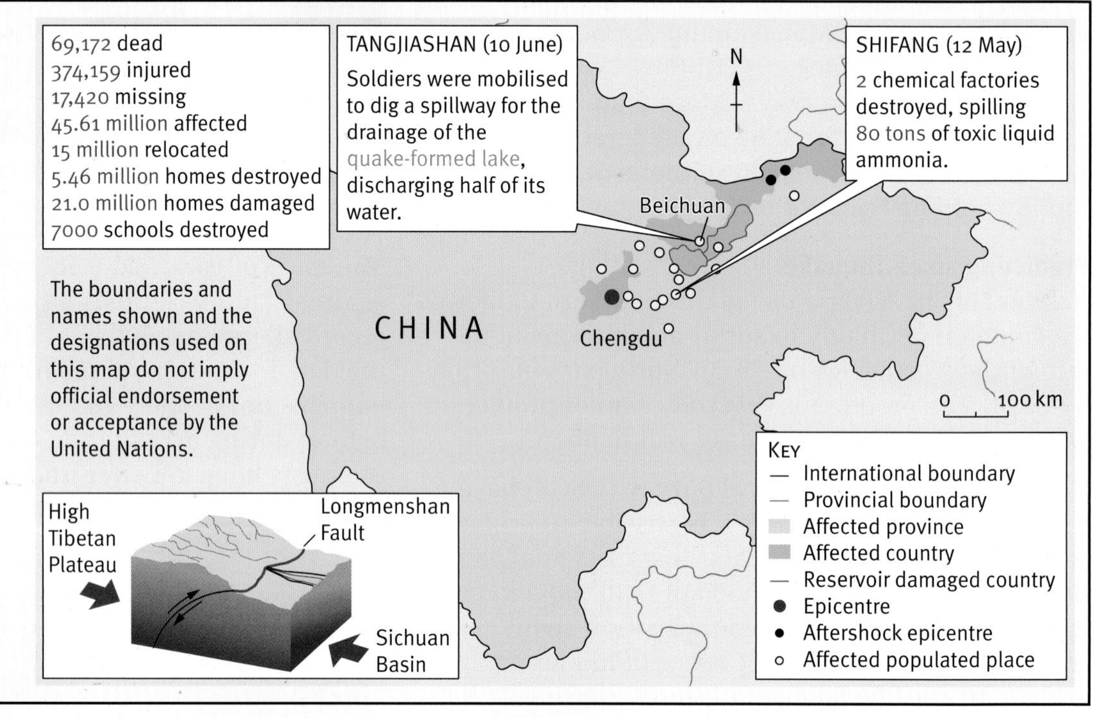

Figure 3.33 ***Over 7000 schoolrooms collapsed in the earthquake.***

Schools built in rush

According to Chinese government experts, the collapse of so many schools in the Sichuan earthquake was probably due to construction flaws occurring in a rush to build schools. This is the first time that officials have admitted that building standards may have contributed to the deaths of many children.

'Since the houses in the quake-hit areas were generally built of mud, many collapsed or were damaged,' said Wu Jingping, the secretary of the Communist Party in Sichuan's Liangshan prefecture. The government has now allocated 27 million yuan (£2 million) from its emergency relief fund for the disaster in Sichuan Province. 'The most important and most urgent problem that needs to be solved is how to help the earthquake victims rebuild their homes,' Mr Wu said. With coal mining areas badly damaged by the tremors, the region is facing a severe fuel shortage ahead of Sichuan's bitter winter.

Gao Li Qiang – *Writing from Sichaun, 4 June 2008*

As usual I was working in my office. Suddenly, the desk where I put my laptop began shaking. After several seconds my laptop started to jump up and down; it was like there were magic hands moving it. We must get out of the building as soon as we can – this is the only idea I had. When we got to the stairs, we found that we could not move anymore because the building, was shaking so strongly that we could not even stand.

One person suddenly yelled, ' Run!' Then everyone started to run down to the ground from 16th floor.

On the first few floors on our way down, the stairs were clear. Nothing was on them. But then we found socks, broken heels, shoes and even shirts all over the stairway.

The street had already been filled up with scared people. Mobile phones did not work and we could not find any ground lines near us. We realized we had lost contact with our family, friends.

The streets were full of people and vehicles. Traffic was so bad at that time and the radio did not give us any information of what was going on. I drove through red lights, crossed the forbidden dual yellow lines. Anyway I reached my family and took them to the outskirts of Chengdu, where there are no high buildings. The night of May 12th we stayed in our car listening to the radio. I thought we would be able to go back to work after few days but it turns out this was just the beginning; the worst was yet to come.

(Source: Threshold website)

Figure 3.34 ***Emergency tents at a stadium in Chengdu.***

Figure 3.35 ***Landslides can dam rivers to form 'Quake lakes'. As the pressure from the flowing water begins to build, the dams become unstable and can fail, resulting in flooding further down the valley.***

ACTIVITIES

1. Make a fact file of the Sichaun earthquake. Include information about its causes and the primary and secondary effects that were caused.
2. Locate Sichuan on an atlas map of China. Use the information in this section to explain why people suffered so badly after this earthquake. Write sections about homes, jobs, health.
3. What made it difficult for Gao Li Qiang to find out what had happened to his family?

THINK ABOUT IT

How might China's one-child policy cause particular hardship for people in Sichuan Province?

PLENARY ACTIVITY

If you had been in charge of the relief work in Sichuan Province, what would have been your priorities and how would you have put them into action?

Putting down roots in earthquake country

GET STARTED

Why do people who live in big cities such as San Francisco and Chengdu continue to believe that they will not die in the next big earthquake?

THINK ABOUT IT

'Earthquakes don't kill people, buildings do.' How true do you think this statement is? What other factors will it depend on?

Many people who live in areas affected by earthquakes know about the risks but they still choose to stay. The way people think about the possible consequences of a major disaster is called hazard perception. The way people respond to an earthquake depends on how they think it threatens them and how well they think they are protected from its impacts. It is not possible to prevent earthquakes. They usually occur with no warning and all people can do is:

- try to predict when an earthquake will occur
- prepare for them in order to reduce the damage
- ensure that rescue procedures are well planned.

Figure 3.36 *A tsunami early warning system.* **(Source: GFZ, German Research Centre for Geosciences)**

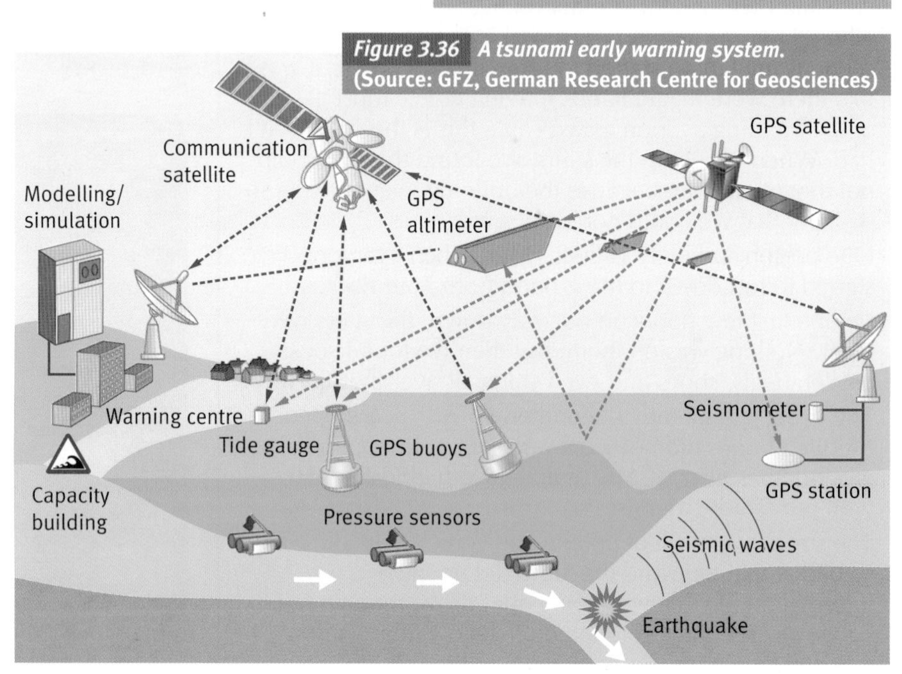

Can earthquakes be predicted?

If scientists can predict when and where an earthquake will happen, people can be warned and lives could be saved, so scientists monitor active zones.

- Sensitive instruments measure earth movements and check the stress building up in rocks. Maps can then be produced of the probability of earthquakes occurring. They can also use instruments to give early warning of tsunamis (Figure 3.36).
- Several earthquakes can occur in a short space of time. Foreshocks occur before some large earthquakes, so scientists can predict the chances of a major shock following.
- The number of recent earthquakes can be plotted to show if a major earthquake can be expected.

If a prediction is wrong then there can be consequences. Many people cannot afford to stop working, even for a few days. In 1986, 56,000 people were evacuated from the towns of Lucca and Modena in Italy. Shops and businesses closed for two days. No earthquake happened and local people were angry about the loss of business and the inconvenience of the evacuation. The mayor was forced to resign.

How can people prepare for earthquakes?

- **Earthquake-proof buildings**: most deaths in cities in earthquake-prone areas are caused when people are crushed inside collapsing buildings. Scientists can test how different building materials respond to shaking and improve their ability to survive. They can also design stronger buildings to add to their chance of remaining upright during an earthquake. Another major danger is the risk of fire from severed power lines and ruptured gas pipes (as seen in the Northridge case study). When buildings are constructed they can provide more flexible pipes to cut down this risk.

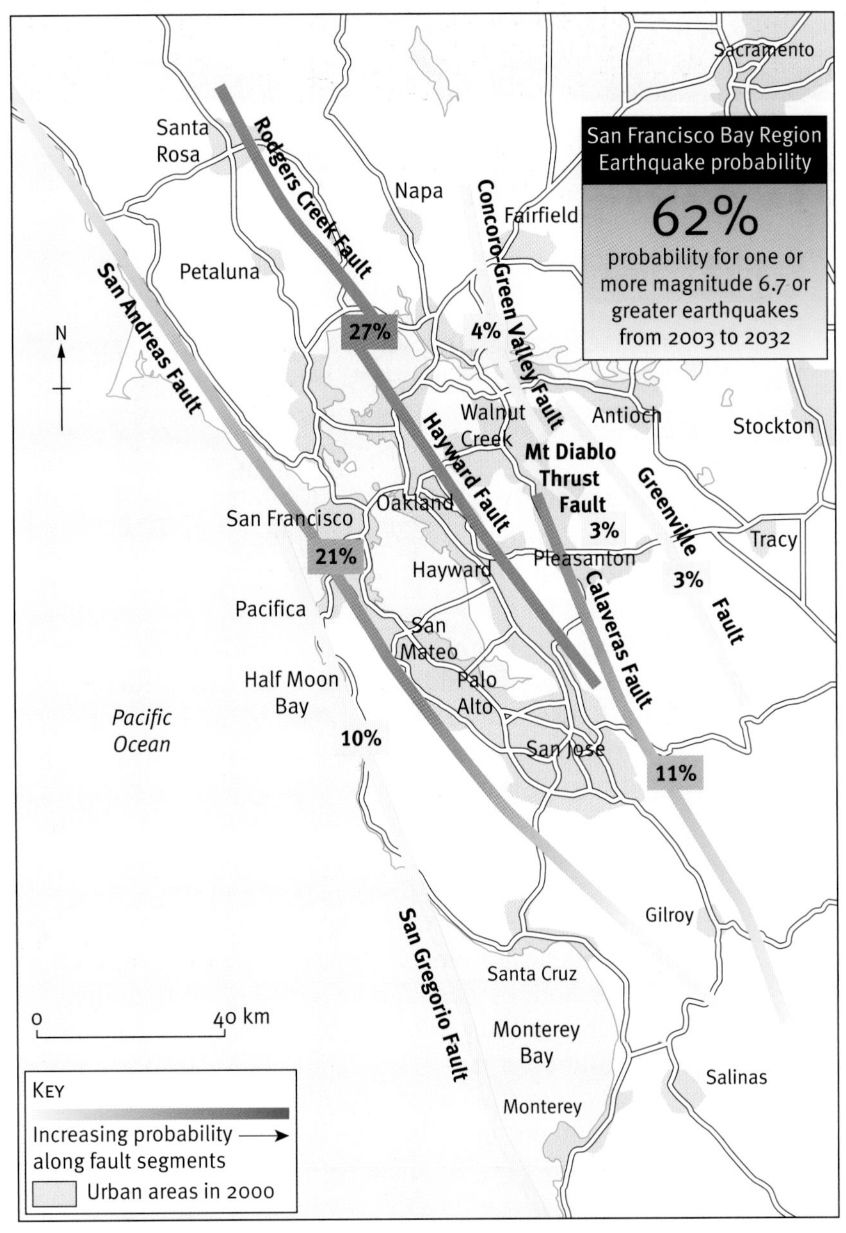

Figure 3.37 ***Probability map of earthquakes in Bay Area, California. (Source: US Geological Survey)***

- **Education**: government agencies and the media provide extensive guidance in earthquake-prone areas to help people prepare for the period before, during and after a tremor. In the USA this is available on the internet: Earthquake Information from the USGS (website). In Japan, 1 September is Disaster Day, the anniversary of the devastating Tokyo earthquake of 1923 that killed 155,000 people. The day is a public holiday when Japanese people practise earthquake drill.
- **Emergency services**: governments must plan carefully to make sure that the emergency services are prepared for possible earthquakes and that relief supplies are ready. If an earthquake is detected as soon as it starts, then better precautions can be taken: for example, power stations can be shut down so that there is less danger of fires from ruptured power lines, hospital generators started, and the emergency services can be alerted. Good communication with earthquake monitoring stations can also save lives. Predicting earthquakes can be expensive, because lots of special scientific equipment is needed. Poorer countries and regions will not be able to afford such equipment.

Figure 3.38 ***The problems of earthquake prediction.***

ACTIVITIES

1 What methods are used to reduce the perception that earthquake zones are dangerous places to live?

2 Study the earthquake probability map (Figure 3.37).

 a What will have influenced the scientists when making their predictions?

 b Which areas area would you choose to live in if you had to move to San Francisco? Explain your answer.

3 Look at the cartoon (Figure 3.38). Do you think scientists and engineers should concentrate on prediction or preparedness to help reduce the impact of earthquakes in the future? Explain your answer.

PLENARY ACTIVITY

What would be the difference between the methods to prepare people for earthquakes in LEDCs and MEDCs?

Does it matter how rich you are?

GET STARTED

Look around your classroom and list five hazards should an earthquake occur. Think of a way to fix each hazard.

THINK ABOUT IT

Which development indicators could help you predict how much an earthquake will affect a country? Use an atlas and your own knowledge to make a list of useful statistics.

A long way from help

In developing world countries, many people still live in rural areas. Even in those countries that are experiencing rapid economic growth, it is often the people who live in the countryside that are the last to see the benefits of increased wealth. This is often particularly so for remote or inaccessible rural regions. The Sichuan area of China has been largely neglected and untouched by China's economic rise and there is a widening gap between the prosperous urban dweller and struggling rural people.

Additional problems brought by inaccessible locations include the following.

- Villages are built in precarious locations that suffer badly in earthquakes. In the Afghanistan earthquake of 1998, 25 villages were completely destroyed.
- The scale of damage to villages and the number of casualties is often not known for several days.
- Local health care is poor.
- People often have little or no health insurance.
- Water supply and sanitation is often poor and there is a great risk of disease after the event.
- The majority of the buildings are poorly constructed and do not stand up to strong shaking.
- Damage is often worsened by other events such as landslides, which hinder rescue attempts and cause damming of rivers.
- Relief can take days to arrive and desperate survivors take matters into their own hands. Widespread looting of homes and shops can occur.
- There are often no accurate maps of the area to help rescuers locate settlements and other areas where survivors might be.
- There often no plans for dealing with the impacts of the earthquake and people are not well prepared.

Buying protection?

In more developed countries people are more prepared for an earthquake and they know what to do when one occurs.

The major economic impacts of earthquakes result from damage to buildings and to the infrastructure that are essential to the operation of the economic activities in the affected area. Infrastructure systems are extremely vulnerable and their failure can result in delays in rescuing survivors as well as in restoring quality of life for those affected. Many businesses will have been dependent on transport and communications for workers, customers, supplies shipments, water electricity, gas, telephone and other services. Some areas such as ports are particularly vulnerable because they are located next to harbours, often built on soft soils and land fill. Major ports are centres of commerce and handle regional and international trade. They are linked with other sectors such as transport, finance, banking and insurance. Ports were affected in the earthquakes in Alaska in 1964, in SanFrancisco in 1989 and particularly in Kobe, Japan in 1995. Some businesses will even find an increase in their activity because of demands placed on them by the disaster. Businesses that may benefit are ones for which there is an increased demand for their products or services for relief, clearing and removing debris, repair and restoration, and reconstruction. However, all these activities have a cost and economically developing countries will find it difficult to fund such large-scale development.

Fact file

Suggested list of items for an earthquake survival kit in California:

- Torch
- Water
- Disinfectant
- Radio and batteries
- Shoes
- Money
- Fire extinguisher

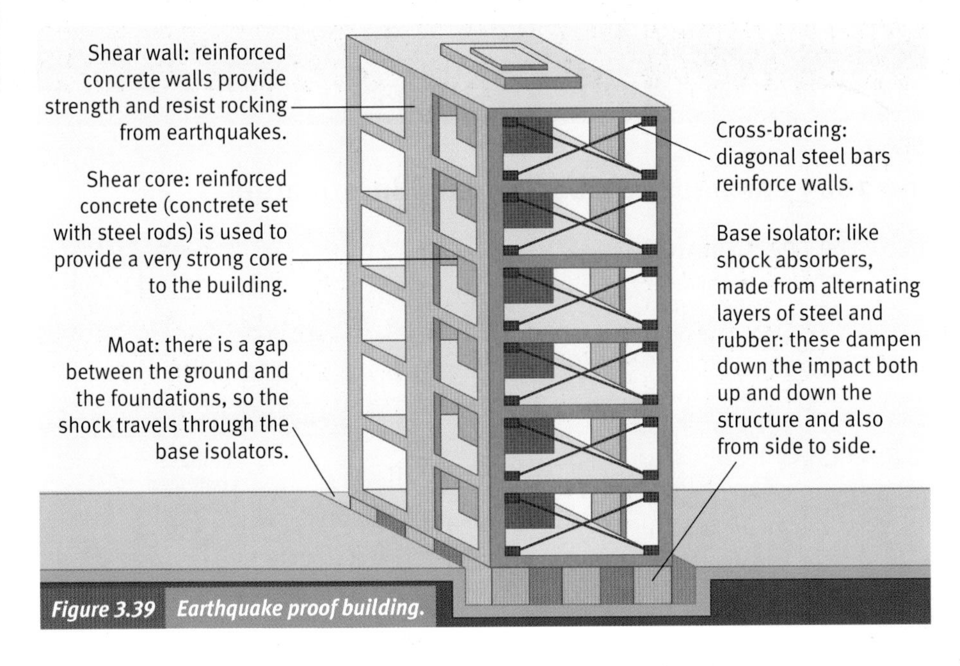

Shear wall: reinforced concrete walls provide strength and resist rocking from earthquakes.
Shear core: reinforced concrete (conctrete set with steel rods) is used to provide a very strong core to the building.
Moat: there is a gap between the ground and the foundations, so the shock travels through the base isolators.
Cross-bracing: diagonal steel bars reinforce walls.
Base isolator: like shock absorbers, made from alternating layers of steel and rubber: these dampen down the impact both up and down the structure and also from side to side.

Figure 3.39 *Earthquake proof building.*

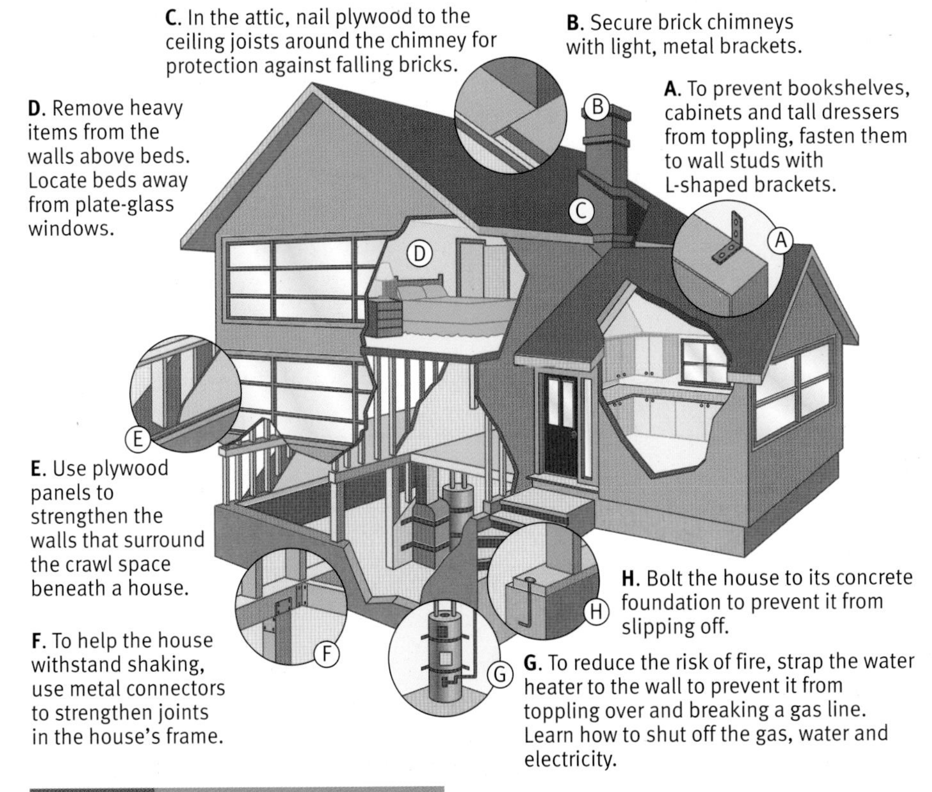

C. In the attic, nail plywood to the ceiling joists around the chimney for protection against falling bricks.

B. Secure brick chimneys with light, metal brackets.

D. Remove heavy items from the walls above beds. Locate beds away from plate-glass windows.

A. To prevent bookshelves, cabinets and tall dressers from toppling, fasten them to wall studs with L-shaped brackets.

E. Use plywood panels to strengthen the walls that surround the crawl space beneath a house.

F. To help the house withstand shaking, use metal connectors to strengthen joints in the house's frame.

H. Bolt the house to its concrete foundation to prevent it from slipping off.

G. To reduce the risk of fire, strap the water heater to the wall to prevent it from toppling over and breaking a gas line. Learn how to shut off the gas, water and electricity.

Figure 3.40 *Hazard spotting in the home.*

Figure 3.41 *Earthquake-proof building in Kobe, Kobe Port Tower.*

RESEARCH LINK

In what ways can a building be earthquake proofed? Give examples of features used in actual structures.

PLENARY ACTIVITY

What can LEDCs countries such as China do to better prepare people for large earthquakes?

ACTIVITIES

1. Explain why earthquakes are more likely to be a disaster for LEDCs such as China rather than MEDCs such as the USA.
2. Describe the ways that people can prepare their homes for a major earthquake.
3. Look at the list of some of the items recommended to make up an earthquake survival kit in California. Explain why each of the items has been included. Add another five items to your list and explain why they might be included.
4. How might your survival kit have been different for people living in a remote area of an LEDC? Explain your answer.

A Decision Making Exercise – Responding to an earthquake: relief or recovery?

In October 2005, Pakistan suffered a devastating earthquake. The magnitude of the event ensured an immediate worldwide response, with international relief agencies rushing aid to the victims. The relief programme lasted months but the effects of the earthquake remain, years after the event. In 2008, a report written on behalf of the UN and many of the NGOs involved in the relief effort suggested that the longer-term recovery programme was in desperate need of further support. Could it be that the world is good at reacting to disaster, providing short-term aid for those most in need, but it lacks the will needed to bring long-term help those who survive? What do you think?

The 2005 earthquake – background information

Figure 3.43 **The Pakistan earthquake.**

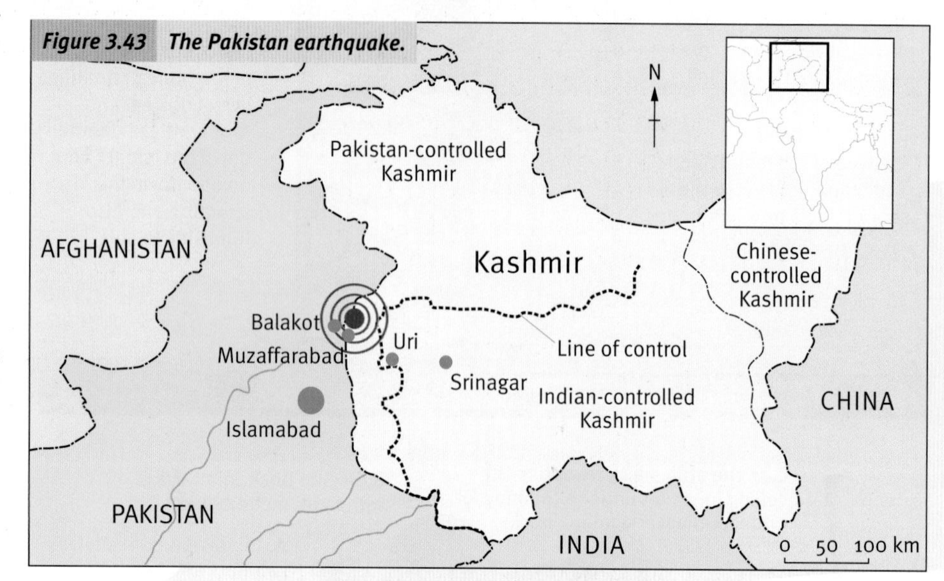

On 8 October 2005 at 03.50 GMT, the quake hit this remote mountainous region of north Pakistan. It measured 7.6 on the Richter scale.

Most estimates suggest that 86,000 people died and hundreds of thousands were made homeless.

The World Bank suggests that it was the worst natural disaster ever to have hit the country.

The immediate relief effort

Figure 3.42 **Survivors gather valuable possessions.**

As with all such disasters the initial relief effort focused on getting short-term aid to the area. This included:

- search and rescue teams to find survivors and pull people from wrecked buildings
- medical teams to save the dying and to administer care for the seriously wounded
- food and water supplies for survivors
- emergency tents for the thousands who had lost their homes.

Earthquake victims

Relief agencies and governments from all over the world pledged US $5.4 billion to help the Pakistan government cope with the overwhelming problems. Immediate aid was more critical as the Himalayan winter was about to produce heavy snowfall and temperatures as low as –15°C. By mid-November the government had distributed 350,000 tents, 3.2 million blankets, 3000 tonnes of medicine and 200,000 tonnes of food. Doctors from all over the world arrived to set up emergency clinics and mobile X-ray facilities. After their initial work to save lives and repair fractures, they concentrated on a vaccination programme to ensure that disease did not spread.

Emergency relief fundraising: within days of the earthquake, the British public raised millions of pounds. All the news bulletins carried pictures and video footage of the suffering. However, the generosity of the British public would be tested even further as one report suggested that an even bigger effort would be needed for the long-term recovery programme. Fundraisers knew that as soon as the images that had provoked such a kind-hearted response disappeared from our screens, the funds would quickly dry up. What are the long-term funds needed for?

Aftershock

MTV journalist Gideon Yago produced a diary after visiting the earthquake-hit zone several months after the event. In the diary he graphically describes scenes that inform the reader of the need for a long-term strategy after aid workers had overcome some of the initial problems. Read his diary at the MTV website. Although the earthquake devastated South Asia and left 3 million people homeless it did not dominate news headlines for a long period.

Figure 3.45 ***Emergency aid reaches the victims.***

The recovery programme – long-term relief

In 2006, the Pakistan government set up ERRA (the Earthquake Reconstruction and Rehabilitation Authority) to oversee the recovery programme. It desperately seeks funds from all over the world to help the area fully recover from the disaster. Some of its aims are outlined in Figure 3.44.

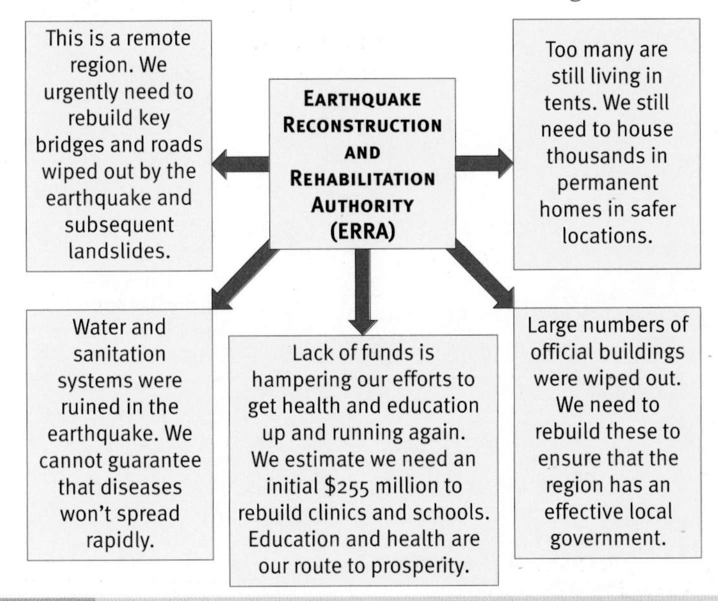

Earthquake Reconstruction and Rehabilitation Authority (ERRA)
This is a remote region. We urgently need to rebuild key bridges and roads wiped out by the earthquake and subsequent landslides.
Too many are still living in tents. We still need to house thousands in permanent homes in safer locations.
Water and sanitation systems were ruined in the earthquake. We cannot guarantee that diseases won't spread rapidly.
Lack of funds is hampering our efforts to get health and education up and running again. We estimate we need an initial $255 million to rebuild clinics and schools. Education and health are our route to prosperity.
Large numbers of official buildings were wiped out. We need to rebuild these to ensure that the region has an effective local government.

Figure 3.44 ***Aims of the Earthquake Reconstruction and Rehabilitation Authority.***
(Source: Earthquake Reconstruction and Rehabilitation Authority)

YOUR TASK

In May 2008, a school raised £1200 towards the Save the Children appeal for further funding to support the work of ERRA. They had recognised the need for long-term support for the victims of the 2005 quake after viewing a BBC documentary on the ongoing problems faced in Pakistan and after visiting the Save the Children website.

12 May 2008, a breaking news story gives brief details of a major earthquake in Sichuan, China. Thousands have been killed with even greater numbers trapped. An appeal has been made for emergency aid. What does the school do? Argue the case for diverting the funds to the China appeal or for donating the funds to ERRA. Use the evidence on these two pages and the resources on the supporting CD.

Tropical storms: where and how?

GET STARTED

The UK has been experiencing more extreme weather in recent years. What sort of damage occurs in very strong winds? List some of the effects you have seen in news reports and on the television.

RESEARCH LINK

How do tropical cyclones get their name?

Tropical storms are amongst the most powerful and destructive meteorological systems on earth. Globally, 80 to 100 develop over tropical oceans each year. Many of these can cause considerable damage to property and loss of life.

A tropical storm is a large depression or **cyclone** that forms over tropical oceans and moves away from the equator. Ocean temperatures must exceed 27°C for a tropical storm to form and the water needs to be at least 60 m deep. The wind speeds are usually in excess of 74 mph. Tropical storms happen during the hottest times of the year – between May and November in the northern hemisphere and between November and April in the southern hemisphere.

Areas that are susceptible to tropical storms are the Caribbean and Central America, South-East Asia and the Indian subcontinent.

KEY TERMS

Cyclone – a system of winds rotating inwards to an area of low pressure.

Hurricane – a violent tropical storm in the Caribbean region.

Typhoon – a violent tropical storm over the Indian and Pacific oceans.

A tropical storm is typically 700 km in diameter and can exceed 13 km in height. In the northern hemisphere the winds rotate in an anti-clockwise direction. Fast-rising warm air creates an area of very low pressure – the eye of the storm – which sucks in air. The rotating winds cause low pressure close to the equator and absorb the moisture from the oceans, creating **hurricanes**, tropical cyclones or **typhoons**.

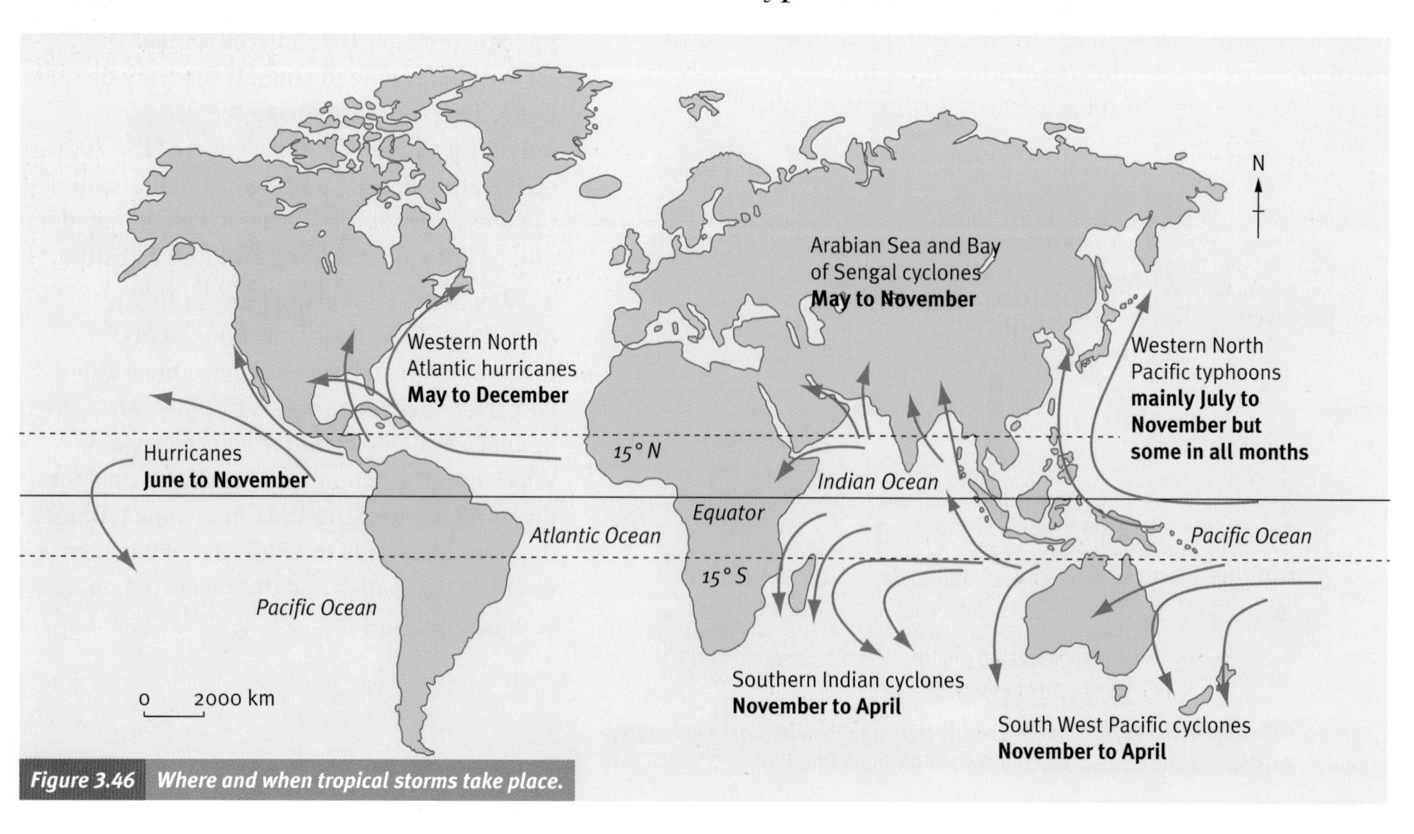

Figure 3.46 Where and when tropical storms take place.

How tropical storms affect people and places

Tropical storms can cause great damage when they reach land. They can damage buildings, power cables and crops. Flooding can be caused by the heavy rainfall brought by a tropical storm and also through storm surges.

An MEDC is better placed to deal with the effects of tropical storms. They have technology, such as weather satellites, to enable them to monitor and predict storms and have well-equipped infrastructure to deal with them when they do strike. This infrastructure includes things such as emergency procedures, resources, robust transport and food supplies. Often in LEDCs people don't have the means to move inland away from the storm and therefore the impact is usually greater.

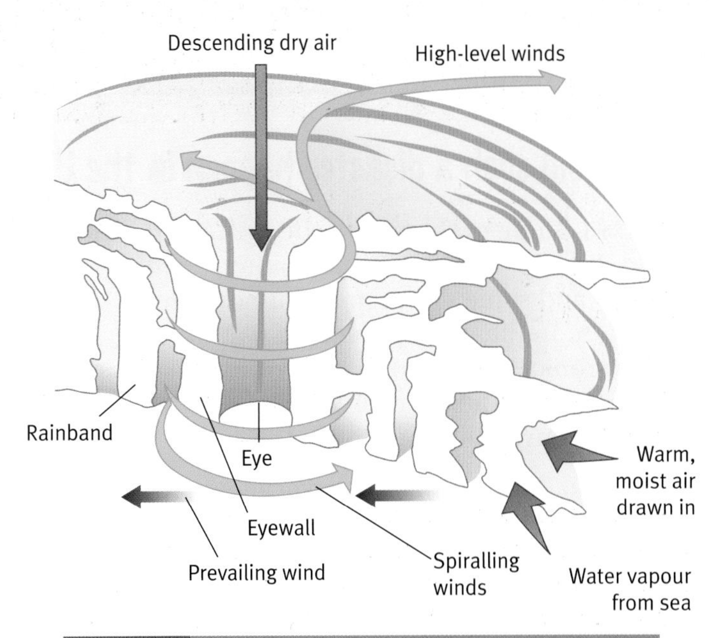

Figure 3.47 *A section through a typical cyclone or hurricane.*

Is the frequency of natural hazards increasing?

There is an ongoing scientific debate about the link between increased North Atlantic hurricane activity and global warming. The 2007 report of the Intergovernmental Panel on Climate Change rates the probability of such a link as 'more likely than not.'

Globally, there is an average of about 90 tropical storms every year. According to an international report in 2007 there is no clear trend in the frequency of cyclones in the Pacific. However, in the North Atlantic there has been a clear increase in the frequency of tropical storms and major hurricanes. From 1850 to 1990, the long-term average number of tropical storms was about 10, including about 5 hurricanes. For the period of 1998–2007, the average is about 15 tropical storms per year, including about 8 hurricanes. This increase in frequency correlates strongly with the rise in North Atlantic sea surface temperature, and recent scientific studies link this temperature increase to global warming.

Fact file

- The deadliest hurricane on record was the Martinique Hurricane of 1780 that killed 20,000 people.
- Hurricanes Camille (1969) and Allen (1980) share the record for the maximum sustained windspeed for an Atlantic hurricane of 165 knots (306 kph).

Impacts: If the projected rise in sea level due to global warming occurs, then the vulnerability to tropical cyclone storm surge flooding would increase causing increased coastal flooding in low lying areas. The rise in sea level temperature could also cause storms of greater intensity to form in the oceans, causing even greater damage when they reach the land.

THINK ABOUT IT

How will climate change affect tropical cyclones?

PLENARY ACTIVITY

Why do MEDCs suffer billions of pounds worth of property damage but LEDCs lose hundreds of lives in tropical storms?

ACTIVITIES

1 How are tropical storms formed?

2 Describe the features of a tropical storm. Use a diagram to illustrate your answer.

How could such a disaster happen in the USA?

The USA is one of the most economically developed countries in the world. It has many scientists monitoring the risk of hazards across the country and some of the most sophisticated levels of technology anywhere in the world. The Gulf of Mexico coast is famous as an area where tropical storms threaten for five months of the year. How then, did Hurricane Katrina cause such major and long-lasting devastation in the area around New Orleans?

Hurricane Katrina began as a very-low-pressure weather system, which strengthened to become a tropical storm and eventually a hurricane as it moved west and neared the Florida coast on the evening of 25 August 2005.

GET STARTED

Measure a height of 3 m in your classroom. Make a list of all the damage that would be done if your school or college was flooded to this depth. What would be destroyed?

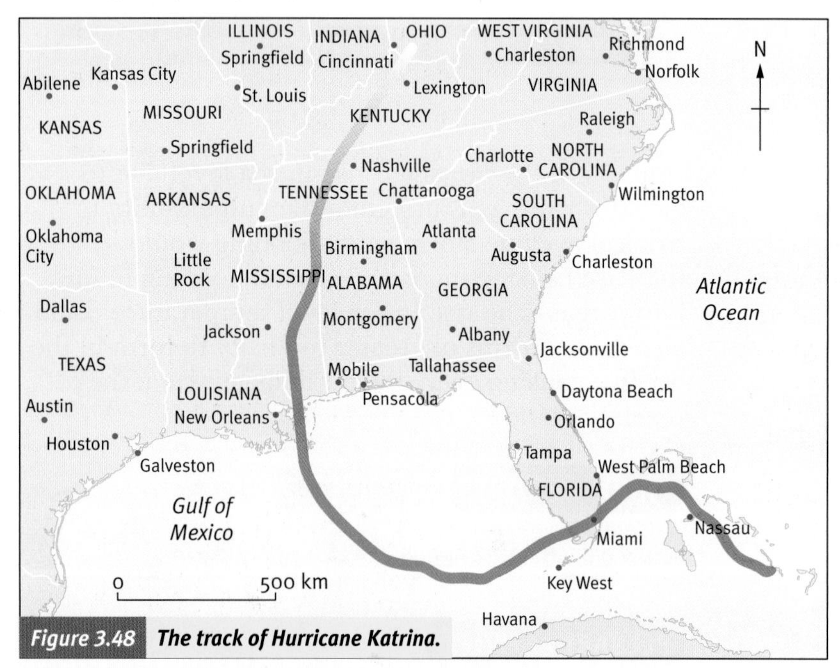

Figure 3.48 *The track of Hurricane Katrina.*

Table 3.5 Timetable of Katrina's approach

23 August	Tropical depression forms off the south-eastern Bahamas.
24 August	Upgraded to tropical storm and given name Katrina.
25 August	Strengthened to become a hurricane and made landfall in the morning north of Aventura in Florida.
26 August	Storm intensified as it crossed the warm waters of the Gulf of Mexico reaching Category 2 status.
27 August	Storm has doubled in size and reaches Category 3 intensity as it crosses the Gulf towards Mexico.
28 August	Reaches its peak strength of Category 5, with maximum winds of 282 kph and a central pressure of 902 mb. Turns north and approaches the US coast.
29 August	Makes its second landfall at 6.10 a.m. at the eastern edge of New Orleans. Hurricane-force winds extend outwards for 193 km from the centre.
30 August	Hurricane follows the course of the Mississippi, being downgraded to a tropical depression near Clarksville, Tennessee.
31 August	Remnants of the storm reach the Great Lakes, causing heavy rain and high winds.

Figure 3.49 *Heavy rain during hurricane.*

Weren't there warnings?

On 27 August, the National Hurricane Center (NHC) ordered a hurricane watch along the Gulf Coast states of the USA. The United States Coast Guard began to gather resources in areas outside the expected impact zone and called up more than 400 reserve members. Aircraft from the Aviation Training Centre were recalled to their base in Mobile so that they could take part in rescue operations.

President Bush declared a state of emergency in areas of Louisiana, Mississippi and Alabama on Saturday 27 August but he did not include any of Louisiana's coastal parishes. The director of the NHC expressed concern that a storm surge produced by the hurricane might overtop New Orleans' levées and flood walls. On Sunday 28 August the National Weather

Service issued a bulletin predicting that the New Orleans area would be 'uninhabitable for weeks' after 'devastating damage' caused by Katrina.

On Sunday 28 August about 1.2 million people were issued with evacuation orders. Fifty-seven emergency shelters were established in coastal areas. Motorways jammed as people obeyed the order to leave. The Louisiana Superdome in the centre of New Orleans was set up for people who could not leave the city. There were not enough buses and coaches to evacuate all the vulnerable communities from the city and public transport, including train services, had shut down.

Impacts on people

- The levées and flood walls that protected New Orleans were breached in 53 different places, allowing water to flood the city up to 3 m deep.
- Flooding was made worse by the heavy rain (250 mm during the storm). It is estimated that over 80 per cent of the city was under water.

Looters take advantage of New Orleans mess

'IT'S INSANE,' says tourist watching theft in the French Quarter.

In the flooded streets of New Orleans, looters floated garbage cans filled with clothing and jewelry down the street in a dash to grab what they could. In some cases, looting on Tuesday took place in full view of police and National Guard troops. At a Walgreen's drug store in the French Quarter, people were running out with grocery baskets and coolers filled with soft drinks, chips and diapers.

(Source: Associated Press)

ACTIVITIES

1. Explain why the hurricane lost strength as it travelled north through the central USA.
2. Draw up a timetable to show the warnings that were given to the people of New Orleans.
3. Explain the primary and secondary effects of the hurricane. What reasons were given to explain the high number of deaths from this hurricane?
4. Many critics of the US federal administration said federal authorities would have made a better job of responding to the disaster if the majority of the victims had been rich white people rather than poor black people. Why is it that poor people are often hit hardest by natural hazards?

'Katrina exposed serious problems in our response capability at all levels of government, and to the extent that the federal government didn't fully do its job right, I take responsibility', President George W. Bush

'I hate the way they portray us in the media. You see a black family, it says, "They're looting." You see a white family, it says, "They're looking for food." And, you know, it's been five days waiting for help because most of the people are black. President Bush doesn't care about black people.' Kanye West, at a relief concert for the victims of New Orleans

- Deaths totalled 1836 – many drowned in the floodwaters and lay in the streets or floated in the water for days before their bodies were recovered.
- Over 10,000 people were made homeless. Most of these were from the poor inner city areas of St Bernard's Parish and the Lower Ninth Ward.
- Over 3 million people were without electricity.
- Most of the roads in and out of the city were damaged and two major road bridges collapsed.
- People were stranded in the flooded city during a summer of record temperatures reaching 85°F during the day. People struggled to get to safe havens such as the Superdome.
- There was shortage of food and no access to a clean water supply for survivors, which raised problems of contaminated water and health risks.

Figure 3.50 **Devastation caused to homes in New Orleans.**

RESEARCH LINK

Flood control barriers can sometimes cause more flooding than they prevent. Use the internet to find out why this might be so.

CASE STUDY Cyclone Nargis, Myanmar 2008

Are the deaths from tropical storms caused by being in the wrong place at the wrong time?

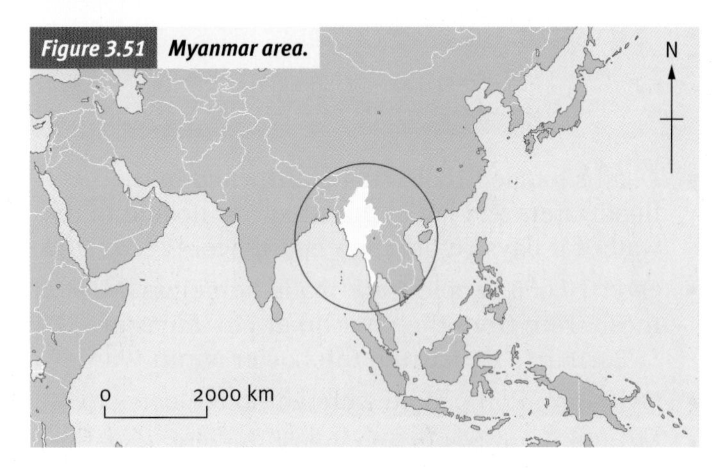

Figure 3.51 **Myanmar area.**

Any country not used to being struck by winds of over 160 kph and torrential rain is going to suffer, but when that country is already poor the effects are much worse. Factor in a government suspicious of help and long-term destruction of natural defences and a disaster now appears to have been waiting to happen.

In late April 2008, an area of very low pressure formed in the Bay of Bengal. The American Typhoon Warning Centre in Hawaii monitored the storm as it got worse and moved north towards land. Satellite images on 29 April showed that a tropical storm had formed. Forecasters predicted that the storm would hit Bangladesh but as its strength ebbed and flowed due to changes to the sea water temperature its course changed towards the coast of Myanmar.

Myanmar

Myanmar is an LEDC with a population of about 55 million. It is one of the poorest countries in South-East Asia with restricted development, mainly due to issues with the ruling military government and a desire to be isolated from the world.

Myanmar relied heavily on primary exports like oil and expensive timber but a lack of roads and railways mean that the country has been unable to fully exploit its natural wealth. However, Myanmar is still the world's largest producer of rubies, although the average Burmese citizen doesn't benefit much from this trade, with a GDP (Gross Domestic Product) per capita of £1900 and a Human Development Index (HDI) of 0.578 (see page 161).

GET STARTED

Using an atlas, find five facts about the physical geography of Myanmar. What is this country's former name?

In a country with the same size population as the UK there is only one person in a thousand with access to the internet. Myanmar ranks 32nd in a list of the world's 50 poorest countries and the last thing it needed was a disaster the size of Cyclone Nargis.

Environmental damage in Myanmar

As is often the case with the world's poorest countries, the environment suffers in order for money to be earned. In Myanmar's case this has meant the destruction of mangrove swamps. In a 75-year period Myanmar has destroyed 83 per cent of its mangroves in the area affected by Cyclone Nargis. These mangroves should have served as natural barriers to stop storm surges. Their root systems slow water flow, catch material and absorb wave energy. In the 2004 Boxing Day tsunami, villages with intact mangrove swamps had suffered significantly less death and destruction than those that had lost theirs.

However mangrove swamps get converted into money-making areas such as shrimp farms, tourist resorts, land for agriculture and re-housing.

The effect of the storm

By 2 May the storm's winds had reached speeds of over 217 kph and it hit the vast delta of the Irrawaddy River in Myanmar, bringing torrential rain and storm surges to the very low-lying and agricultural delta.

The UN estimated that 1.5 million people were severely affected by Cyclone Nargis, with official figures estimating the death toll at over 125,000 people. Some reports suggest 95 per cent of buildings were destroyed, and all suffered significant damage, including roofs being ripped off. Millions of people were made homeless as a result of the storm. The local infrastructure was unable to withstand the storm, with reports that sewerage systems were overwhelmed, causing leakage which contaminated rice fields and caused disease. Electricity lines were also destroyed, roads were swept away and stagnant, dirty water encouraged vast numbers of mosquitoes to breed.

The biggest impact was not from the high winds but from the flooding. Low pressure meant that the sea was no longer depressed by the atmosphere and so the sea level rose. As a result the low-lying coastal areas of the Irrawaddy River were flooded. The short-term effects were that people drowned and many were made homeless. In the longer term fields were destroyed, families were devastated and development was put back by years.

The response to Cyclone Nargis

As the scale of the disaster became clear it also became apparent that the military government of Myanmar was not doing as much as the rest of world thought it could. Stories leaked out of aid being kept in the major cities and not spreading out to rural areas. Foreign aid workers were either not allowed in to help or were restricted in where they could go.

More than a week after the cyclone first hit only one in ten of those affected had received some sort of aid. MEDC governments loudly accused the government of Myanmar of making the situation even worse. The army harassed volunteers, check points on rivers stopped journalists from covering the disaster for television, leaflets were distributed to inform people that aid deliveries are only attracting children and others who do not need help. What lies behind these tactics is hard to say, but experts suggest that it is an attempt to preserve the prestige of Myanmar within its own boundaries and across the world.

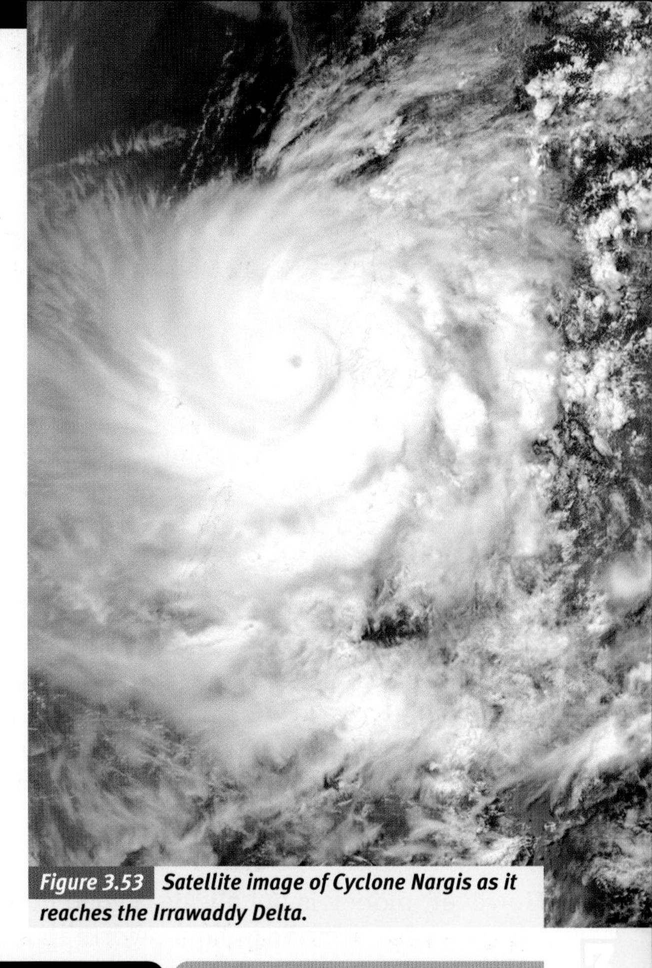

Figure 3.53 *Satellite image of Cyclone Nargis as it reaches the Irrawaddy Delta.*

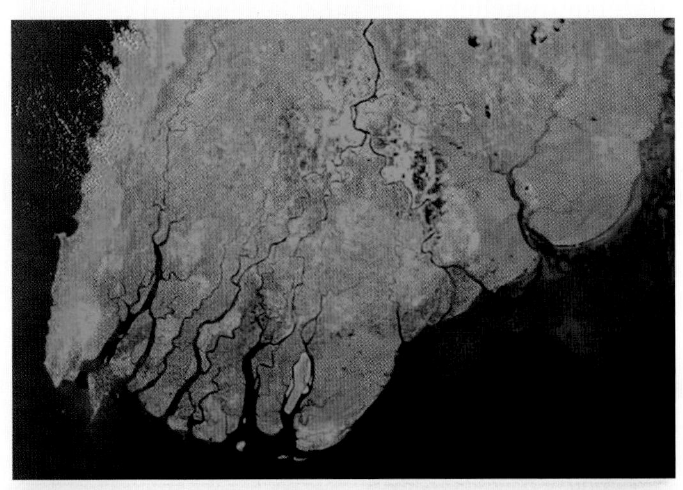

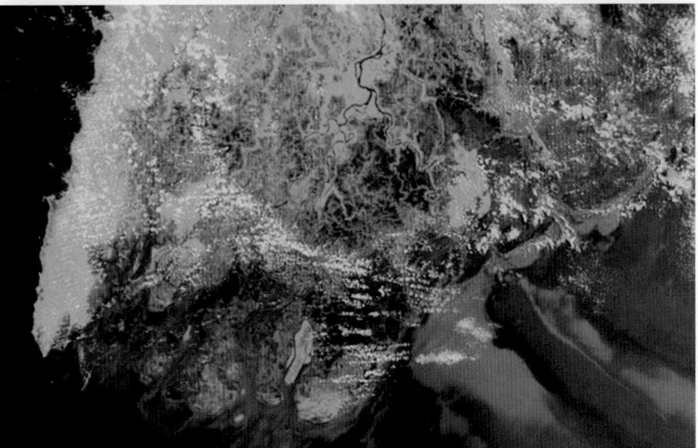

Figure 3.52 *The Irrawaddy Delta before and immediately after Cyclone Nargis. Notice how much is now under water.*

THINK ABOUT IT

Much of the anger of the western governments has been used to highlight previous human rights abuses in Myanmar and the lack of democracy in the country. This has meant that the rulers of Myanmar have been even slower to let foreigners in.

Should the UK government have done anything other than send in supplies to help?

ACTIVITIES

1 What were the physical factors that made Cyclone Nargis so destructive in Myanmar?
2 Why did so many people choose to live in such a potentially dangerous area as the Irrawaddy Delta?
4 What things would Myanmar need after a disaster like Nargis? How would they change as time went by?

PLENARY ACTIVITY

If you were in charge of defending Myanmar against future cyclones what would be your priorities? How would you go about putting them into action?

What features of a tropical storm affect people?

MEDCs and LEDCs are equally at risk from the passage of tropical storms and both suffer many of the same effects. Damage to buildings, roads, power cables and agricultural land can be caused by winds. Extensive flooding can result from storm surges as well as from the heavy rains that accompany these storms.

GET STARTED

List five things you think you need to survive in the world. Add five things that you would want to have with you should a disaster strike. Would these lists be different for someone living in the economically developing world? If so, how?

A wide range of factors contribute to an area's resistance to hazards. LEDCs suffer because building codes, warning systems, defences, emergency services and communications are poor, resulting in high death tolls. Wealthier countries are more likely to be able to evacuate people, but the value of their infrastructure means they suffer higher losses in economic terms.

However, money is not always a good measure of the damage caused. The loss of an American home worth thousands of dollars, but covered by insurance, may be less important than the loss of a wooden shelter and a few cows in the Ganges delta that represents a family's entire wealth and future – and is not insured.

In the Irrawaddy delta the winds from Cyclone Nargis reached 210 kph and produced a 4-m storm surge. Villages along the banks of the river made their living from fishing, but the storm surge destroyed their boats, smashing them against the river banks. More than 20 families had their homes destroyed and clean water supplies were contaminated. The following newspaper extract describes some of the impacts on people.

Figure 3.54 **Fishermen mending nets in Mya Yar Kone.**

Jumpstarting livelihoods in a cyclone-hit village

'We used to get a good price for our fish in the local market,' says U Thin Nu, the 44-year-old leader of Mya Yar Kone village, a community of almost 300 people in Myanmar's Ayeyarwady [Irrawaddy] delta. 'I would like to see regular business come back.'

Almost four months after Cyclone Nargis, economic life is at a standstill in Mya Yar Kone. The village, which depended for most of its livelihoods on catching fish and crabs and sending them to the township centre of Labutta, is still struggling with the after-effects of the storm.

'These communities are now well covered in terms of shelter and emergency relief,' said U Htun Tin, UNDP Township Coordinator in Labutta. 'But reviving livelihoods has been more of a challenge.'

UNDP has been actively supporting the village since soon after the cyclone. Grants have been given to farmers to hire casual labour so they can plant their crops on time. Twenty-four households qualified for shelter assistance, and the entire community has received grants to clean up common infrastructure such as village ponds.

Now the challenge is to give the community members the support they need to get long-term income-generating activities off the ground. Grants have also been provided to fishing families for boats and nets. And UNDP recently introduced the concept of Self-Reliance Groups (SRGs) to the village. These groups, which have been successful in creating sustainable livelihoods in other parts of Myanmar, allow members to start up small-scale business activities while building up a common fund.

Still, Mya Yar Kone has a long way to go before it is fully back on its feet. U Than Swe, at 81 the oldest man in the village, remembers a time when boats from the community regularly shuttled down the river to Labutta, full of fresh fish and fat crabs.

'Now, we need help,' he said. 'We lost almost everything. We need boats and we need nets.'

And one other thing, he added: 'Can you help us make our crabs bigger?'

(Source: United Nations Development Programme)

Katrina floods wipe out years of research

WORK ON HEART DISEASE, cancer, AIDS, other ailments may be lost forever

Important work on heart disease, cancer, AIDS and a host of other ailments may be lost forever to scientists at Tulane and Louisiana State universities' medical schools in New Orleans.

LSU lost all of its 8,000 lab animals, including mice, rats, dogs and monkeys.

(Source: Associated Press)

Figure 3.55 *Damaged coastal bridge from New Orleans.*

RESEARCH LINK

Search the online American newspapers for the latest news of the effects of Katrina.

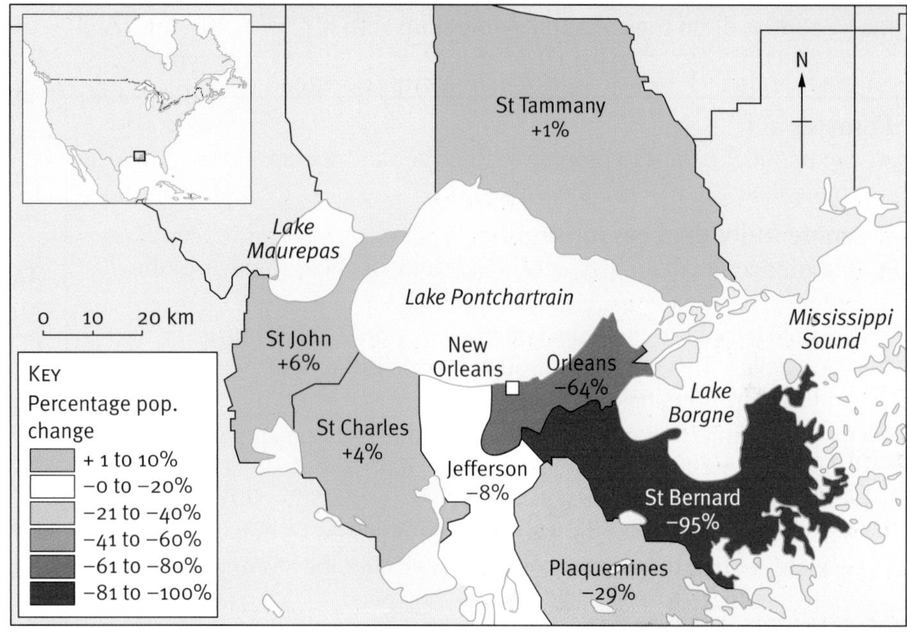

Figure 3.56 *Population change in the Mississippi delta area. (Source: US Census Bureau)*

THINK ABOUT IT

What political issues in a country might affect the amount of aid one government is willing to give to another?

ACTIVITIES

1 Using the article 'Jumpstarting livelihoods in a cyclone-hit village', complete the table below to show the responses to the cyclone in Myanmar.

Short-term effect	Response	Long-term effect	Response
More than 20 homes destroyed	Emergency shelters provided	Lack of income from fishing	Grants from UN Development Programme enabled villagers to buy new boats and nets

2 Study the map showing population change in the Mississippi delta over a six-month period.

  a Describe the changes it shows.

  b Why do you think people had still not returned to their homes?

  c Some people will never go back. Explain how that will affect New Orleans and how the areas that have taken in the refugees will be affected.

3 How do tropical storms affect people in MEDCs differently from those in LEDCs?

PLENARY ACTIVITY

One local newspaper described the aftermath of Hurricane Katrina as being 'like a Third World disaster'. Choose whether you agree or disagree with this statement and prepare your list of arguments with actual examples. Take part in a debate with the rest of class.

How can people be saved from the impacts of tropical storms?

GET STARTED

Can people ever be protected from the devastation that occurs when a tropical storm has hit their country, or do they 'just have to put up with it'?

There are many different measures that can be introduced to reduce the impacts of cyclones; Figure 3.57 lists just some.

Table 3.6 Comparing impacts of tropical storms

	Storm	
	Hurricane Gustav (USA)	**Cyclone Sidr (Bangladesh)**
Date	August 2008	November 2007
Windspeed	241 kph	257 kph
Death toll	10	3447
Damage	US $15 billion	US $1.7 billion

Disaster aid

- Set up relief camps to provide food, water and shelter in the immediate aftermath of the storm.

Promote individual responsibility

- Inform people about insurance against the possible impacts of the storm.
- Educate people so that they know what to do when a hurricane approaches. This might be through broadcasting or providing leaflets and holding meetings.
- Offer residents low-cost loans that must be spent on strengthening existing properties with storm shutters. Hurricane shelters can be built.
- Set up co-operative banks with low interest rates, so that when people want to rebuild their homes or businesses after the storm they can easily get loans to buy materials that they need.

Hazard-resistant design

- Build embankments or levées around settlements so that storm surges are less likely to flood coastal areas.
- Strengthen houses and roofs so that they are not damaged during the high winds.
- Build houses on stilts so that they are less likely to flood.
- Improve sanitation so that there is less danger of sewage leaking into drinking water.
- Build better roads so that evacuation can be more effective, and rescue and relief workers can reach affected areas more quickly.
- Strengthen and raise the banks of rivers so that the increased flow caused by the heavy rain and storm surges will not cause flooding.
- Build sea walls to protect densely populated areas near the coast.

Warning and forecasting

- Improve weather stations so that they can give early warning of storms.
- Get better at warning people and evacuating them from the danger area. This can be done through the use of satellites.
- Improve land use planning.
- Plant shelter belts of trees along the coast that will shield homes from the effects of the strong winds.
- Zone land use so that housing is not near the coast. Agricultural land can be allowed to flood.
- Replant and reinstate lost swamplands on coastlines.

Figure 3.57 Reducing the impact of cyclones.

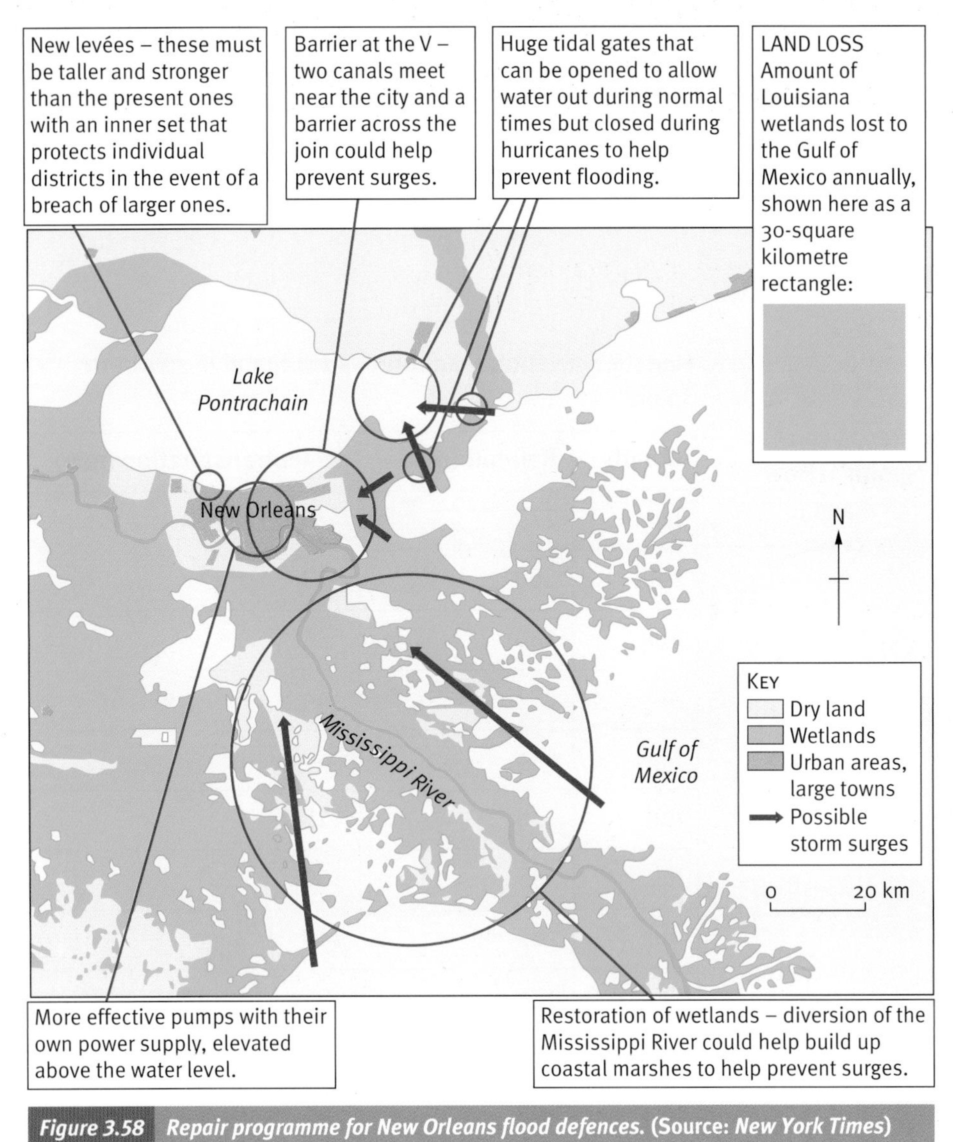
New levées – these must be taller and stronger than the present ones with an inner set that protects individual districts in the event of a breach of larger ones.
Barrier at the V – two canals meet near the city and a barrier across the join could help prevent surges.
Huge tidal gates that can be opened to allow water out during normal times but closed during hurricanes to help prevent flooding.
LAND LOSS Amount of Louisiana wetlands lost to the Gulf of Mexico annually, shown here as a 30-square kilometre rectangle:
Lake Pontrachain
New Orleans
N
Mississippi River
Gulf of Mexico
KEY
Dry land
Wetlands
Urban areas, large towns
Possible storm surges
0
20 km
More effective pumps with their own power supply, elevated above the water level.
Restoration of wetlands – diversion of the Mississippi River could help build up coastal marshes to help prevent surges.

Figure 3.58 *Repair programme for New Orleans flood defences.* (Source: *New York Times*)

Figure 3.59 House on stilts.

PLENARY ACTIVITY

Carry out a survey of tropical storms that have occurred over the past 12 months. You can use the US National Hurricane Center website for this. Choose two storms that have affected LEDCs and two that have affected MEDCs. Enter the names of the storms into your favourite search engine. Make a note of the number of references that come up. Will there be any differences in the numbers that you find? Can you explain this?

ACTIVITIES

1 Copy and complete the following table by adding each of the possible measures from Figure 3.58 explaining how each measure will reduce the impact of the hurricane and indicating whether this measure would be more sustainable, expensive and appropriate for LEDCs, MEDCs or both.

	Tackles the cause by	Tackles the impact by	Sustainable?	Expensive?	Appropriate for LEDC or MEDC or both?
Educate people					
Plant trees along the coast as a shelter belt					

2 Look back at the photographs in Figure 3.52. Why would building a coastal sea wall be impractical in Myanmar?

3 Select either New Orleans or Myanmar.

  a Choose three measures that you think would be most effective in reducing the impact of tropical storms.

  b Share your answers with a partner and then decide on the three you think would be most effective between you.

  c Are there any other steps which you think could be taken to improve the situation?

What is a drought and where do droughts occur?

GET STARTED

Make a list of as many different uses of water as you can think of. Compare your list with another person's. How many of the uses do you think you could not manage without? For which ones do you think you could cut down on the amount you use?

What causes droughts?

A drought occurs when there is not enough rain to support people or crops. The cause of droughts is easily understood, but hard to prevent. Depending on the location, crop failures, famine, high food prices and deaths can occur. Unlike other forms of severe weather or natural disasters, droughts often develop slowly.

Different parts of the world have different climates so each country has its own definition of drought. Drought is a normal, recurrent feature of climate. It occurs almost everywhere, although its features vary from region to region. Defining drought is therefore difficult; it depends on differences in regions and needs. Drought in Libya is said to occur when rainfall is less than 180 mm, but in Bali, drought is considered to occur after a period of only 6 days without rain.

In the most general sense, drought originates from a reduction of precipitation over an extended period of time, resulting in a water shortage for some activity, group or environmental sector. Knowing the amount of precipitation in an area is not enough to predict whether drought will occur. It is necessary to consider the *effective* precipitation – the difference between the amount of rain that falls and the amount of evapotranspiration.

Fact file

- A drought lasted over 400 years in the Atacama Desert in Chile, where the average annual rainfall is just 1.5 cm.

Table 3.7 Evapotranspiration and rainfall in southern Spain

Month	Rainfall (mm)	Evapotranspiration (mm)
Jan	275	20
Feb	200	23
Mar	180	34
Apr	277	44
May	344	80
Jun	138	115
Jul	85	150
Aug	85	140
Sept	174	95
Oct	471	60
Nov	387	30
Dec	368	20

Droughts are one of the top three threats to population in the world (along with famine and flooding). They can be devastating and their impact far-reaching and severe. Sometimes a drought takes decades to develop fully and predicting droughts is difficult. Because of the slow onset of droughts, it is difficult to estimate their cost.

In areas where a dry season is a common feature, such as much of Africa, droughts are defined as periods of two years or more with below-average rainfall. In some countries, droughts can continue over months, years or even decades.

RESEARCH LINK

Use the Global Drought Monitor website to discover where droughts are occurring now.

Factors affecting the severity of drought

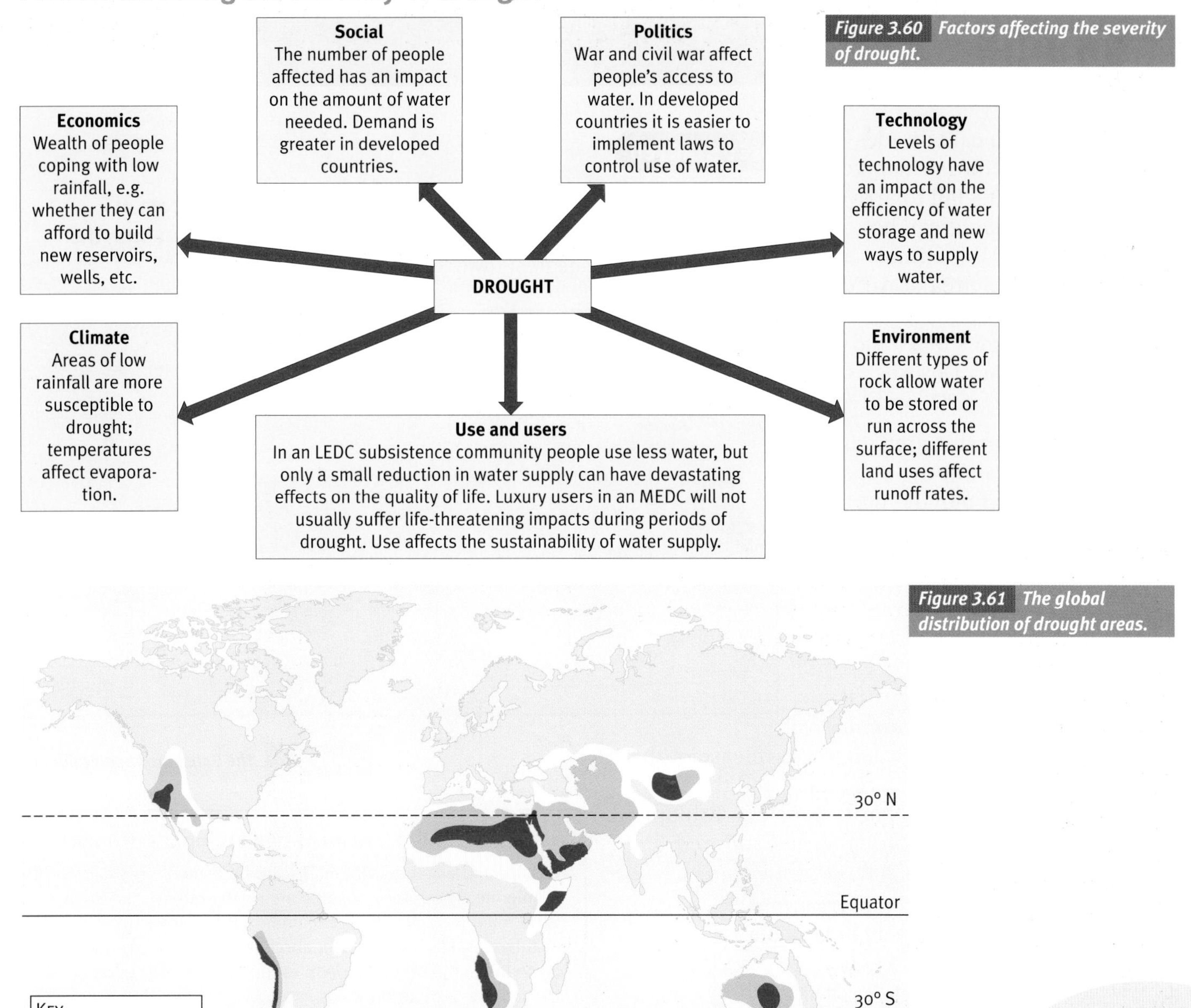

DROUGHT
Social
The number of people affected has an impact on the amount of water needed. Demand is greater in developed countries.
Politics
War and civil war affect people's access to water. In developed countries it is easier to implement laws to control use of water.
Economics
Wealth of people coping with low rainfall, e.g. whether they can afford to build new reservoirs, wells, etc.
Technology
Levels of technology have an impact on the efficiency of water storage and new ways to supply water.
Climate
Areas of low rainfall are more susceptible to drought; temperatures affect evaporation.
Environment
Different types of rock allow water to be stored or run across the surface; different land uses affect runoff rates.
Use and users
In an LEDC subsistence community people use less water, but only a small reduction in water supply can have devastating effects on the quality of life. Luxury users in an MEDC will not usually suffer life-threatening impacts during periods of drought. Use affects the sustainability of water supply.
30° N
Equator
30° S
Key
Extreme aridity
Arid
Semi-arid

Figure 3.60 **Factors affecting the severity of drought.**

Figure 3.61 **The global distribution of drought areas.**

ACTIVITIES

1. From the map in Figure 3.61, describe the distribution of areas affected by drought in the world. In what ways does the map not represent the true picture of drought as a hazard?
2. Look at Table 3.7, which shows rainfall and potential evapotranspiration for southern Spain. Work out the effective rainfall for each month. Draw two graphs on the same axis (rainfall as a bar graph and evapotranspiration as a line graph). In which months is southern Spain in danger of drought? Explain your answer.
3. For each of the six factors affecting the severity of drought, give an example of how it may cause hardship for people in (a) LEDCs and (b) MEDCs.
4. List four groups of people who benefit and four groups who suffer from drought.

PLENARY ACTIVITY

'Land as mere land has no value. What makes it valuable is its access to water.' Is this true for LEDCs and MEDCs?

CASE STUDY Ethiopia in the Sahel – drought in an LEDC

Countries in tropical latitudes experience a distinct wet and dry season. If the rains fail to arrive, the resulting drought can be disastrous for crops, animals and people. The Sahel area of Africa is an area particularly prone to drought. It stretches in a band south of the Sahara Desert between latitudes 12°N and 17°N.

Drought occurs in this area when the moist, rainy air at the Equator is prevented from moving north and reaching the countries of the Sahel. Higher sea temperatures are also thought to be partly responsible.

GET STARTED

Using an atlas, work out the longest distance from the west coast of Mauritania to the east coast of Somalia. How does this compare with the distance from your nearest town to London?

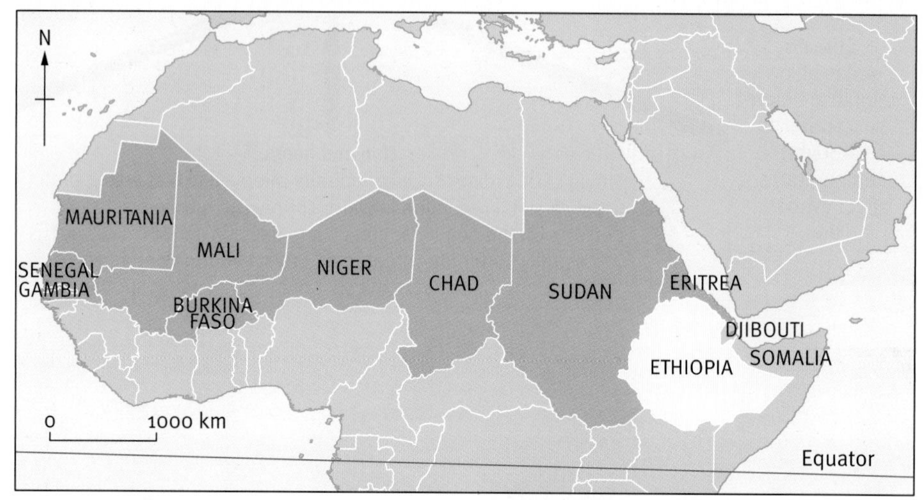

Figure 3.62 *The Sahel area and Ethiopia.*

Ethiopia

The drought of 1984–85 in Ethiopia resulted in over a million people dying. This disaster led to the release of the song 'Do They Know It's Christmas?' and the founding of the Live Aid charity concerts. Ethiopia continues to suffer from droughts and suffered another major disaster in 2006.

June 2006: Rains fail again in Ethiopia

This year the short rains, which begin in February, and the long rains starting in June both failed to supply enough water for the country's needs.

August 2006: Herdsmen flee Ethiopia's drought

Pasture and water are drying up and herdsmen have started to migrate. Thousands of cattle are on the move – but there is no relief in sight. All people can do is helplessly look on as their animals die one after another. Under temperatures of up to 50°C, the nomadic peoples trek hundreds of kilometres through the arid desert lands in search of water and grazing pastures. There are reports of clashes along the road between Dire Dawa and Djibouti, as Afar and Issa tribes fight for scarce grazing.

More aid needed

Most years, Ethiopia has to depend on some level of food aid as it rarely grows enough to feed the whole population. The initial estimates were that 5.2 million people would need food aid, but the extent of the drought and therefore of the food shortfall has meant that the numbers affected are rising. The country's prime minister has said that six million people are in need of immediate aid and that number could rise to 15 million if international donors do not help Ethiopia.

Disease

There are also increasing fears of an outbreak of cholera due to the huge number of carcasses of dead animals in the Awash River, the main water source for people in the area.

Aid agencies say that people who are already weak due to lack of food are especially vulnerable to diseases.

Agriculture

Farmers have planted once, twice and then a third time. Each time the crop has withered. The agricultural practices that served them so well are now well and truly exhausted. All over the Ethiopian highlands, farms have been divided and subdivided so often they are little more than scraps of land. Farmers cultivate tiny plots, some of which are very steep.

Figure 3.63 *Ethiopian landscape showing an animal skeleton.*

RESEARCH LINK

Which other countries of the Sahel are experiencing drought this year? Make a list of the impacts of the drought. Use the internet to research your answers.

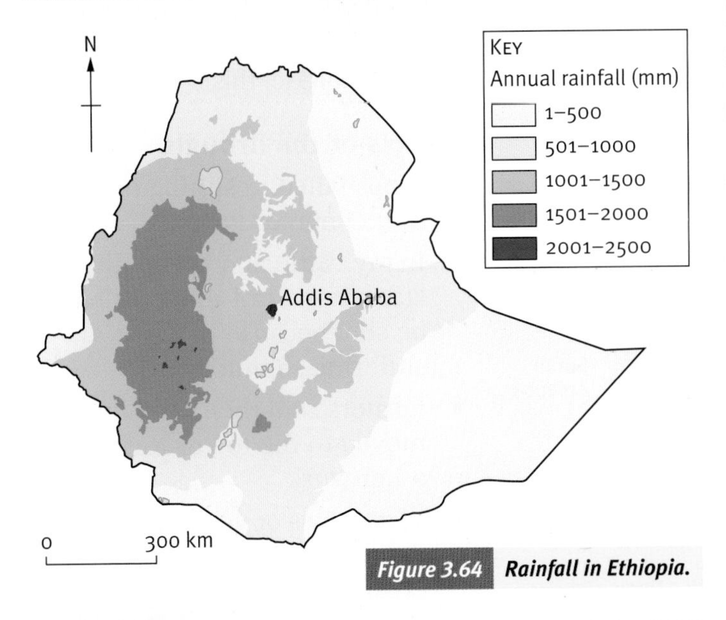

Figure 3.64 *Rainfall in Ethiopia.*

ACTIVITIES

1. What are the natural causes of drought in the Sahel region?
2. Using the map compare the areas of drought and rainfall.
3. Use the Fact file and cuttings to make a bubble map of the causes and effects of drought in Ethiopia.

Figure 3.65 *Primitive farming methods.*

Fact file

Ethiopia

- Population: 77.7 million – doubled from 33.5 million in the 1984 famine. Only 1 in 8 Ethiopians lives in a town – the lowest proportion on Earth. Families average 5.4 children.
- Size: 1.1 million km^2 (twice the size of France).
- Climate: Hot and humid in the lowlands, cool in the uplands. Average daily temperatures 28–30°C.
- Land use: Due to demands for fuel, construction and fencing, at least 77 per cent of the country's tree cover has been cut down in the last 25 years. These have been replaced by plantations of eucalyptus which are soil-depleting.
- Employment: 85 per cent of the population rely on farming for a living.
- Conflict: From 1998 to 2000 Ethiopia was in the grip of a violent civil war between two neighbouring areas. Millions of dollars were spent on both sides to support their armies.
- Aid: Ethiopia gets the most relief aid and the least development aid of any poor country in the world.
- Health: In 2004, there were 2–3 million HIV/AIDS sufferers. One million children had lost their parents to AIDS and more than 250,000 children under five live with the disease.

PLENARY ACTIVITY

Identify different people's perceptions of the landscape shown in the photographs in Figures 3.63 and 3.65 – as grazing land; as an area for agriculture; as an area receiving food aid.

CASE STUDY Australia 2005–2006 – drought in an MEDC

GET STARTED

Elghty per cent of all agricultural production is exported from Australia. Use an atlas to find out what the products are in terms of crops and livestock. How might a drought affect each of the agricultural types you have mentioned?

KEY TERMS

El Niño – large climatic disturbances in the southern Pacific Ocean that occur every 3–7 years.

Australia is the world's driest inhabited continent and often suffers from droughts. The drought of 2005–6 took its toll on the lives and livelihoods of farmers as well as livestock and crops. The 2005 winter was the fourth driest for more than a century. Reservoirs emptied and there were severe water restrictions in the populated coastal areas of Queensland.

Fact file

The number of sheep in Australia fell by 6 million between 2004 and 2006.

What caused the drought?

Australia's droughts are often caused by a meteorological event called **El Niño**. A strong El Niño effect in 2002 caused a reversal in the normal winter ocean-current patterns in the Pacific Ocean. Normally there is warm water in the western ocean off the coast of Australia and cooler water in the east off Peru and Chile. Strong Trade Winds carry moist air across the ocean and convection currents form above the warm water and result in heavy rain and thunderstorms along the Australian coast. Occasionally this situation is reversed and the cooler water occurs off this stretch of coast. The Trade Winds are weaker and this results in drought conditions.

RESEARCH LINK

Are droughts occurring more frequently and over larger areas in Australia? Do a web search to find out how many droughts the continent has suffered over the past 100 years. What might be the explanations for this?

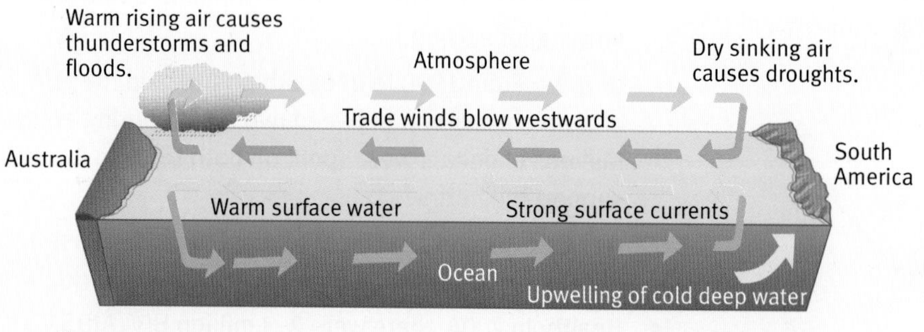

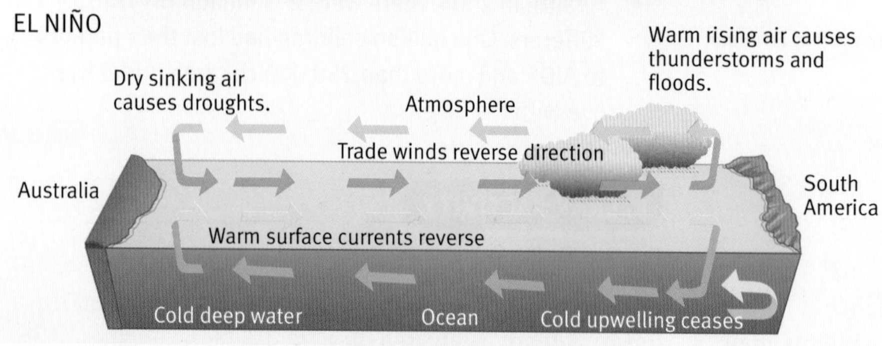

Figure 3.66 ***El Niño effects.*** **(Source: Jacaranda Project)**

Effects of the drought:

- crops failed, resulting in a loss of income for farmers
- many cattle and sheep died of starvation or of thirst, or had to be shot because they were suffering so much
- farmers sold their land and moved to the towns to find other work
- severe loss of vegetation due to the drought resulted in soil erosion
- farmers had to borrow large amounts to buy feed for their animals
- there were several large bushfires and dust storms during the drought
- water quality declined as water stores ran out, leading to the formation of toxic algae.

Can Australia be drought-proof?

Figure 3.67 ***Drought regulations in Queensland, Australia.*** **(Source: Ryebuck Media Pty Ltd, Australia)**

ACTIVITIES

1 Which of the diagrams below (**a** or **b**) represents normal weather patterns for Australia, and which represents abnormal ones?

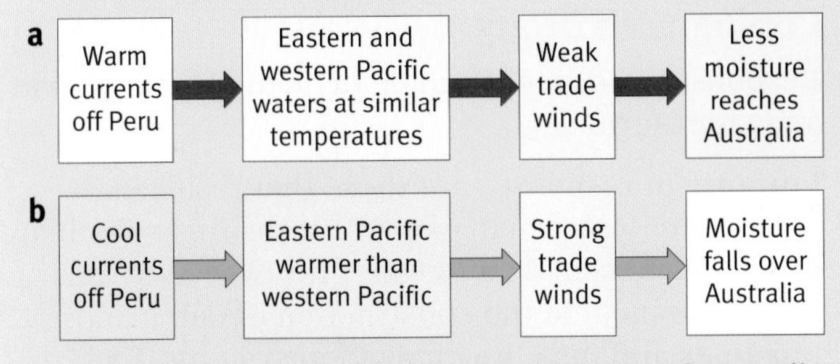

2 Look at Figure 3.62 and describe the areas of Australia most prone to drought. Using an atlas map of population distribution in Australia, name three areas that are densely populated and three areas that are sparsely populated that suffer from drought.

3 Imagine you lived in the house shown in Figure 3.67. How would water conservation measures affect your family in a week?

4 Copy and complete the table below to show how each of the following are affected by drought in Australia.

Aspect	Likely impact	Explanation
Agricultural products		
Livestock		
Bushfires		
Cost of living		
Erosion		
Duststorms		
Debt – personal		
Debt – national		
Gardens and parks		
Water quality		
Health		

Coping with drought: LEDCs and MEDCs

GET STARTED

From what you have learnt so far about drought, explain how drought can be caused by both natural and human factors.

How does drought affect LEDCs?

Each drought that occurs in the semi-arid lands of LEDCs results in less grazing land and fewer crops in areas where there are some of the highest birth rates in the world.

Over 900 million people are affected by the process of land degradation or **desertification** that is occurring in areas of unreliable rainfall across the world. The main cause is the human mismanagement of a very fragile environment that was previously productive land. Figure 3.70 shows some of the causes of desertification.

Population increase and poor agricultural practices result in the soil losing its protective vegetation cover. The land then becomes at risk from soil erosion as it becomes exposed to the wind and the occasional heavy rainstorm.

Figure 3.68 *Dried up river bed in Bedfordshire.*

How does drought affect MEDCs?

Compared with LEDCs, the impact of droughts on people in MEDCs is relatively small.

The impacts of drought in Spain, 2005

The year 2005 saw Spain's worst drought since records began. The water shortages hit a range of economic sectors very hard.

- **Tourism:** Spain's biggest earner. Fountains and beach showers switched off. Golf courses told to reduce water use.
- **Farming:** half the cereal crops lost in the centre and the south of the country; some regions lost everything. Little water to irrigate tomato and flower crops – a major earner. No grass to feed livestock so farmers depended on buying in animal feeds, leading to feed prices shooting up – a farming crisis.
- **Domestic water use:** water is rationed across half of the country to ensure people have water in their homes.
- **Forest fires** swept over large areas of Spain.
- **Tension in relations with Portugal:** many rivers in Portugal have their sources in Spain and Portuguese people felt Spain was stealing their water.

Coping with drought

Responses to drought can be divided into long term and short term.

Long-term responses are those that result in permanent water management measures, which are designed to secure more water for everyday use. Strategies include the construction of dams and pipelines and installing irrigation systems. In developed countries, these will often carry water from areas of surplus to areas of demand. For example, Wales provides a lot of the water for the urban areas of Birmingham and the West Midlands.

KEY TERMS

Desertification – the degradation of land in arid and semi-arid areas resulting primarily from human activities and influenced by climatic variations.

These strategies are expensive, and do not usually include any measures that will improve efficiency.

Short-term responses include the more temporary methods used by people to overcome specific drought events. They concentrate on reducing water demand and include water rationing, cash and/or food aid to the affected areas. These are generally less sustainable than the long-term measures and do not help people to be better prepared for the next drought.

Figure 3.69 Crops destroyed due to lack of rain in Ethiopia.

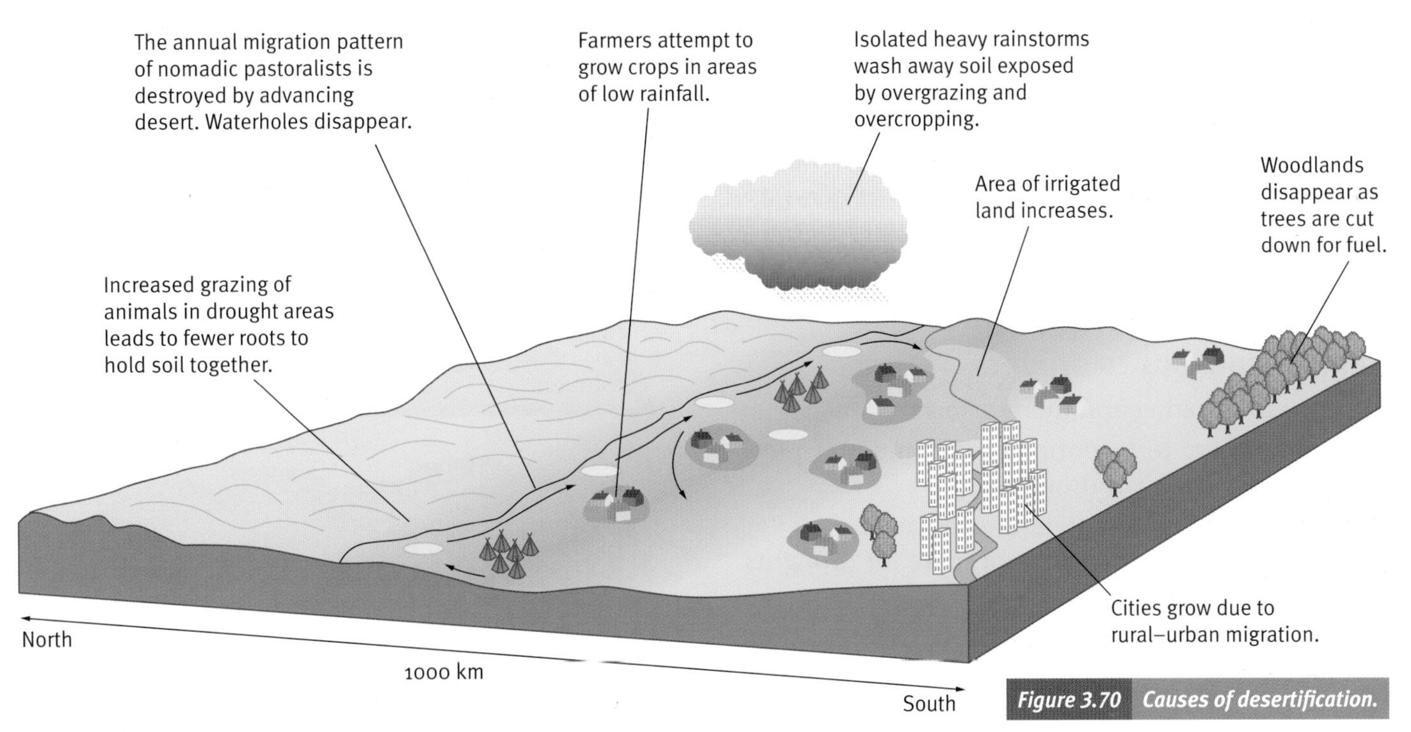

The annual migration pattern of nomadic pastoralists is destroyed by advancing desert. Waterholes disappear.
Farmers attempt to grow crops in areas of low rainfall.
Isolated heavy rainstorms wash away soil exposed by overgrazing and overcropping.
Area of irrigated land increases.
Woodlands disappear as trees are cut down for fuel.
Increased grazing of animals in drought areas leads to fewer roots to hold soil together.
Cities grow due to rural–urban migration.
North
1000 km
South

Figure 3.70 Causes of desertification.

ACTIVITIES

1. What is desertification? Make a table to show the different causes of desertification and show how they result in degradation of the land.
2. Explain the difference in sustainability between long-term and short-term responses to drought. Give an example of each type for MEDCs and LEDCs.
3. Compare the two photographs in Figures 3.68 and 3.69 by putting each in turn in the centre of a Development Compass Rose (right) and considering the questions at the four compass points.
4. Using the internet, research the work of two aid agencies in areas suffering from drought. Make a fact file of the work that each does. Include information about where they work; what type of aid they offer; how much they spend; which people are leading on the support.

Natural
These are questions about the environment, energy, air, water, soil, living things and their relationship to each other. These questions are about the built as well as the 'natural' environment.

Who decides?
These are questions about power, who makes choices and decides what is to happen; who benefits and loses as a result of these decisions and at what cost.

Economic
These are questions about money, trading, aid, ownership, buying and selling.

Social
These are questions about people, their relationships, their traditions, culture and the way they live. They include questions about how, for example, gender, race, disability, class and age affect social relationships.

(Source: Development Education Centre, Tide Global Learning)

Living with drought: how can the impact of drought be controlled?

GET STARTED

Think of five ways that the use of water in your house could be controlled.

Few developing countries have the technology to build big dams to store water. In rural areas it is often a long walk to the local well or river several hours away. More economically developed countries can afford to buy their way out of the consequences of poor water conservation at the moment.

Some responses to drought

Improve forecasting and monitoring:

- monitoring of temperatures over time may give advance warning of droughts
- use of satellites to spot early changes in vegetation can help people prepare for droughts.

Improve water conservation and farming methods; these responses can either be low tech or high tech.

Low-tech responses include:

- low walls built across fields to reduce run-off
- areas created for tree planting to conserve moisture in the soil
- using micro-dams to store water that can then be used for irrigation
- using stone piles to collect water by allowing water to condense onto the cold stones
- planting drought-resistant crops that can withstand a lack of water.

Hi-tech responses include:

- drip-and-sprinkle irrigation
- use of concrete water coolers to utilise hot groundwater
- improving water supply and irrigation by building reservoirs and new wells
- cloud seeding – injecting clouds with 'seeds' of silver iodide, salt or dry ice to make the clouds' water or ice particles bigger and yield more rain
- seawater greenhouses that use solar power to evaporate seawater and produce fresh water for irrigation.

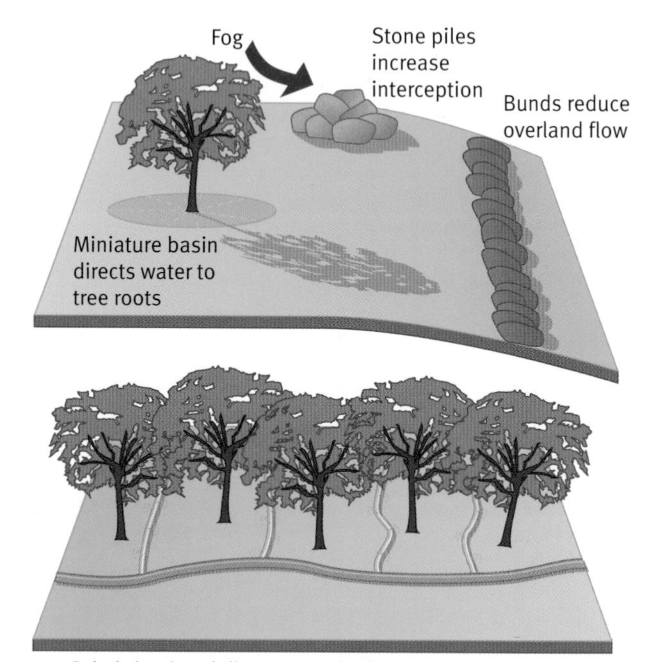

Figure 3.71 *Some 'low-tech' responses to drought.*

Expert guidance on fighting drought in Australia

'A lot of the water that's used in agriculture is lost in evaporation, and clearly, the more we can cap bores and we can pipe water to troughs, that way we're losing less and we're using water far more efficiently. Landscapes that had a lot more native vegetation in them first of all, will have improved water quality. They have plants in them, the native pastures, naturally grassy woodlands, respond much better to recovery from drought. They've been here for millions of years. Some of the systems we try and replace them with are not as resilient. So I consider that retaining as much native vegetation as possible, really gives you the resilience and robustness to accommodate some, if not all of these difficult things.'

(Source: Professor Hugh Possingham, Director of Ecology at the University of Queensland)

Figure 3.72 Wells in Ethiopia.

New water sources for the desert

Vast greenhouses that use seawater for crop cultivation could be combined with solar power plants to provide food, fresh water and clean energy in deserts, under an ambitious proposal from a team of architects and engineers.

The Sahara Forest Project envisages huge greenhouses with concentrated solar power (CSP), a technology that uses mirrors to focus the sun's rays, creating steam to drive turbines to generate electricity.

The installations would turn deserts into lush patches of vegetation, according to its designers, and do away with the need to dig wells for fresh water, an activity that has depleted aquifers across the world.

Virtually any vegetables could be grown in the greenhouses, according to the designers. The nutrients to grow the plants could come from local seaweed or be extracted from the seawater.

ACTIVITIES

1 Identify some of the responses to drought from the resources on these pages. Make notes on their advantages and disadvantages, using a table with three columns: Suggested management strategy; Advantages; Disadvantages.
2 Give at least one example of a suggested drought management strategy for each of these scales: personal, local, regional, national, international.
3 Research one type of strategy in detail and create a presentation or poster that includes an explanation of how it works. Include arguments for and against its use.

PLENARY ACTIVITY

'Droughts are a normal and natural part of the environment. They cause such great problems now because of the way we have developed our land use practices over time – crop farming and raising animals require regular and reliable access to water. In the future drought relief should be tied to ecologically sustainable production, ending of land clearance and urban expansion in arid areas.'

Read the statement above. Do you agree with the views expressed? What would be the social, economic and environmental impacts of this change for LEDCs and for MEDCs?

Foundation:

Key Geographical Themes 8-mark case study question

Case study – A climatic hazard event.

i Name a type of climatic hazard and give the location where it took place.
ii Explain the natural processes that caused this event.
iii Explain how human activities affected the impact of the climatic hazard. **[8 marks]**

Mark scheme

Level 1: The student names a valid type of hazard and a valid location, but shows only limited relevant knowledge and information in explanation. Meaning may not be communicated very clearly because of mistakes in writing. **[1–3 marks]**

Level 2: Names a valid type of hazard and a valid location, with relevant development of the explanation in either (ii) or (iii). Demonstrates some relevant knowledge, based on a range of factual information and evidence. Meaning is communicated clearly. **[4–6 marks]**

Level 3: Names a valid type of hazard and a valid location, with relevant development of the explanation in either (ii) or (iii). Demonstrates thorough knowledge, based on a full range of relevant factual information and evidence. Meaning is communicated very clearly. **[7–8 marks]**

i Hurricane Katrina, New Orleans, USA.

ii This was caused by warm air rising quickly over the Atlantic because of the warm sea temperatures.
Water is evaporated and this sucks in more warm air to cause high wind speeds.

iii New Orleans is a city on low-lying land.
The people thought they were safe.
However, the storms destroyed the levées and large areas were flooded, thousands were drowned or made homeless.

Examiner's comment

A well-known example has been given in (i) with a clear, accurate location.
Without the location this answer would only be able to achieve seven marks maximum.

The candidate has done well to explain the complex processes which cause tropical storms. The warm air rising because of high sea temperatures is the basic idea. The Atlantic fits the given case study example. The high speed winds resulting from the rapid evaporation is the development of the explanation.

The low-lying land idea is a basic way in which human activities made the hazard worse. The failure of the flood protection levées is the development of the idea.
Explaining how human activities reduced the impact of a hazard would also be valid.
Full marks have been achieved by this case study answer.

End of unit activities

Thinking about geography – living graphs and reasoning

A *living graph* is a visual thinking strategy that can help us make connections between geographical patterns and processes and their impact on people's lives.

Using living graphs allows us to *develop theories* based on what we think the graphs show. In this way we are further developing our ability to *explain* and *justify* our thinking through informed *reasoning*.

Natural hazards – living graph

The line graph in Figure 3.73 shows the intensity of a volcanic eruption in Indonesia over the course of a month. At its most intense, the eruption generated earthquakes that reached 7.2 on the Richter Scale. Below the graph, there are twelve numbered statements.

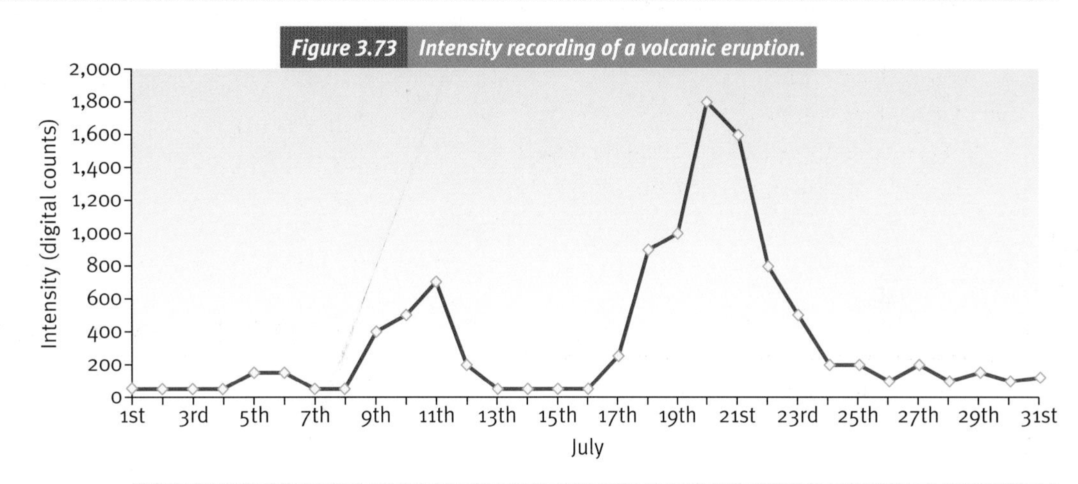

Figure 3.73 *Intensity recording of a volcanic eruption.*

1 Rescue services stand down	2 Grandfather says prayers	3 Atin goes to school	4 Gunter queues for rice
5 Coffin makers rest	6 The helicopter flies away	7 Dogs start howling	8 The TV goes blank
9 The farmer loads his truck with food	10 The aide wakes the President	11 Umbrellas sell well	12 The airport closes

ACTIVITIES

1. Working with a partner, look at the statements and think about the sequence in which these events are likely to take place in relation to the eruption intensity pattern shown in the graph. There may be several possibilities!
2. Match up each statement to a date when they would be likely to occur.

Developing your reasoning skills

3. Join up with another pair and compare your sequences – get the other pair to justify to you their reasoning for placing their statements. Do you agree?
4. Think of some statements of your own that *describe the eruption sequence* and match them to a date when they would be likely to occur.
5. Finally, think up some more statements like the ones in the table and swap them with another pair – can they match up a date for these and justify why?

THINKING ABOUT REASONING

- From Activity 1, how did you decide the order in which the statements would go?
- Did you get any visual images in your head? Where were they from?
- Was it easy or hard to justify your decisions to the other pair? Explain why.
- In what other subjects do you have to justify and reason your decisions? How could you use the *living graph* strategy to help you?

Chapter 4

Economic development

There are many contrasts between rich and poor countries in development and economic activity. A major contrast is in farming. In the rich world, most countries have less than 5 per cent of their working populations employed in agriculture. This is because farming is capital intensive, with a high investment in machinery and other inputs such as fertiliser. In many poor countries, more than 60 per cent of the working population are employed in agriculture. Here, farming is very labour intensive, with little money available for the inputs that would make farming more efficient. While rich countries are either able to export significant amounts of food or afford to buy the food they cannot produce, many poor countries struggle to feed their populations. Having enough to eat is probably the most important aspect of a person's quality of life.

Consider this

In the last few years food prices have been rising rapidly. This has caused difficulties for people on low incomes in rich countries. However, the problems for people in poorer countries have been much greater, with millions of people unable to afford the most basic diet. In some countries food protests and riots have occurred. The media have referred to this issue as 'the global food crisis'. What impact will this have on development? Will rising food prices cause living standards to fall substantially in some countries? Will development be more difficult in the future than it has been in the past?

QUESTIONS FOR INVESTIGATION

- **a** What is meant by 'development'?
- **b** How and why are there variations between the employment structures of different countries?
- **c** What determines the location of different economic activities?
- **d** How do multinational companies affect development?
- **e** How can economic activity affect the physical environment at a variety of scales, including global?

Is there more to development than wealth?

GET STARTED

In small groups, make a list of countries and individual people you think of as wealthy. What are the signs of this wealth for both individuals and countries?

What is wealth?

When we think of our own wealth or 'economic well-being' we consider how much money we have and the value of our possessions. People with a lot of money are considered to be rich, and people with little money are said to be poor. Wealth is a fairly narrow idea or concept because it refers to money alone, and does not include other things that may be important to our happiness or **quality of life**.

What is development?

Development is a much more wide-ranging concept than wealth. It includes wealth, but it also includes other important aspects of our lives. For example, many people would consider good health to be more important than wealth. People who live in countries which are not democracies, where freedom of speech cannot be taken for granted, often envy those who do live in democratic countries. Figure 4.1 presents one view of the factors that make up the quality of life (with an image of considerable wealth in the background).

Development occurs when there are improvements to individual factors making up the quality of life. For example, development occurs in LEDCs when:

- a new well is dug and a pump supplied to bring clean water to a village for the first time. Benefits will include improvements to health, and the fact that people will no longer have to walk long distances to collect water
- local food supply improves due to new investment in machinery and fertilisers
- the electricity grid extends outwards from the main urban areas to rural areas
- a new road or railway improves the accessibility of a remote province.

Figure 4.1 ***The quality of life.***

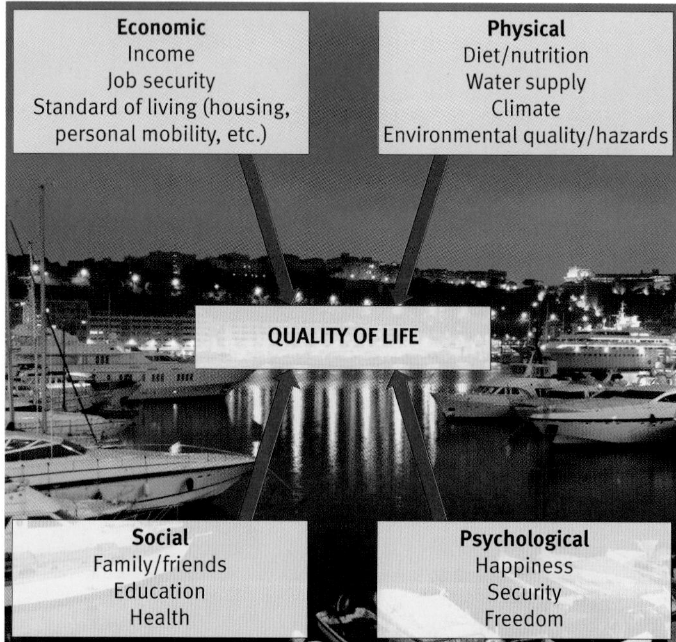

THINK ABOUT IT

Look at Figure 4.1. Select what you think are the four most important aspects of the quality of life. Justify your selections.

KEY TERMS

Development – the use of resources and the application of available technology to improve the standard of living within a country.

Environmental impact assessment – a document required by law in many countries, detailing all the impacts on the environment of a project above a certain size.

Quality of life – this term sums up all the factors that affect a person's general well-being and happiness.

Sustainable development – development that meets the needs of the present without harming the ability of future generations to meet their own needs.

Why does development need to be sustainable?

Most people would agree that development is a good thing for the population of a country, at least in the short term. However, for many examples of development the longer-term costs may outweigh the benefits. As a result, future generations may have a lower quality of life because the development that occurred was unsustainable. For example:

- modern farming methods may increase soil erosion and reduce soil fertility
- rapid industrialisation may lead to large-scale pollution of land, air and water
- rising incomes and commercialisation may undermine long-standing traditions, leading to social breakdown.

As more and more countries have suffered from such problems, our understanding of **sustainable development** has improved. Now careful planning can avoid many of the mistakes of the past. For example, new construction projects above a certain size require an **environmental impact assessment** in most countries. Environmental impact assessments look not only at the problems the new development might create immediately, but also the disadvantages it might bring in the future.

THINK ABOUT IT

Why is development a good thing – at least in the short term?

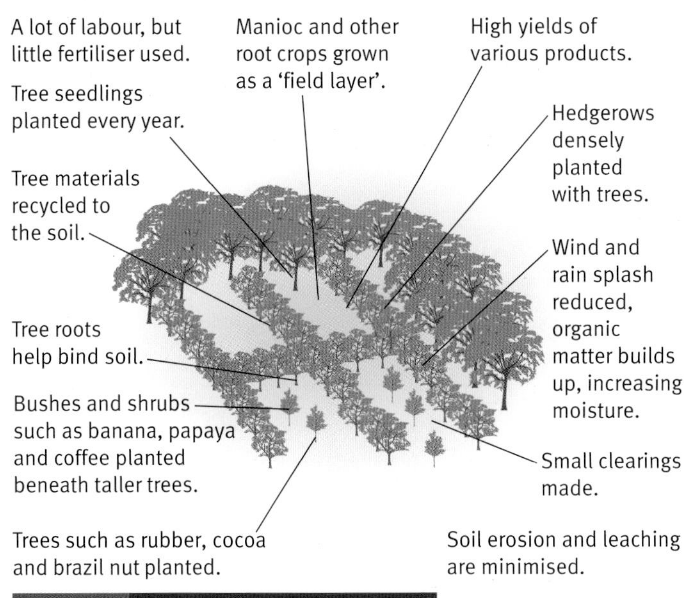

Figure 4.2 ***Sustainable farming methods.***

Examples of sustainable development include:

- organic farming to ensure that the long-term fertility of the soil is not reduced by the over-use of chemical fertilisers
- fishing at a level that does not reduce fish stocks and endanger the supply of this source of food in the future.

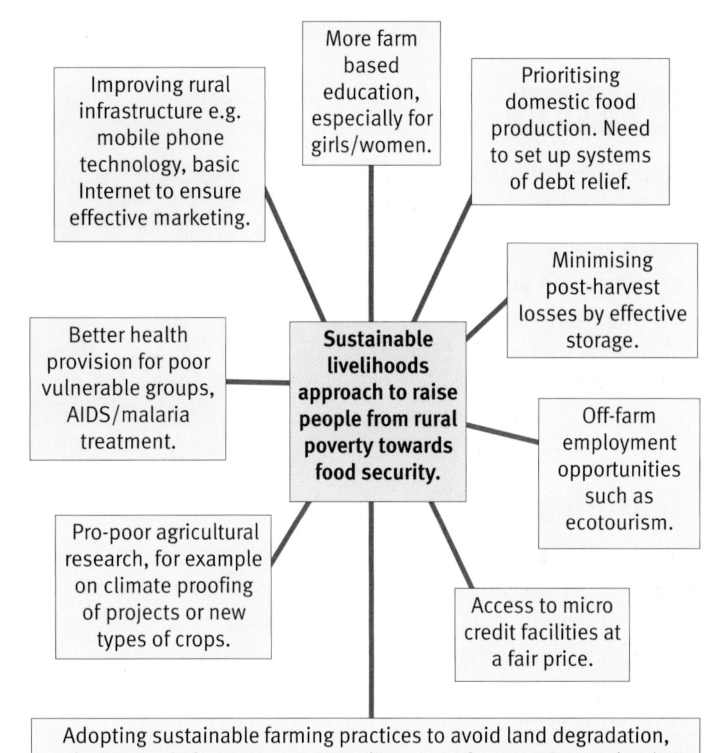

Figure 4.3 ***How rural poverty can be tackled by sustainable development practices.*** **(Source: GeoFactsheet no 228)**

ACTIVITIES

1. What is wealth?
2. List three differences you would expect to find between a wealthy country and a poor country.
3. What is sustainable development?
4. **a** Describe the differences in wealth that you can see in the region in which you live.

   b Suggest reasons for these differences.
5. Why has so much development over the past 50 years proved to be unsustainable?
6. Describe and explain the sustainable development practices shown in Figures 4.2 and 4.3.

PLENARY ACTIVITY

Should we look after ourselves or should we think about future generations? Explain your response to this question.

How can economic well-being and the quality of life be measured?

GET STARTED

Look at the photos in Figures 4.4 and 4.5 of life in a very rich country and a very poor country. What evidence can you find in the photos for one being rich and the other poor?

Figure 4.4 *Doha is the capital city of Qatar, which has the highest GDP per capita (per person) in the world.*

Measuring wealth

The most common indicator of a country's wealth is the Gross Domestic Product (GDP). The Gross Domestic Product is the total value of goods and services produced by a country in a year. To take account of the different populations of countries, the Gross Domestic Product per capita is often used, in which case the total GDP of a country is divided by the total population.

Organisations such as the UN now publish GDP data at **purchasing power parity (PPP)**. This takes account of differences in the cost of living between countries.

Table 4.1 shows the extent to which GDP per capita varies around the world. The data shown by the world map in Figure 4.6 do not take into account the way in which the cost of living can vary between countries. Because of this, they are referred to as 'nominal' GDP data.

Figure 4.5 *Slums near the Taj Mahal, India.*

Measuring the quality of life

The way that the quality of life has been measured has changed over time. In the 1980s the Physical Quality of Life Index (PQLI) was devised.

Table 4.1 Top four and bottom four countries in GDP per capita (PPP)

Top 4 countries	
Qatar	80,900
Luxembourg	80,500
Malta	53,400
Norway	53,000
Bottom 4 countries	
Ethiopia	800
Eritrea	800
Sierra Leone	700
Niger	700

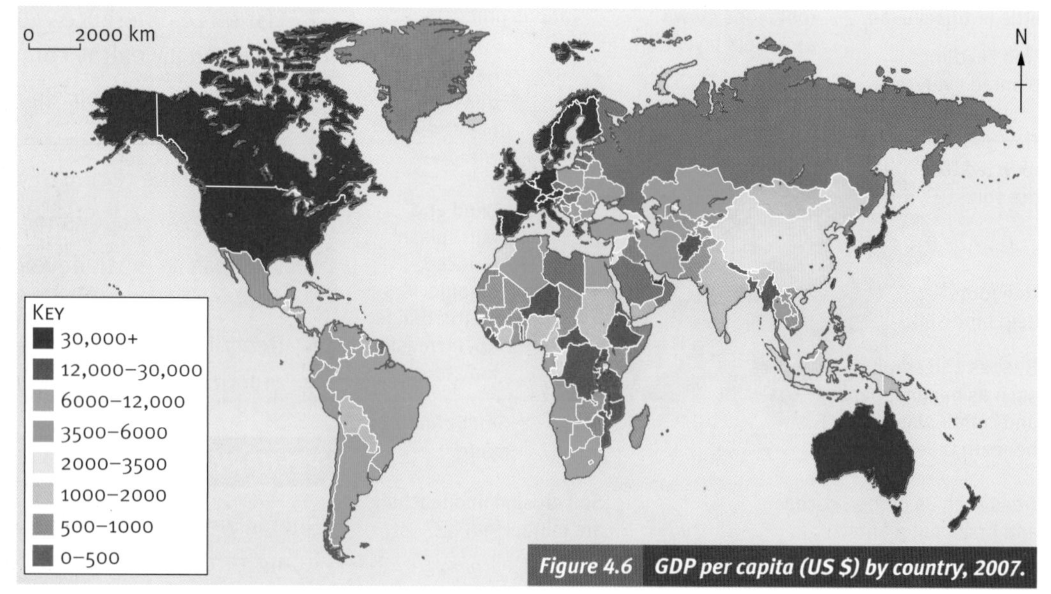

Figure 4.6 *GDP per capita (US $) by country, 2007.*

The PQLI was the average of three development factors: literacy, life expectancy and infant mortality. However, in 1990 the Human Development Index (HDI) was devised as a better measure. It contains three variables:

- life expectancy
- educational attainment (how many children go to primary, secondary and tertiary education, plus adult literacy)
- GDP per capita (PPP$).

The actual figures for each of these three measures are converted into an index, which has a maximum value of 1.0 in each case. The three index values are then averaged to give an overall Human Development Index value. This also has a maximum value of 1.0. Every year, the UN publishes the Human Development Report, which uses the HDI to rank all the countries of the world in their level of development.

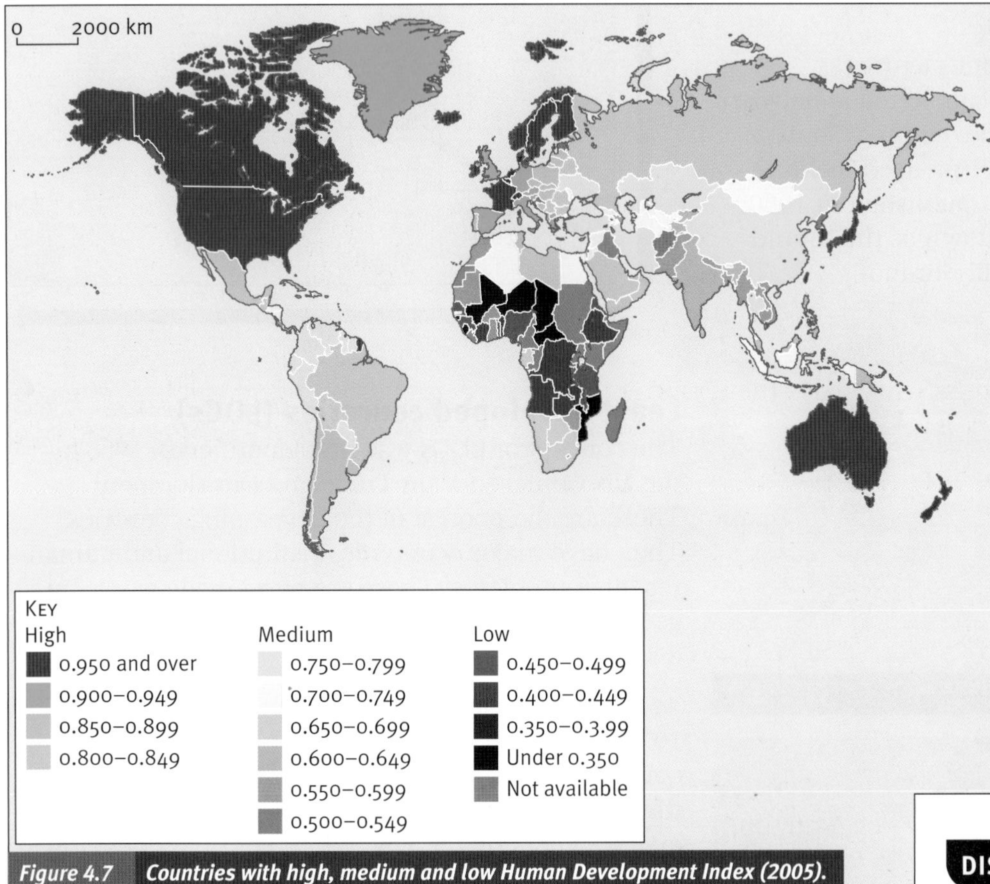

Figure 4.7 *Countries with high, medium and low Human Development Index (2005).*

ACTIVITIES

1. Define Gross Domestic Product per capita.
2. Look at Figure 4.6. Use an atlas to identify two countries in each of the eight GDP per capita classes shown in the key.
3. Look at Table 4.1. Work out the difference in GDP per capita (PPP) between the wealthiest and the poorest country in the world. How would this difference affect somebody's life?
4. Why are organisations such as the UN increasingly using GDP data at purchasing power parity (PPP)?
5. Describe the global distribution of GDP per capita shown in Figure 4.6. How does this compare with the map of human development (Figure 4.7)?

KEY TERMS

GDP at purchasing power parity (PPP$) – the GDP of a country converted into US dollars on the basis of the purchasing power parity of the country's currency. This is assessed by calculating the number of units of a currency required to purchase the same representative basket of goods and services that a US dollar would buy in the USA.

Fact file

Poverty

Poverty can be viewed in a number of ways. The most important are:

- **Absolute poverty:** a person is absolutely poor if his or her income is less than a particular level. This will vary from country to country and from region to region, depending on the cost of living.
- **Relative poverty:** refers to falling behind most others in the community, determined by inclusion in a bottom-income group, such as the poorest 10 per cent.

DISCUSSION POINT

Imagine that the UN has decided to add one more measure of development to the HDI. What do you think it should be and why? Use your atlas for ideas.

PLENARY ACTIVITY

Which countries do you consider to be poor? What are the signs of poverty in these countries? Are people in the UK ever poor?

Identifying and explaining why countries are at different stages of development

GET STARTED

Work in pairs for a few minutes to consider how you would expect the UK to be at a noticeably higher level of development than India.

The North–South divide

Although the global development picture is complex, a general distinction can be made between the developed 'North' and the developing 'South' (Figure 4.8). These terms were first used in *North–South: A Programme for Survival* published in 1980. This publication is generally known as the 'Brandt Report' after its chairperson Willy Brandt.

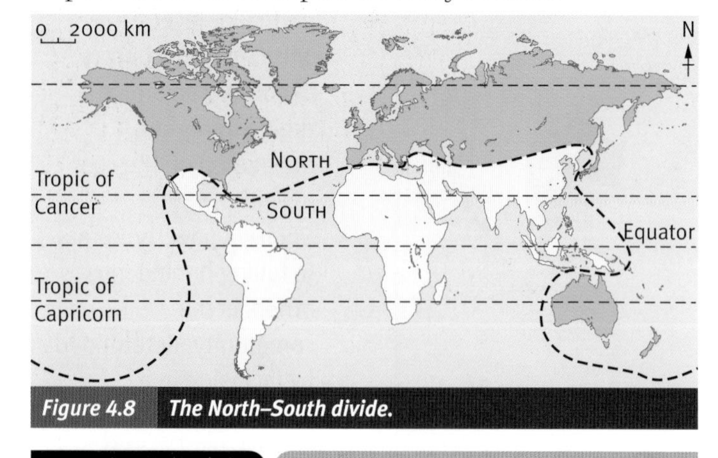

Figure 4.8 The North–South divide.

THINK ABOUT IT

Is the **North–South divide** still a valid division nearly 30 years after publication?

Other terms used to distinguish between the richer and poorer nations are:

- Developed and developing
- More economically developed countries (MEDCs) and less economically developed countries (LEDCs).

Stages of development

Over the years, there have been a number of descriptions and explanations of how countries moved from one level of development to another. A reasonable division of the world in terms of stages of economic development is shown in Figure 4.9.

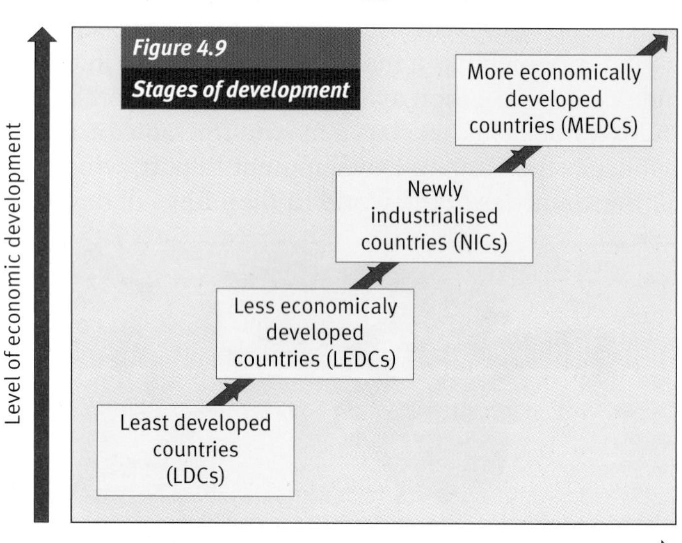

Figure 4.9 Stages of development

Least developed countries (LDCs)

The concept of LDCs was first identified in 1968 by the UN Conference on Trade and Development. These are the poorest of the developing countries. They have major economic, institutional and human resource problems. These are often made worse by geographical handicaps and natural and man-made disasters.

At present 49 countries are identified as LDCs. With 10.5 per cent of the world's population, these countries generate only one-tenth of 1 per cent of the global income. The list of LDCs is reviewed every 3 years by the UN. When countries develop beyond a certain point they are no longer considered to be LDCs.

Newly industrialised countries

The first countries to become newly industrialised countries (NICs) were South Korea, Singapore, Taiwan and Hong Kong. The media referred to them as the 'four Asian tigers'. A 'tiger economy' is one that grows very rapidly. The reasons for the success of these countries were:

- a good level of infrastructure
- a skilled but relatively low-cost workforce
- cultural traditions that revere education and achievement

- governments welcoming foreign direct investment (FDI) from transnational corporations
- distinct advantages in terms of geographical location
- governments encouraging banks to lend to companies at low interest rates.

KEY TERMS

North–South divide – the simple division between MEDCs (mainly in the north) and LEDCs (mainly in the south).

The success of these four countries provided a model for others to follow such as Malaysia, Brazil, China and India. In the last 15 years the growth of China has been particularly impressive. South Korea and Singapore have developed so much that many people now consider them to be MEDCs.

Explaining the development gap

There has been much debate about the causes of development. Detailed studies have shown that variations between countries are due to the following factors.

Physical geography:

- landlocked countries have generally developed more slowly than coastal ones
- small island countries face considerable disadvantages in development
- tropical countries have grown more slowly than those in cooler latitudes, reflecting the cost of poor health and unproductive farming. However, richer non-agricultural tropical countries such as Singapore do not suffer a geographical deficit of this kind
- some countries have a wealth of natural resources that have boosted economic growth.

Economic policies:

- economies that welcomed and encouraged foreign investment have developed faster than closed economies
- fast-growing countries tend to have high rates of saving and low spending
- good government, law and order and lack of corruption generally result in a high rate of growth.

Demography:

- progress through the demographic transition is a significant factor, with the highest rates of growth experienced by those nations where the birth rate had fallen the most.

Development differences: more advanced developing countries

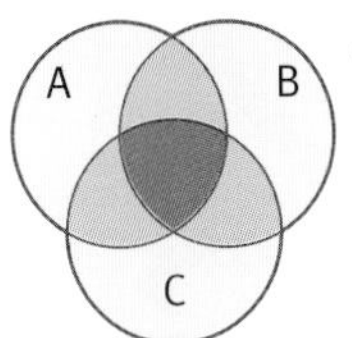

A: Largest countries in region
B: Countries with abundant natural resources
C: Newly industrialised countries

■ Highest level of development

Development differences: least advanced developing countries

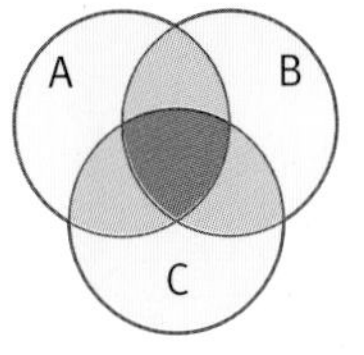

A: Land-locked or island developing countries
B: Countries with few natural resources
C: Countries seriously affected by natural hazards

■ Lowest level of development

Figure 4.10 ***Fast and slow development LEDCs.***

Figure 4.11 ***Seoul, South Korea – a newly industrialised country.***

Table 4.2 Energy consumption per capita for high, middle and low-income countries

Total energy consumption per capita 2003 (kg of oil equivalent per person)	
High-income countries	5435
Middle-income countries	1390
Low-income countries	494

PLENARY ACTIVITY

How do you think economic well-being and quality of life in the UK might change over the next 20 years?

ACTIVITIES

1. Look at Figure 4.8. Describe the locations of the developed North and the less developed South.
2. Look at Figure 4.9, then:
   a. draw your own version of the four stages of development shown in the model
   b. with the help of Figures 4.6 and 4.7 in particular, suggest a country for each stage of the model.
3. a. What is an LEDC?
   b. Describe the characteristics of an LEDC.
   c. The least developed countries are sometimes called LDCs. Suggest and explain some possible countries for this category.
4. a. Name four NICs.
   b. With reference to Figure 4.10, suggest why some LEDCs have been able to develop into NICs while many have not.

How can development be affected by aid?

GET STARTED

Should rich countries help poorer countries by donating aid? If so, how much should they give as a proportion of the whole country's earnings a year?

Figure 4.12 *UK aid project in Bangladesh.*

When did international aid begin?

The first time aid was given by one country to help develop another was when the USA set out to reconstruct the war-torn economies of Western Europe and Japan after the Second World War. The success of this aid programme, called the Marshall Plan, meant that similar approaches were used to improve development in LEDCs from the 1950s.

Why is international aid necessary?

Most LEDCs have been keen to accept **international aid** because of the:

- **'foreign exchange gap'**, whereby many LEDCs lack the hard currency to pay for imports such as oil and machinery that are vital to development
- **'savings gap'**, where population pressures and other drains on expenditure prevent the accumulation of enough capital to invest in industry and infrastructure
- **'technical gap'**, caused by a shortage of skills needed for development.

KEY TERMS

International aid – the giving of resources (money, food, goods, technology, etc.) by one country or organisation to another poorer country. The primary objective is to improve the economy and quality of life in the poorer country.

The different types of international aid

Figure 4.13 shows the different types of international aid. The basic division is between the following.

- **Official government aid**, where the amount of aid given and who it is given to is decided by the government of an individual country. The Department for International Development (DfID) runs the UK's international aid programme.
- **Voluntary aid** is run by Non-Governmental Organisations (NGOs) or charities such as Oxfam, ActionAid and CAFOD. NGOs collect money from individuals and organisations. However, an increasing amount of government money goes to NGOs because of their expertise in running aid efficiently.

Official government aid can be divided into the following:

- **Bilateral aid**: given directly from one country to another. This is often tied to other commercial deals between the involved countries
- **Multilateral aid**: provided by many countries and organised by an international body such as the UN.

Aid supplied to poorer countries is of two types:

- **short-term emergency aid**: provided to help cope with unexpected disasters such as earthquakes, volcanic eruptions and tropical cyclones
- **long-term development aid**: directed towards continuous improvement in the quality of life in a poorer country.

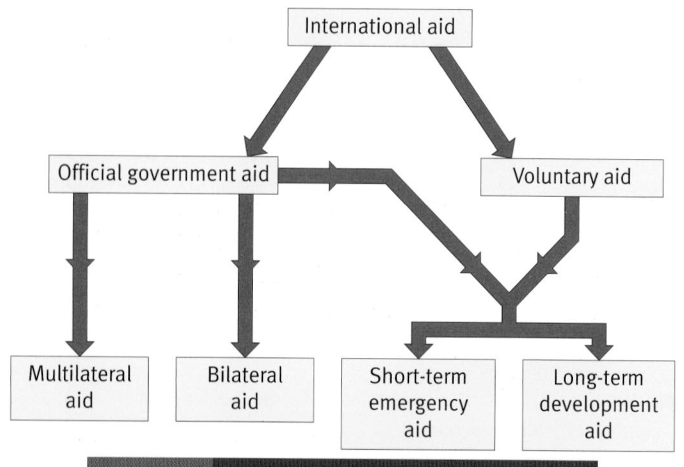

Figure 4.13 *The different types of international aid.*

THINK ABOUT IT

Why would one country want to help another?

Can aid speed up development?

In its best form, there can be little doubt that aid can combat poverty. For example it can:

- provide vital investment in agriculture and water supply. Clean water and an adequate diet are essential for health and development
- buy machinery to set up new industries. This then creates new job opportunities
- finance new infrastructure projects such as roads, railways and electricity grids, which are important for the development of new economic activities and raising the quality of life.

There is no doubt that many countries have benefited from international aid. All the countries that have developed into NICs from LEDCs had received international aid. However, their development has been due to other reasons too. It is a complicated business and so it is difficult to be precise about the contribution of international aid to the development of each country. Critics of international aid argue that:

- too often aid fails to reach the very poorest people and when it does the benefits are frequently short-lived
- a significant proportion of foreign aid is 'tied' to the purchase of goods and services from the donor country and often given for use only on jointly agreed projects. Some of these ties have involved the LEDC buying large amounts of arms and other military equipment
- aid is often used on large capital-intensive projects, which may actually worsen the conditions of the poorest people
- aid may delay the introduction of reforms; for example, the substitution of food aid for land reform
- areas can become dependent on aid.

Many geographers argue there are two issues more important to development than aid: first, changing the terms of trade so that developing nations get a fairer share of the benefits of world trade (Figure 4.14); second, writing off the debts of the poorest countries.

RESEARCH LINK

What have been the effects of sending second-hand clothes from the UK to developing countries?

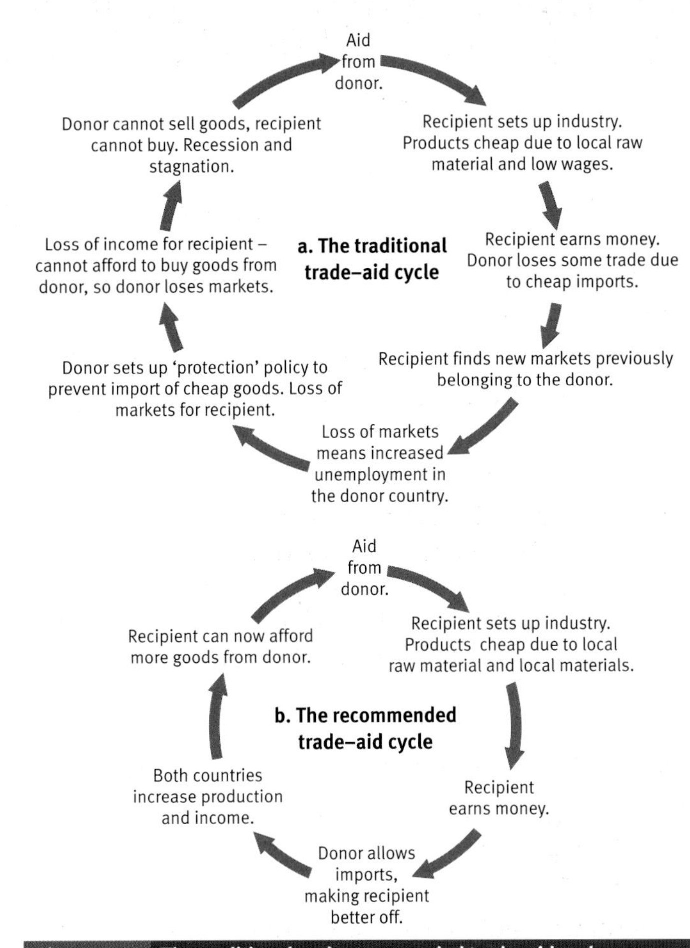

Aid from donor.
Recipient sets up industry. Products cheap due to local raw material and low wages.
Recipient earns money. Donor loses some trade due to cheap imports.
Recipient finds new markets previously belonging to the donor.
Loss of markets means increased unemployment in the donor country.
Donor sets up 'protection' policy to prevent import of cheap goods. Loss of markets for recipient.
Loss of income for recipient – cannot afford to buy goods from donor, so donor loses markets.
Donor cannot sell goods, recipient cannot buy. Recession and stagnation.
a. The traditional trade–aid cycle
Aid from donor.
Recipient sets up industry. Products cheap due to local raw material and local materials.
Recipient earns money.
Donor allows imports, making recipient better off.
Both countries increase production and income.
Recipient can now afford more goods from donor.
b. The recommended trade–aid cycle

Figure 4.14 The traditional and recommended trade–aid cycles. **(Source: Reproduced with the permission of Nelson Thornes Ltd from *The New Wider World* by David Waugh, isbn 978-0-7487-7376-3 first published in 1998)**

ACTIVITIES

1. Define international aid.
2. Explain the difference between official government aid and voluntary aid.
3. **a** What is meant by the initials 'NGO'?
   b Make a list of all the NGOs you have heard of.
   c Compare your list with those of other people in your class.
   d Produce a 'top five' class ranking of NGO popularity.
   e What do they have in common?
4. Draw cartoons to show the three types of gap that stop countries developing.
5. Produce a flow diagram to show how aid can speed up development.
6. Explain at least two possible disadvantages of international aid.
7. Why do LEDCs agree to tied aid deals? Are there times when they should not?
8. How would you address each of the gaps that work against development?

CASE STUDY WaterAid in Mali – sustainable aid?

GET STARTED

Make a log of how you have used water in your day so far, and roughly how many litres you think you have used.

NGOs: leading sustainable development

NGOs have often been much better at directing aid towards sustainable development than government agencies. The selective nature of such aid has targeted the poorest communities using **appropriate technology** and involving local people in decision-making.

WaterAid in Mali

WaterAid was established in 1981. Its first project was in Zambia, but its operations spread quickly to other countries. Mali is one of the countries in which WaterAid currently operates.

Mali, in West Africa, is one of the world's poorest nations (Figure 4.15). The natural environment is harsh and deteriorating. Rainfall levels, which are already low, are falling further and desertification is spreading. Currently 65 per cent of the country is desert or semi-desert. Mali has a population of around 12 million. The total percentage of the population with sustainable access to improved water supply is about 50 per cent. WaterAid has been active in the country since 2000.

WaterAid's main concern is that the fully privatised water industry frequently fails to provide services to the poorest urban and rural areas. It is running a pilot scheme in the slums surrounding Mali's capital Bamako, providing clean water and sanitation services to the poorest people. Its objective is to demonstrate to both government and other donors that projects in slums can be successful, both socially and economically.

WaterAid has financed the construction of the area's water network. It is training local people to manage and maintain the system, and to raise the money needed to keep it operational. Encouraging the community to invest in its own infrastructure is an important part of the philosophy of the project. According to Idrissa Doucoure, WaterAid's West Africa Regional Manager, 'We are now putting our energy into education programmes and empowering the communities to continue their own development into the future. This will allow WaterAid to move on and help others.'

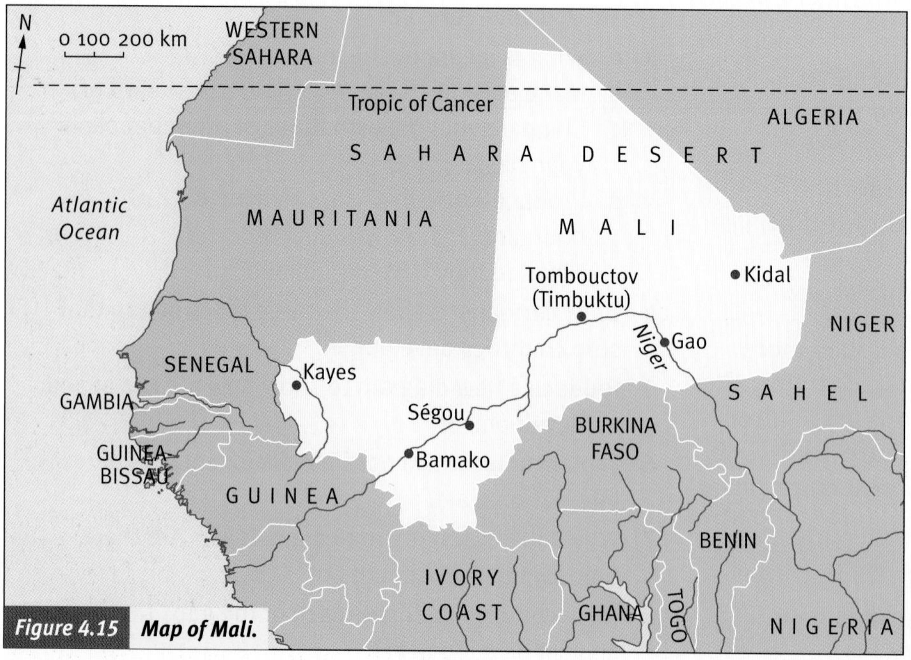

KEY TERMS

Appropriate technology – aid supplied by a donor country whereby the level of technology and the skills required to service it are properly suited to the conditions in the receiving country.

Figure 4.16 **WaterAid in Mali.**

Figure 4.15 **Map of Mali.**

THINK ABOUT IT

What aspects of the WaterAid programme could be described as sustainable?

Already significant improvements in the general health of the community have occurred.

The combination of safe water, sanitation and hygiene education maximises health benefits and promotes development. The combined benefits of safe water, sanitation and hygiene education can reduce the number of deaths caused by diarrhoeal diseases by an average of 65 per cent (WHO). A child dies every 17 seconds from diseases associated with lack of access to safe water and adequate sanitation.

In the longer term, communities are able to plan and build infrastructure, which enables them to cope better in times of hardship. In areas with WaterAid projects, life in times of drought is eased by the factors shown in Figure 4.18.

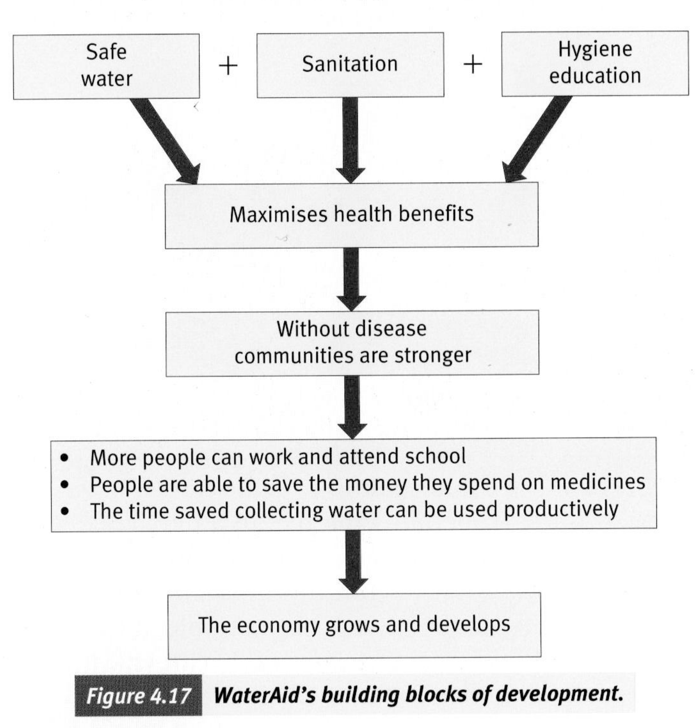

Figure 4.17 ***WaterAid's building blocks of development.***

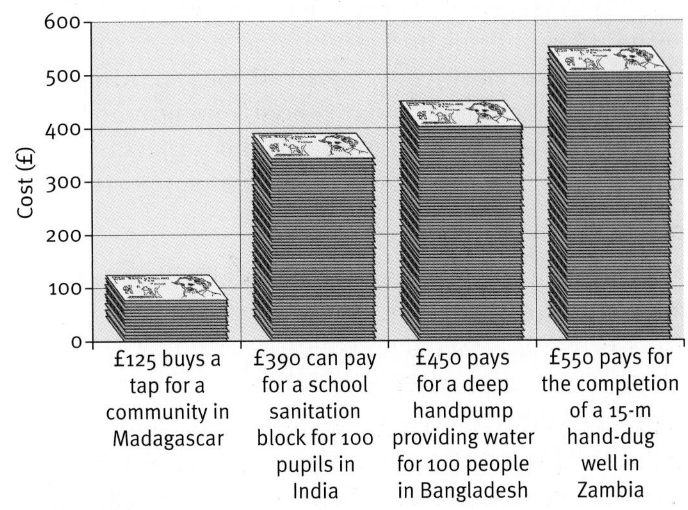

Figure 4.18 ***WaterAid: what donations can buy.*** **(Source: WaterAid)**

ACTIVITIES

1 a Write your own easily remembered definition of appropriate technology.
  b Why is it so important for sustainable development?
2 Describe the location of Mali.
3 a How is WaterAid helping in Mali?
  b How sustainable do you think the WaterAid project is?
4 Look at Figure 4.17.
  a What do you understand by the following terms:
   - Safe water
   - Sanitation
   - Hygiene education?

   How does each one help a community?
  b Why is it so important to combine these three factors to maximise the health benefits to a community?
  c Explain why healthier communities are more likely to be able to improve their own living standards.
  d For each component of sustainability, explain how providing clean water can make a difference.
5 Why do you think NGOs have been able to invest in sustainable projects much more than national governments?
6 Is short-term sustainable aid possible? Is it desirable?

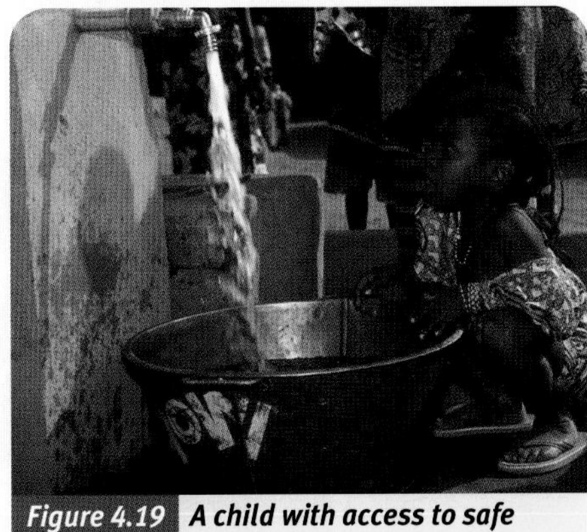

Figure 4.19 ***A child with access to safe water in Mali.***

RESEARCH LINK

Look at WaterAid's website and find the section 'Sustainable technology in action'. Produce a summary of no more than 150 words using this section.

PLENARY ACTIVITY

Why is water such an important issue in Mali?

A Decision Making Exercise – Goat Aid or Computer Aid?

Goat Aid and Computer Aid projects have been launched all over the world by a wide range of different charitable organisations. Much of the work is carried out in Africa. Explore the merits of supporting each project. Consider some of the criticisms aimed at both. Whilst each project has very different goals and aspirations, both receive significant support from the British public. After a fundraising activity in your school, enough money has been raised to support one of the projects. Which project would you recommend supporting?

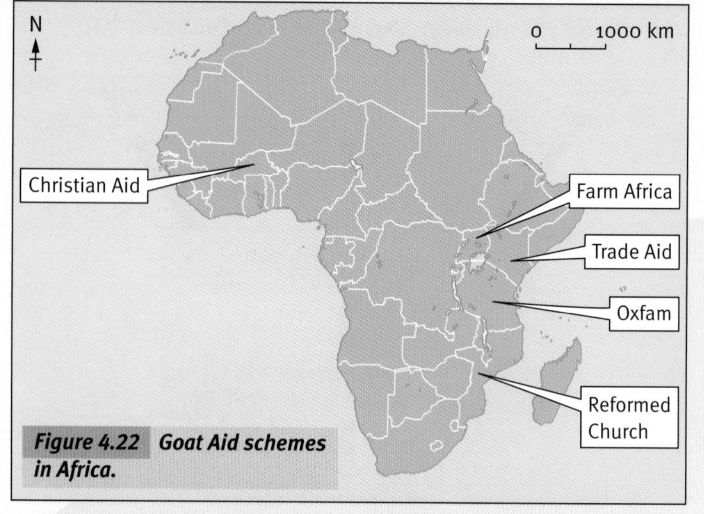

Figure 4.22 ***Goat Aid schemes in Africa.***

Goat Aid

The simple flow diagram in Figure 4.21 illustrates how people can use the gift of a small herd to increase their social and economic standing, whilst living in a challenging environment.

As a Christmas gift, giving a goat or sponsoring a goat for Africa has become very popular in recent years. What do such projects attempt to bring to the people of rural Africa (see map in Figure 4.22)? Whilst there are subtle differences between the projects, all of them seek to provide new opportunities to people who live in a harsh environment. Each charitable organisation provides examples of how rural communities can flourish after receiving their goats. This example is taken from an Oxfam project in Tanzania. You will find more details on the Oxfam website.

Figure 4.20 ***Noolarami Koolanga from Tanzania explains what her goats mean to her and her family.***

'With these goats I am now somebody. I won't miss food again. I'm sure that my small children will now have milk. It was very difficult before because we used to have to go to the bush and carry [and sell] firewood so that we could buy some milk in the shop. Maybe when they multiply, I can sell one to cater for the needs of my family. I'll sell a goat to buy beadwork, because the beads have a lot of profit. And then with the profits I will educate my children. I will also buy good food, and eventually I hope I can build a tin-roofed house, because I have seen others do that.'

(Source: Oxfam GB)

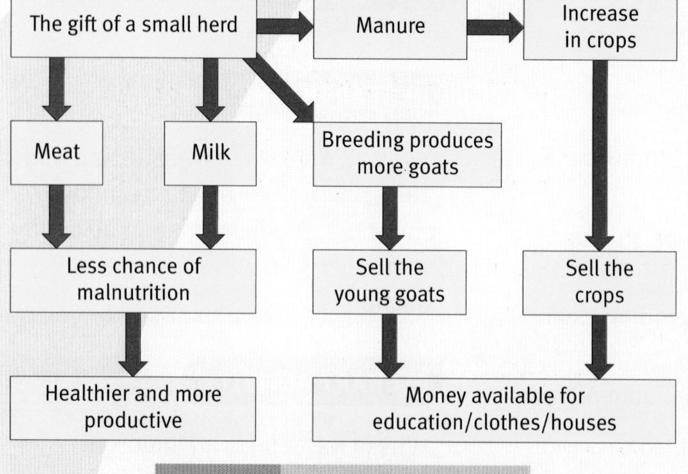

Figure 4.21 ***The benefits of Goat Aid.***

Goat Aid attracts some controversy

In recent years, projects like this have received some criticism. Some experts believe that:

- goats actually create more problems for local people in that they need grazing land, water and veterinary care, all of which are in short supply in sub-Saharan Africa
- whilst goats will survive on almost any vegetation, their feet and their constant grazing adds to the very real and increasing problem of desertification
- it is estimated that the number of hoofed animals has increased from 300 million in 1965 to over 650 million in 2005. In that time, the same experts point out that poverty has not decreased!

Computer Aid

Computer Aid is a fast-developing alternative to some of the more traditional aid programmes worldwide. The primary aim of all the charities involved in this initiative is to help close the 'digital divide' between the less developed world and the more developed world.

Computer Aid International (UK) is one of the leading charities involved in this work. It asks UK businesses and organisations to donate unwanted computers and peripherals to support education and health projects across the world, particularly in Africa. Figure 4.25 shows how it works.

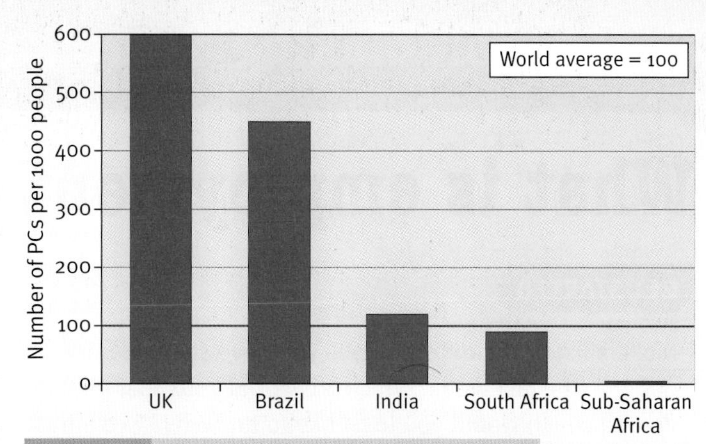

Figure 4.24 **Number of PCs in different populations.**

Companies donate unwanted IT equipment in the UK to Computer Aid.

↓

Computer Aid International checks and digitally cleans the equipment in the UK.

↓

The equipment is shipped to Africa and is distributed to partner organisations.

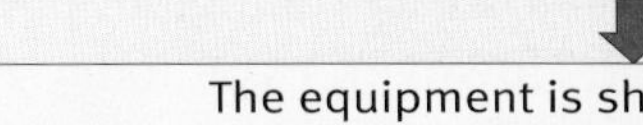

↓ ↓ ↓

Health Care
Remote medical centres in the Kenyan bush can now tap into the expertise of doctors in Nairobi (the capital) using PCs, digital cameras and scanners.

Using the internet, the same remote centre can access the World Health Organization's website for up to date information and treatment for the big killers in Kenya, i.e. Malaria, HIV/AIDS and TB.

Farmers
The Kenyan Met Office has computerised its forecast services. Farmers can now get crucial, accurate advice on when the rains are likely to fall in their area and which variety of crop will grow best.

Muthoka, a farmer from the remote west, has reported bumper harvests, providing enough to eat for his family and enough surplus to sell for profit.

Students
Our Lady Fatimah School in one of Nairobi's poorest slums has taken delivery of 25 PCs. Linda Otieno has developed her computer skills and now looks forward to finding a well paid professional job to support her extended family.

Information Technology has been recognised by the UN as a driving force for development. It will have long-term benefits.

Figure 4.23 ***How Computer Aid International (UK) works.***
(Source: Simon Birch, *Independent*, 2006)

Figure 4.25 ***Enjoying the benefits of Computer Aid.***

Computer Aid attracts some controversy

Not all Computer Aid projects receive the same positive reports. There are examples across Africa where software has been unreliable and incompatible. Often, there are insufficient staff to provide technical support when needed. Many projects fail to consider the running costs of the new equipment and users end up frustrated and disappointed! Some consider Africa to be the dumping ground for unwanted hardware.

YOUR TASK

Consider the evidence on these two pages and the additional information on the supporting CD. If your school raised £500 to donate to an African development charity, which of the two programmes would you recommend? In your report, you should identify the strengths and the weaknesses of each scheme before making your decision. You should comment on how far you think both projects are sustainable.

What is employment structure?

GET STARTED

Make a list of the different jobs you have seen people doing in the last week. Which job would you most like to do and which would you least like to do? Why? Explain someone else's choices using geographical terms and ideas.

Employment structure

There are hundreds of different jobs in the UK. All of these can be placed into four broad employment sectors:

- **the primary sector** collects raw materials from the land and the sea. Farming, forestry, mining, quarrying and fishing make up most of the jobs in this sector. Some primary products are sold directly to the consumer, but most go to secondary industries for processing
- **the secondary sector** manufactures primary materials into finished products. Activities in this sector include the production of processed food, furniture and motor vehicles. Secondary products are classed either as consumer goods (produced for sale to the public) or capital goods (produced for sale to other industries)
- **the tertiary sector** provides services to businesses and to people. Retail employees, drivers, architects, teachers and nurses are examples of jobs in this sector
- **the quaternary sector** uses high technology to provide information and expertise. Research and development is an important part of this sector. Jobs in this sector include aerospace engineers, research scientists, computer scientists and biotechnology workers. Quaternary industries have only recently been recognised as a separate group. Before then, jobs now classed as quaternary were placed in either the secondary or tertiary sectors depending on whether a tangible product was produced or not. However, even today much of the available information on employment does not consider the quaternary sector.

Figure 4.26 **Formula 1 engineers tuning a car with laptops.**

Table 4.3 Employment structure in the UK 1964–2005

		1964		1973		1979		1983		1990		2005
	(000s)	(% of total employment)	(000s)	(% of total employment)	(000s)	(% of total employment)	(000s)	(% of total employment)	(000s)	(% of total employment)	(000s)	(% of total employment)
Total primary	1201	5.1	773	3.4	692	3.0	672	3.0	476	2.1	297	1.1
Total secondary	10,978	46.9	9573	42.4	8911	38.5	7748	35.4	6093	26.6	4437	16.6
Total tertiary	11,178	47.8	12,320	54.4	13,556	58.5	13,465	61.4	16,351	71.3	21,916	82.2
Total employment	23,357		22,664		23,158		21,891		22,920		26,650	

(Source: Office for National Statistics, Crown copyright, 2006)

The **product chain** can be used to illustrate the four sectors of employment. The food industry provides a good example (Figure 4.27). Some companies are involved in all four stages of the food product chain.

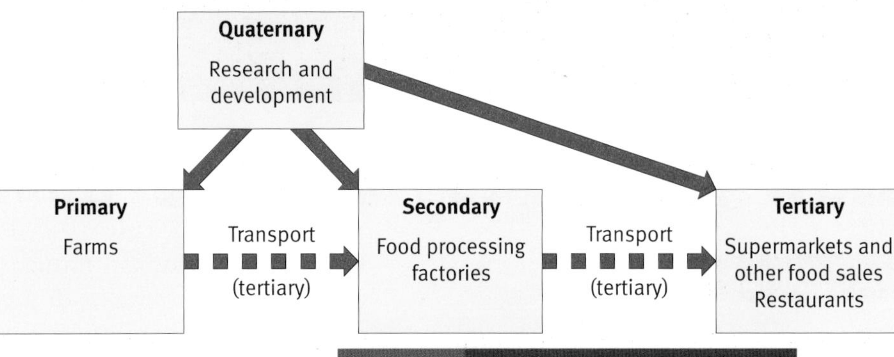

Figure 4.27 The food industry's product chain

The formal and informal sectors of employment

Jobs in the *formal sector* have contracts and pay taxes. Such jobs generally provide better pay and much greater security than jobs in the *informal sector*. Fringe benefits such as holiday and sick pay may also be available. Formal sector employment includes health and education service workers, government workers, and people working in established manufacturing and retail companies.

In contrast, the informal sector is that part of the economy operating outside official recognition. Employment is generally low paid and often temporary and/or part-time in nature. Such employment is outside the tax system and job security will be poor, with an absence of fringe benefits. About three-quarters of those working in the informal sector are employed in services. Typical jobs are street food vendors, messengers, repair workers and market traders. Informal manufacturing tends to include both the workshop sector, making for example cheap furniture, and the traditional craft sector. Many of these goods are sold in bazaars and street markets.

The government estimates that about 5 per cent of all employment in the UK is in the informal sector. This usually occurs when people insist on being paid in cash and do not declare this to the Inland Revenue. In LEDCs the informal sector may account for up to 40 per cent of the total economy.

KEY TERMS

Product chain – the full sequence of activities needed to turn raw materials into a finished product.

DISCUSSION POINT

Class discussion – are there any examples of the informal sector operating in your local area?

Part-time work

More people are working part time than ever before. Some choose part-time work because they like working this way or it fits their current lifestyle, but for many people it is the only work available. The greatest proportion of part-time jobs is in the tertiary sector.

THINK ABOUT IT

Why couldn't we find any examples of informal industry in our research for this book?

ACTIVITIES

1 List two jobs in each of the four sectors of employment.
2 Explain the differences between the formal and informal sectors of employment.
3 Why is informal employment considered a bad thing by governments?
4 a Suggest why part-time work has become more important in the UK in recent decades.
  b Do some research to find out which part-time jobs are available in your area.
5 Draw up a table to show the advantages and disadvantages of the informal sector of employment to both individuals and a country as a whole.
6 Why do migrants tend to start in informal industries?

PLENARY ACTIVITY

How many teachers in your school work part time? Why do they work part time rather than full time? (Some may choose not to answer this question: how would you record this in your survey?)

How do employment structures vary?

GET STARTED

Which important industries in the UK would employ very few people in a very poor country?

LEDCs

People in the poorest countries of the world are heavily dependent on the primary sector for employment. Most of these people will work in agriculture and many will be **subsistence farmers**. In some areas where the population is very high and the amount of land is very limited, there will not be enough work available for everyone to work a full week. The work available is often shared and people are said to be underemployed.

In some regions of LEDCs, primary industry may dominate the economy. Work in mining in LEDCs is often better paid than jobs elsewhere in the sector, but the working conditions are often very harsh. In poor countries, higher paid jobs in the secondary, tertiary and quaternary sectors are usually very few in number. The tertiary jobs that are available are often in the public sector. Public sector jobs such as teaching, nursing and refuse collection are paid for by the government.

Many of the world's poorest countries are primary-product-dependent; in other words, they rely on one or a small number of primary products for most of their export earnings. This makes them very vulnerable to changes in world markets. For example, if a country relies on coffee for most of its export earnings and the price of coffee falls substantially, it will have far less money to pay for the imports it needs and less to invest in health, education and other important aspects of the quality of life. Countries with diverse economies are much more resilient to price changes.

NICs

In NICs, employment in manufacturing has increased rapidly in recent decades. NICs have reached the stage of development whereby they attract **foreign direct investment** from **multinational companies (MNCs)** in both the manufacturing and service sectors. The business environment in NICs is such that they also develop their own domestic companies. Both processes create employment in manufacturing and services.

Figure 4.28 ***The working conditions for a teacher in Kabul.***

KEY TERMS

Foreign direct investment – overseas investments in physical capital by MNCs.

Multinational companies (MNCs) – firms that produce goods in more than one country.

Subsistence farming – the most basic form of agriculture where the produce is consumed entirely or mainly by the family who work the land or tend the livestock. If a small surplus is produced it is sold or traded.

THINK ABOUT IT

Explain what sorts of conditions allow the development of the quaternary sector.

The increasing wealth of NICs allows for greater investment in agriculture. This includes mechanisation, which results in the falling demand for labour. So, as employment in the secondary and tertiary sectors rises, employment in the primary sector falls. Eventually, NICs may become so advanced that the quaternary sector begins to develop. Examples of NICs where this has happened are South Korea, Singapore and Taiwan.

MEDCs

MEDCs are often referred to as 'post-industrial societies' because far fewer people are now employed in manufacturing industries than in the past. Most people work in the tertiary sector, with an increasing number in the quaternary sector. Jobs in manufacturing industries have fallen for two reasons:

- many manufacturing industries have moved to take advantage of lower costs in NICs. Cheaper labour is often the main attraction, but many other costs are also lower
- investment in robotics and other advanced technology has replaced much human labour in many manufacturing industries that remain in MEDCs.

Table 4.4 Employment structure of an LEDC, NIC and MEDC

Country	Primary (%)	Secondary (%)	Tertiary (%)
UK (MEDC)	1.1	16.6	82.2
China (NIC)	43.0	25.0	32.0
Bangladesh (LEDC)	63.0	11.0	26.0

RESEARCH LINK

Study the employment pages of your local newspaper's website. How many jobs are available in each of the four sectors of employment?

A graphical method often used to compare the employment structure of a large number of countries is the triangular graph. One side (axis) of the triangle is used to show the data for each of the primary, secondary and tertiary sectors. Each axis is scaled from 0 to 100 per cent. The indicators on the graph show how the data for the UK can be read.

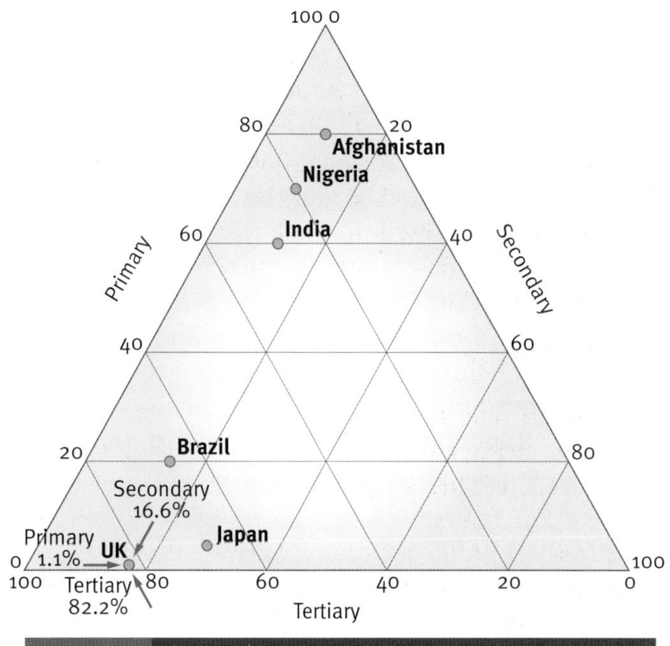

Figure 4.29 ***Comparison of employment sectors in six countries.***

ACTIVITIES

1. Define:
   a. subsistence agriculture
   b. primary product dependence.

   What is considered bad about each concept for a country's prospects of development?
2. Why do so few people in LEDCs work in the:
   a. secondary
   b. tertiary
   c. quaternary sectors?
3. Why does the primary sector dominate employment in the poorest countries of the world?
4. Explain the changes in employment structure that have occurred in NICs.
5. On a copy of the triangular graph, plot the positions of China and Bangladesh using the data in Table 4.4. Can you find a country for each area of the triangle?

THINK ABOUT IT

Some geographers propose another label of development for the poorer countries: LLEDCs. What sort of employment structure would they have?

PLENARY ACTIVITY

How have employment structures changed in the UK? Why are some economists concerned that there are so few manufacturing jobs left in the UK?

How have employment structures changed over time and how may they change in the future?

GET STARTED

What jobs in the UK did not exist 50 years ago?

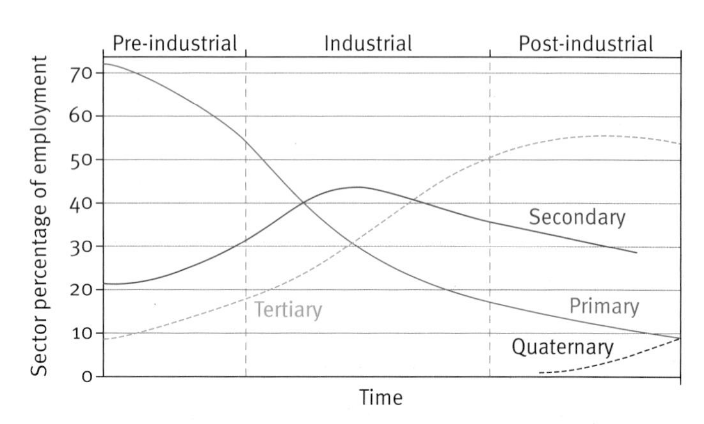

Figure 4.31 ***The sector model.***

As an economy advances, the proportion of people employed in each sector changes (Figure 4.32). Countries such as the UK and the USA are 'post-industrial societies' where most people work in the tertiary sector. Yet, in 1900, 40 per cent of employment in the USA was in the primary sector. However, the mechanisation of farming, mining, forestry and fishing drastically reduced the need for labour in these industries. As these jobs disappeared, people were forced to move to urban areas where jobs in the secondary and tertiary sectors were expanding. Less than 4 per cent of employment in the USA is now in the primary sector.

Human labour has been replaced in manufacturing too. In more and more factories, robots and other advanced machinery handle assembly-line jobs that once employed large numbers of people. Also, many manufacturing jobs once performed in MEDCs are now done in NICs. In 1950, the same number of Americans was employed in the secondary and tertiary sectors. By 1980, two-thirds were working in services. Today, 78 per cent of Americans work in the tertiary sector.

The tertiary sector is also changing. In banking, insurance and many other types of business, computer networks have reduced the number of people required. But elsewhere service employment is rising, such as in health, education and tourism.

Figure 4.30 ***Labour-intensive farming in the early twentieth century.***

Outsourcing

A major change in employment has been the increase in **outsourcing**. Companies do this to save money. Work can be outsourced to companies in the same country or it can go abroad where labour and other costs are much lower. For example, many British and American companies have outsourced their call centres to India.

KEY TERMS

Outsourcing – where a company contracts out some of its work to another company. This usually happens because a company can save money. However, it can also happen if a company lacks certain skills.

Figure 4.32 An Indian call centre outsourced by a British company.

RESEARCH LINK

Many British companies outsource tasks to India. Bangalore is one of the major outsourcing locations. Investigate the advantages and disadvantages of outsourcing to Bangalore.

Employment structure: the future

The nature of work in the UK has changed markedly over the last 50 years. It will continue to change in the future as the process of globalisation continues. The key questions are:

- Will even fewer people work in the primary sector and which tasks will be performed by those that remain?
- How much further will manufacturing employment fall and which products will Britain still produce?
- Which service sector jobs will decline and which will increase in importance?
- Which totally new services will begin to provide employment in the future?
- How many people will be unemployed at various stages in the future and what status and standard of living will they have? Should they be supported?
- What changes will occur in (a) the working week, (b) paid holidays, (c) retirement age, (d) pensions, (e) the school leaving age, (f) working conditions, (g) the location of employment?
- How much control can the British government have over these issues?

Employment is one of the most important factors in most people's lives. It is the income from employment that influences so many aspects of an individual's quality of life. The UK of 2025 is likely to be very different from the present state of the country.

With further advances in ICT there will be a greater opportunity for more people to work from home. Telemobility will allow many people to perform the same tasks from home that they now do in their office. However, a decade ago it was thought that higher technology home-working would be more important now than it has actually turned out to be. It seems that the physical clustering of people in organisations has proved more difficult to break down than many commentators thought. People like to talk!

It seems likely that international commuting and employment migration (geographical mobility) within the EU will increase as economic and psychological barriers to movement recede. The degree of mobility should also increase as the pace of change quickens.

ACTIVITIES

1. Describe and explain the changes shown in Figure 4.31.
2. Briefly describe how employment in the USA has changed since 1900.
3. What is 'outsourcing'?
4. Why does outsourcing occur on such a large scale?
5. Explain the changes that are likely to occur in employment in the UK in the future.
6. For three of the questions about the future of the UK's employment, draw a flow map to show what might happen.

PLENARY ACTIVITY

Ask your headteacher how the school is planning to tackle the challenges of the next 10 years: what skills do the school's senior management think GCSE students will need for the jobs you will all be doing in 10 years' time?

What determines the location of different economic activities? Primary industry

GET STARTED

What are the new primary industries that are appearing in the UK?

The location of primary industry is influenced by both physical and human factors. The physical environment sets out the possibilities. For example, iron ore mining can only occur at a location if the mineral is physically present. However, the presence of a mineral or fuel does not guarantee it will be mined. A mining company will carefully work out all the costs involved to ensure that it will make a profit if it decides to exploit a mineral resource.

Figure 4.33 **Wheat cultivation in the Canadian prairies.**

Example: Wheat farming in the Canadian Prairies

Extensive farming

Three-quarters of Canada's farmland lies in the Prairie Provinces of Alberta, Saskatchewan and Manitoba. It is one of the world's largest areas of **extensive** commercial cereal farming. Here, farms averaging 300 hectares in size stretch for almost 1500 km from east to west.

The development of wheat cultivation

The hot, dry summers of the Prairies favour wheat production. Wheat was first grown here in the early 1600s. Since then, many changes have occurred:

- the family farm has grown from a few hand-worked hectares to a large, highly mechanised unit

THINK ABOUT IT

What are the physical and human factors that influence the type of farming that takes place in a region?

KEY TERMS

Extensive farming – when one type of farming and large farms dominate a very large area. Inputs per hectare are low compared with intensive farming, such as market gardening.

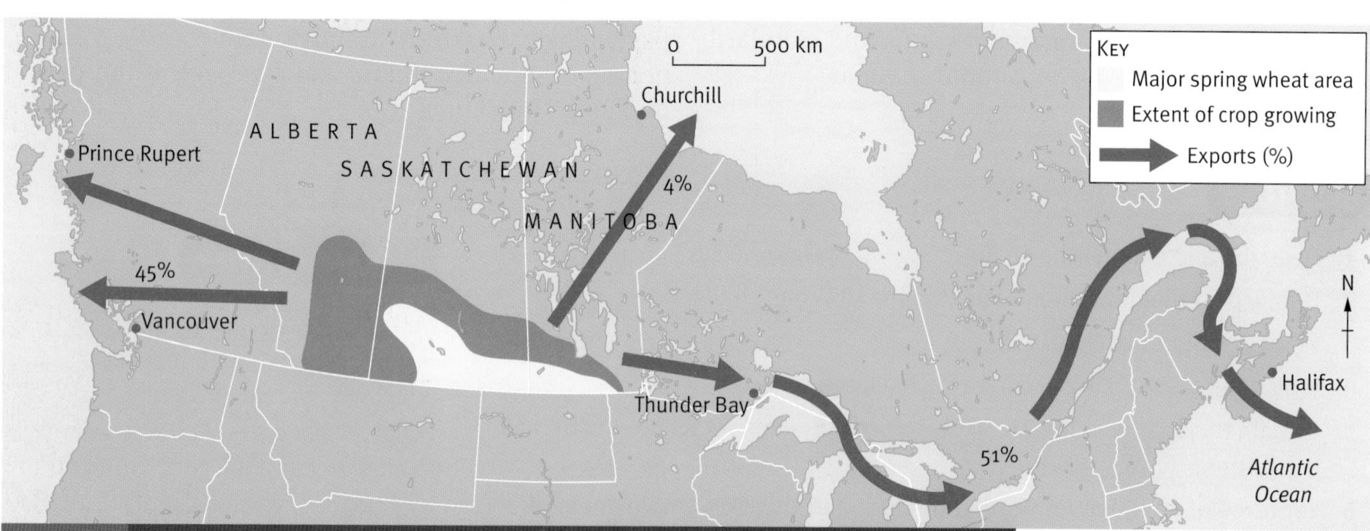

Figure 4.34 **Grain export routes in the Prairies. (Source: *North America in Focus*, Hodder Education, 1990)**

- new strains of wheat can withstand colder temperatures and a shorter growing season. This is 20 days less than in the early 19th century
- today's wheat is resistant to many diseases that once destroyed huge areas of crops
- inputs of fertiliser, herbicides and pesticides have steadily increased
- ploughing is now little used, as it increases soil erosion by making the topsoil more likely to drift. Instead, cultivators cut weeds below the surface without turning the soil
- a modern combine harvester allows one person to harvest more than 2000 bushels (a measure equivalent to 8 gallons) a day.

Transport to markets

Most of the grain produced in the Prairies is sold in other parts of Canada or abroad. The transport of grains has a set routine:

- grain is taken by road to a local elevator. This is a wooden structure that can hold around 70,000 bushels. There are 5000 elevators in the Prairies
- the amount and quality of grain required at the shipping terminals is decided six weeks before delivery.
- trains pick up the grain and deliver to port. Here it is either loaded on ships or placed in the huge terminal storage elevators.

Fact file

Top six wheat-producing countries, 2006–2007 (million metric tonnes)

1. EU (124.7)
2. China (104.0)
3. India (69.3)
4. USA (49.3)
5. Russia (44.9)
6. Canada (25.2)

(Source: Washington Grain Alliance)

Figure 4.35 **Grain terminal at Thunder Bay.**

Physical hazards and sustainability

At times, severe droughts occur in the Prairies when farmers lose part or all of their crops. Dry soil crumbles into fine particles that are easily picked up by the wind. In fact, the main long-term problem is soil erosion. Prairie farmers try to protect the soil by:

- **crop-fallow rotation**: fields are left fallow (uncultivated) every other year to replenish soil moisture. The crop stubble is left to further reduce the effect of the wind
- **ripping**: a Caterpillar tractor is used in winter to cut the frozen soil and mould it into 30-cm chunks. These chunks break the effect of the wind close to the soil surface.
- **strip farming**: the farmer leaves narrow strips of fallow land, perpendicular to the prevailing wind, between seeded fields.

Export competition and global stocks

Export sales of wheat are important to Prairie farmers. Without exports farm incomes would fall. There is often intense competition between sellers on the world food market.

THINK ABOUT IT

After decades of abuse of the land all the big changes in prairie farming are towards sustainability. Why do you think this is?

ACTIVITIES

1. What is extensive commercial cereal farming?
2. Describe the location of the Prairies.
3. Discuss three ways in which wheat farming has changed over time.
4. Look at Figure 4.34. Describe the locations of the main shipping terminals. Suggest reasons for these different locations.
5. Draw flow maps to show how the new farming methods are sustainable.
6. Give examples of how tertiary and quaternary employment are linked to wheat.

PLENARY ACTIVITY

A major issue in world farming is the subsidies that rich countries such as Canada, the USA and the EU give to their farmers. Why do you think this is such a big issue?

What determines the location of different economic activities? Secondary industry

GET STARTED

Working in pairs, list the secondary industries that are nearest to your school.

Decision making: choosing the best location

Every day decisions are made about where to locate factories. The factors affecting industrial location vary from industry to industry and their relative importance may change over time. Ideally, a company wants to choose the **optimum location** for its factory. However, this does not always happen because companies may not have all the information needed to make such a choice, or they may lack the skills to analyse all the information available correctly.

KEY TERMS

Optimum location – the location that best satisfies the objectives of the company. For most, this will be the maximum profit location.

Example: Slovakia – the changing location of EU car manufacturing

Car manufacturing is one of the world's largest industries. Within the EU, investment in car manufacturing has shifted from Western to Eastern Europe in recent years, as countries like Slovakia have joined the EU. This is because eastern EU countries such as Slovakia can manufacture cars at a lower cost than western EU countries.

The location factors that have attracted the car industry to Slovakia are:

- relatively low labour costs
- low company taxation rates
- a highly skilled workforce, particularly in areas that were once important for heavy industry
- a strong work ethic, resulting in high levels of productivity
- low transport costs because of proximity to Western European markets
- very low political risk because of the stable nature of the country

Table 4.5 The factors that influence industrial location

Physical factors	Human factors
Site: the availability and cost of land. Large factories in particular will need flat, well-drained land on solid bedrock. An adjacent water supply may be essential. *Raw materials:* industries requiring heavy and bulky raw materials that are expensive to transport will generally locate as close to these raw materials as possible. *Energy:* at times in the past, industry needed to be located near fast-flowing rivers or coal mines. Today, electricity can be transmitted to most locations. However, energy-hungry industries, such as metal smelting, may be drawn to countries with relatively cheap hydro-electricity. *Natural routeways and harbours:* essential factors in the past, but still important today as many modern roads and railways still follow natural routeways. Natural harbours provide good locations for ports and the industrial complexes often found at ports. *Climate:* some industries such as aerospace and film benefit directly from a sunny climate. Indirect benefits such as lower heating bills and a more favourable quality of life may also be apparent.	*Capital (money):* business people, banks and governments are more likely to invest money in some areas than others. *Labour:* increasingly it is the quality and cost of labour rather than the quantity that are the key factors here. The reputation, turnover and mobility of labour can also be important. *Transport and communications:* transport costs are lower in real terms than ever before but remain important for heavy, bulky items. Accessibility to airports, ports, motorways and key railway terminals may be crucial factors for some industries. *Markets:* the location and size of markets is a major influence for some industries. *Government influence:* government policies and decisions can have a big direct and indirect impact. Governments can encourage industries to locate in certain areas and deny them planning permission in others. *Quality of life:* highly skilled personnel who have a choice about where they work will favour areas where the quality of life is high.

Figure 4.36 Slovakia's location in Europe.

- attractive government incentives due to competition between Slovakia and other potential receiving countries
- good infrastructure in and around Bratislava (the capital) and other selected locations
- an expanding regional market for cars as per capita incomes increase.

Volkswagen expands

Prior to EU membership, Slovakia already boasted a Volkswagen (VW) plant with an output of 250,000 cars a year. The Bratislava plant is one of the top three VW factories in the world, producing the Polo, the Touareg and the SEAT Ibiza.

In addition to its car manufacturing plant in Bratislava which was founded in 1991, VW also has a plant manufacturing components in Martin. The latter was opened in 2000. Between the two plants, VW employs 8700 workers. A number of companies supplying parts to VW have also opened up in Slovakia.

Figure 4.37 Car manufacturing in Slovakia.

Other recent investment

In 2006, Hyundai opened a major car factory in Slovakia. The location of the factory is near Žilina, 200 km north-east of Bratislava. As with other large car plants, it is attractive for some of its main suppliers to locate nearby. With its seven suppliers the total investment is estimated to be US $1.4 billion.

In 2006, Peugeot opened a large new car plant in Trnava, 50 km from Bratislava. When it reaches maximum production this state-of-the-art plant will export 300,000 cars a year to Western Europe and to other parts of the world.

ACTIVITIES

1. List two physical and two human factors that can affect industrial location.
2. Describe the geographical location of Slovakia.
3. Explain the reasons for such a high level of investment in car manufacturing in Slovakia by foreign multinational companies (MNCs).
4. With the help of a sketch map, describe the locations of VW, Hyundai and Peugeot in Slovakia.
5. Look again at the list of qualities that Slovakia has that attracts car factories. Draw up a table that sorts them into qualities the UK does and does not have
6. Which of the qualities that the UK lacks could the government help with? Do you think it would help?

PLENARY ACTIVITY

Why have Slovakia and other eastern EU countries been so keen to attract foreign car manufacturers?

What determines the location of different economic activities? Tertiary industry

GET STARTED

Write an A-to-Z acrostic for service industries.

Example: business and financial services in the City of London

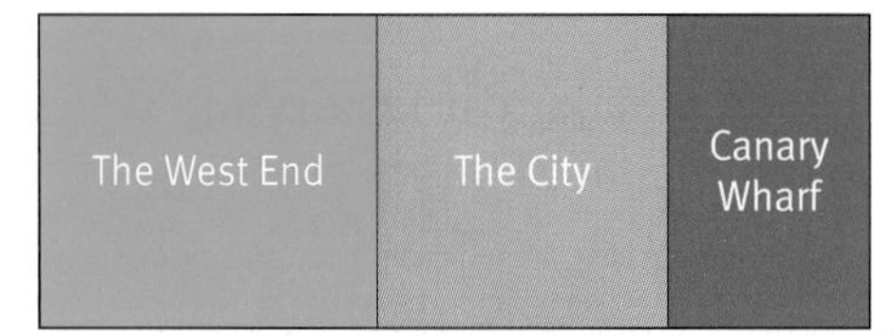

Figure 4.38 London's CBD.

Figure 4.39 The Bank of England.

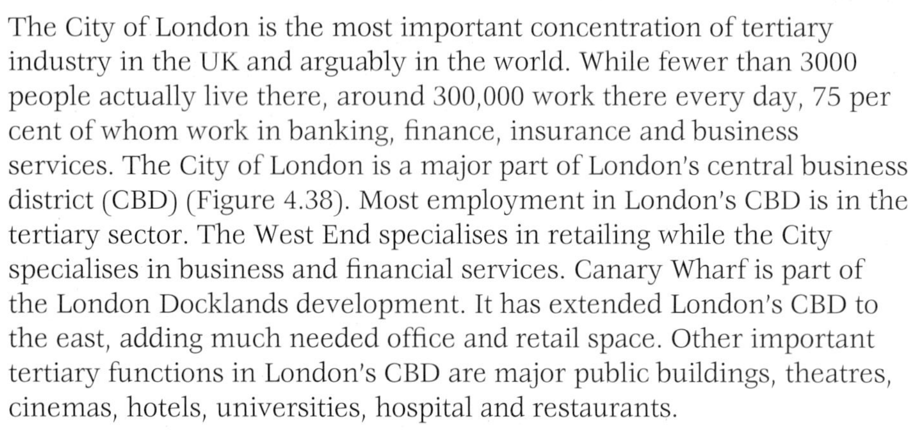

The City of London is the most important concentration of tertiary industry in the UK and arguably in the world. While fewer than 3000 people actually live there, around 300,000 work there every day, 75 per cent of whom work in banking, finance, insurance and business services. The City of London is a major part of London's central business district (CBD) (Figure 4.38). Most employment in London's CBD is in the tertiary sector. The West End specialises in retailing while the City specialises in business and financial services. Canary Wharf is part of the London Docklands development. It has extended London's CBD to the east, adding much needed office and retail space. Other important tertiary functions in London's CBD are major public buildings, theatres, cinemas, hotels, universities, hospital and restaurants.

Among the important buildings in the City are the Bank of England, the London Stock Exchange and Lloyd's of London (insurance).

The City of London is the major reason why London is classed as a '**global city**'. It is one of the big three financial centres in the world, along with New York and Tokyo.

Continuing investment is vital if London is to retain its position. The Corporation of London's Economic Development Unit is responsible for maintaining London's global position. The Unit's objectives are:

- to ensure that all the leading companies in global finance and commerce are in the City and that they have the professional support to function efficiently
- to enhance the quality of the working and living environment within the City

Fact file

City of London

- US $1359 billion foreign exchange turnover each day (34 per cent of global share).
- 53 per cent of the global foreign equity market (stocks and shares).
- World's leading market for international insurance.
- US $1686 billion pension fund assets under management
- 75 per cent of the world's largest 500 companies located in London, most in the City.
- 254 foreign banks in London, most in the City.
- 692 foreign companies listed on the London Stock Exchange.

Figure 4.40 ***Lloyd's of London.***

Figure 4.41 ***The London Stock Exchange.***

KEY TERMS

Back offices – offices of a company handling high-volume communications by telephone, electronic transaction or letter. The location of such low-to-medium level functions can be relatively flexible and they have been increasingly decentralised to locations where space, labour and other costs are relatively low.

Global city – major world city supplying financial, business and other significant services to all parts of the world. The world's major stock markets and the headquarters of large MNCs are located in global cities.

- to ensure an efficient infrastructure and a high-quality workforce by working closely with the property and training sectors
- to market the attributes of the City and of London as a whole on a worldwide basis
- to achieve a stable property market by using its influence as both planning authority and landowner.

The decentralisation of 'back office' functions

Over the last 20 years, routine '**back office**' functions have been moved away from core locations such as the City to less expensive sites. This has happened as companies have sought to reduce costs in order to remain competitive. Back-office functions process large volumes of paper, electronic transactions and telephone enquiries. They include international call centres and customer services, such as direct banking and computer support. These routine functions have relocated to:

- elsewhere in large urban areas of the UK (such as Birmingham, Newcastle)
- more distant locations in peripheral regions (such as Northern Ireland, Scotland)
- developing countries with pools of labour able to handle such tasks (such as India).

Although relocation can reduce costs in a number of ways the main savings are in terms of labour and office space costs. Labour accounts for about 70 per cent of total costs in back-office functions and thus considerable savings can be made by moving from London to lower-wage regions within the UK.

Some companies have really sought to slash labour costs by moving back-office functions to the developing world. For example, India and Russia have become major centres for subcontracting computer programming.

ACTIVITIES

1 Draw a bubble map to show the advantages of the location of the City of London.

2 Which types of tertiary employment are concentrated in the City of London?

3 Why is London considered to be a 'global city'?

4 a Explain why many 'back office' functions have been removed from the City of London.

  b Where have such functions relocated?

5 Look at the objectives of the Economic Development Unit. Why are they so important to achieve? How would you go about doing these things?

PLENARY ACTIVITY

Some companies that moved back-office functions overseas are now moving them back to UK. Why do you think this is happening?

What determines the location of different economic activities? Quaternary industry

GET STARTED

Where are the nearest quaternary activities to your school? What are these activities?

Example: quaternary industry in Ottawa, Canada

High-technology industry is the fastest-growing manufacturing industry in the world. It all began in Silicon Valley (the Santa Clara valley), south of San Francisco in the 1960s. Since then it has spread across the world. Virtually all MEDCs and NICs have at least one **high-technology cluster**.

'High tech' companies use or make silicon chips, computers, software, robots, aerospace components and other very technically advanced products. These companies put a great deal of money into scientific research. Their aim is to develop newer, even more advanced products.

'Silicon Valley North': the Ottawa Region of Canada

High technology is the fastest-growing industry in Canada today. Ottawa (Figure 4.42) has the greatest number of electronics and computer companies in Canada. The area is known as Silicon Valley North, after the world-renowned Silicon Valley (South) in California. The other main high-technology clusters in Canada are in Toronto, Vancouver and Montreal. Ottawa is a popular location for high-tech industry because:

- it is the capital city. Decision-makers and experts of all kinds work there

DISCUSSION POINT

How do high-tech companies benefit from clustering together?

KEY TERMS

High-technology cluster – where high-tech companies group together in a region because their location factors are similar and such companies benefit from being in close proximity.

RESEARCH LINK

Where are the main high-tech regions in the UK?

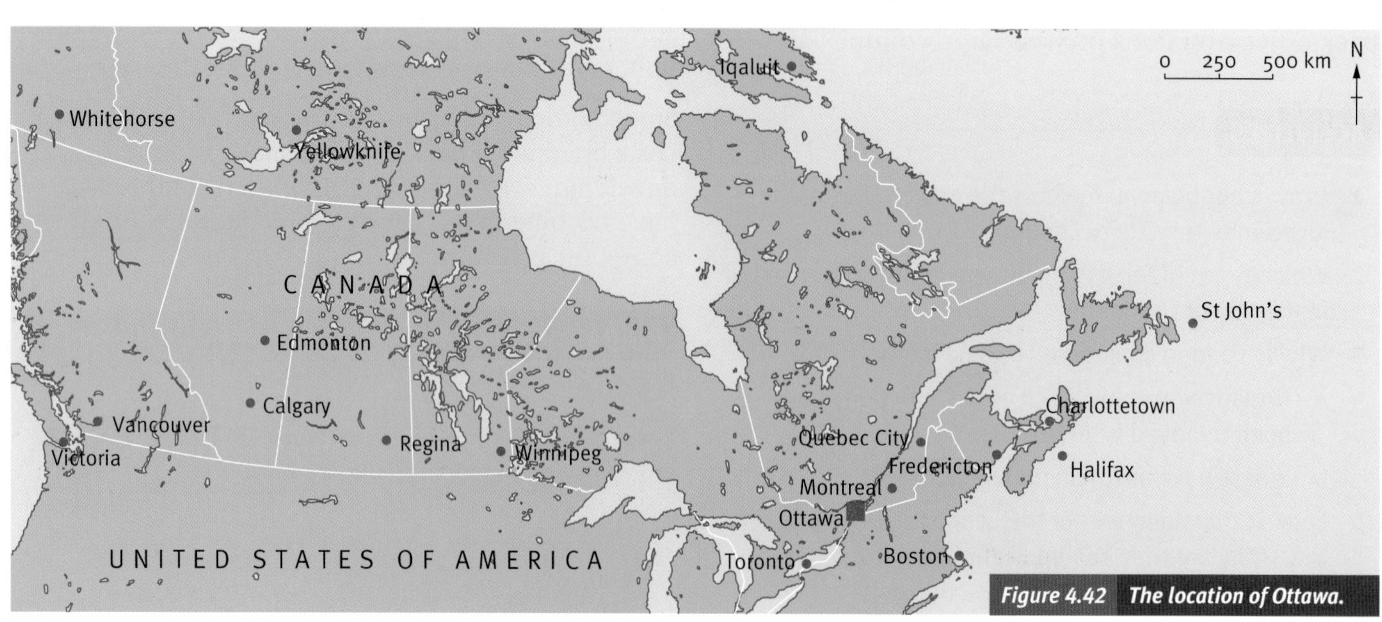

Figure 4.42 The location of Ottawa.

Figure 4.43 A high-tech office in Ottawa.

- the region has benefited from the growth of public scientific research laboratories such as the Atomic Energy of Canada Laboratory
- Ottawa's two universities are famous for science and engineering. There are strong formal and informal links between high-tech firms and the universities
- many high-tech companies began with the help of government grants and many now rely on government contracts for work
- high-tech firms benefit from 'agglomeration economies' (being close together). They 'swap' ideas and staff
- the region has had good access to investment money, which has been vital in providing the start-up costs of many new companies
- it is close to major Canadian and US markets where many of the products are sold
- Ottawa is a very pleasant place to live. The surrounding physical environment is attractive and the city has a high quality of life. This helps firms to attract highly skilled workers
- the city's economic development corporation advertises the advantages of Ottawa internationally.

Figure 4.44 Many high-tech offices have a modern and creative environment.

The high-tech companies in the Ottawa region include: 3M, Cisco Systems, Dell, Hewlett-Packard, IBM, Intel, LogicVision, Mitel and Nortel. A 2007 survey recorded 1841 high-tech companies in the Ottawa region, employing over 78,000 people (Figure 4.45). There is a good spread of large, medium-sized and small high-tech companies in the region.

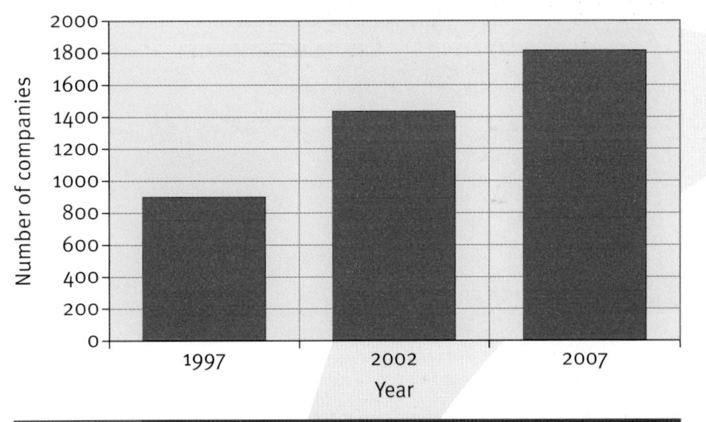

Figure 4.45 The growth of high-tech industries in the Ottawa region 1997–2007. (Source: OCRI)

ACTIVITIES

1. a What is high-tech industry?
   b What is a high-tech cluster?
2. Explain three reasons why high-tech industry has clustered in the Ottawa area.
3. Using an atlas and Figure 4.42, draw an annotated sketch map to show the location of Ottawa
4. Why do governments encourage high-tech clusters in their countries?
5. Explain the reasons for the development of Ottawa's high-tech cluster.
6. Technology clusters have disadvantages too. What do you think they would be?

PLENARY ACTIVITY

Could your local area become a technology cluster? Why? If not, what would your local council need to do for it to happen?

GET STARTED

Make a list of the main cities in south-east Brazil.

South-east Brazil (Figure 4.46) is the **economic core region** of Brazil. Its primary, secondary, tertiary and quaternary industries generate large amounts of money for Brazil. Its economic well-being and quality of life are higher than almost all other parts of the country. The area outside the economic core is known as the **periphery** (Figure 4.47).

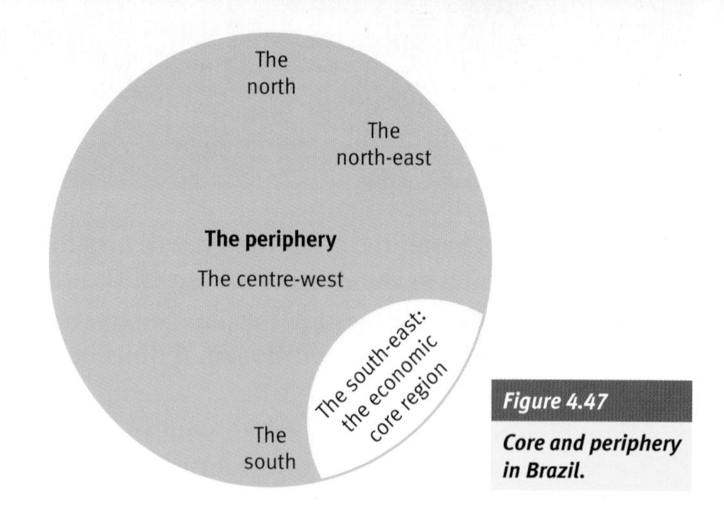

Figure 4.47 ***Core and periphery in Brazil.***

Primary industries

- The warm temperature, adequate rainfall and rich soils (weathered from lava) provide many opportunities for farming. The region is important for coffee, beef, rice, cacao, sugar cane and fruit.

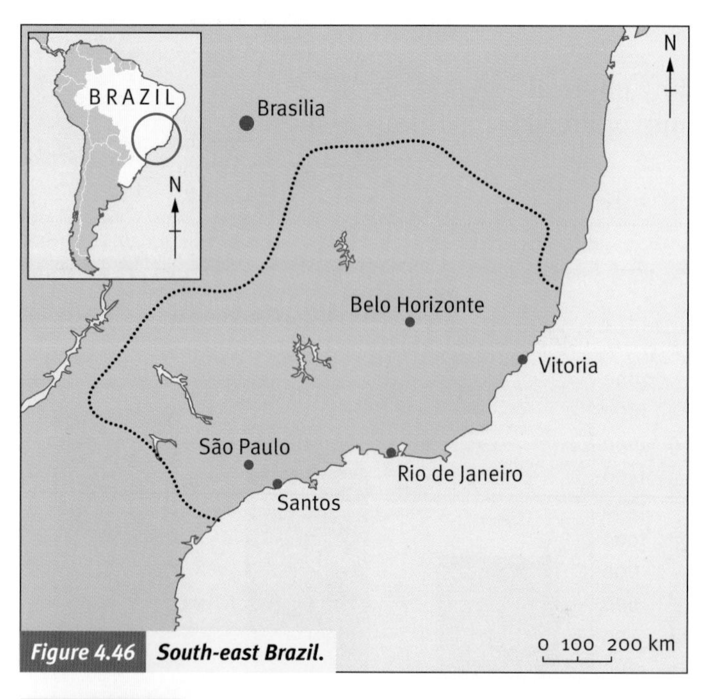

Figure 4.46 ***South-east Brazil.***

- Large deposits of gold, iron ore, manganese and bauxite have made mining a significant industry.
- The region is energy-rich, with large deposits of oil and gas offshore. Hydro-electric power is generated from large rivers flowing over steep slopes.
- The temperate rainforest provides the raw material for forestry.
- Fishing is important for many of the coastal settlements.

Table 4.6 The population of Brazil's five regions (Source: Census 2000, US Census Bureau)

Region	Population (millions)
South-east	72.4
North-east	47.7
South	25.1
North	12.9
Centre-west	11.6

KEY TERMS

Cumulative causation – the process whereby a significant increase in economic growth can lead to even more growth as more money circulates in the economy.

Economic core region – the most highly developed region in a country with advanced systems of infrastructure and high levels of investment, resulting in high average income.

Periphery – the parts of a country outside the economic core region. The level of economic development in the periphery is significantly below that of the core.

DISCUSSION POINT

Suggest why the success of the south-east's primary industries was a big factor in the initial development of secondary industry in the region.

Figure 4.48 *Offshore oilfield, south-east Brazil.*

Secondary industries

The south-east is the centre of both foreign and domestic investment in manufacturing industry. In the 1950s and 1960s the government wanted Brazil to become an NIC. Because the south-east had the best potential of all Brazil's five regions, investment was concentrated here. The region is the focus of the country's road and rail networks. It contains the main airports and seaports. It also has a significant pipeline network for oil and gas. More MNCs are located in the south-east than the rest of Brazil. With the highest population density in Brazil, the labour supply is plentiful. The region also has the highest educational and skill levels in the country.

The car industry is a major activity in the region. Most of the world's large car makers are here, including Ford, GM, Toyota, VW and Fiat. Other manufacturing industries include food processing, textiles, furniture, clothing, printing, brewing and shoemaking. The raw materials located in the region and the large market have provided favourable conditions for many of these industries. However, cheaper imports of shoes, clothes and textiles from Asia have led to a number of companies in the region closing.

Tertiary industries

São Paulo is by far the largest financial centre in South America. The headquarters of most Brazilian banks are in São Paulo. Most major foreign banks are also located there. This is not surprising, as Brazil dominates the economy of South America and São Paulo is the largest city in South America.

ACTIVITIES

1. What is an economic core region?
2. List two primary and two secondary industries in south-east Brazil.
3. Why is the south-east the focus of tertiary and quaternary industry in Brazil?
4. Explain the process of cumulative causation.

Quaternary industries

The south-east is the centre of research and development in both the public and private sectors. Just 83 km from São Paulo is São José dos Campos, where the Aerospace Technical Centre is located. It conducts teaching, research and development in aviation and outer space studies. Many people would be surprised to know that aircraft and aircraft parts make up Brazil's largest export category.

Figure 4.49 *Aircraft manufacture in south-east Brazil.*

The process of cumulative causation

The success of the first large wave of investment by foreign MNCs in the South-east encouraged other MNCs to follow suit. For the last 50 years the south-east has experienced an upward cycle of growth. This is known as the process of **cumulative causation** or the multiplier effect.

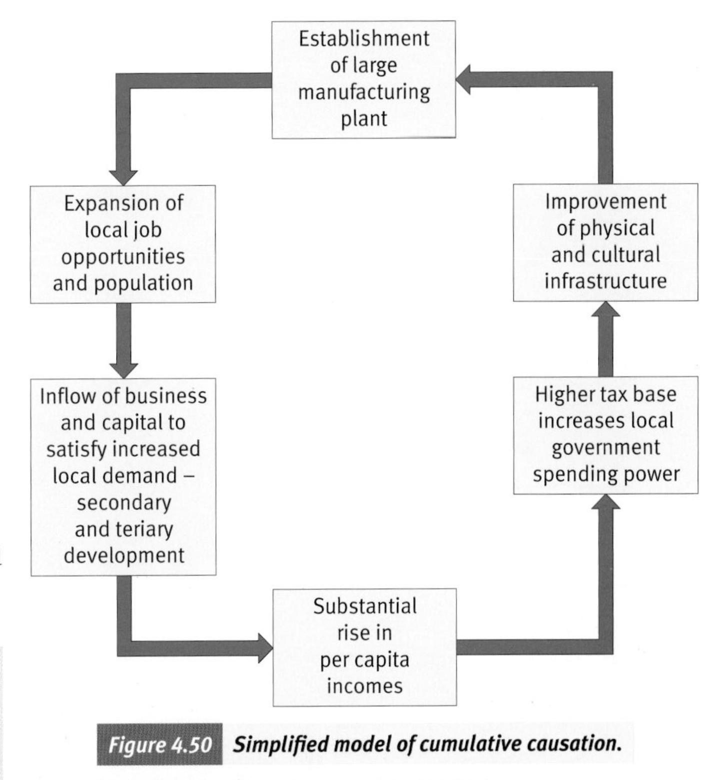

Figure 4.50 *Simplified model of cumulative causation.*

PLENARY ACTIVITY

Explain how the process of cumulative causation has affected the development of economic activities in Brazil.

Multinational companies and globalisation

GET STARTED

Think of something that has happened this month that shows that the world could be thought of as being smaller than it was 30 years ago.

Multinational companies (MNCs) are firms that produce goods or services in two or more countries. The largest MNCs may operate in dozens of different countries. Every year the business magazine *Fortune* publishes a list of the 500 largest corporations in the world. The top 12 for 2007 are shown in Table 4.7.

MNCs have played a key role in the process of **globalisation**. They have been the main drivers of global shift. This is where investment in industry and services has steadily been redirected from MEDCs to selected LEDCs. This has been done to cut costs, particularly the cost of labour. It is this process that has resulted in the emergence of an increasing number of NICs since the 1960s. There are many aspects of globalisation. These are summarised in Table 4.8.

1600 A level playing field which world trade gradually begins to exploit.

1850 Colonisation tips the balance. Primary products exploited by colonial powers to fuel industrialisation. Products sold back at high cost to the colony.

What's the problem? (the North)
Mind the gap!
We need a level playing field! (the South)
N
S

2000 Age of neo-colonisation. TNCs from developed North now exploit resources from the South, supported by the idea of free trade. The South falls further back as it has to borrow money to develop, and pay back the debts from export resources. Labour is now globalised. Technology widens the gap as routine production can be done in LECDs.

Figure 4.51 ***How colonisation turned into globalisation. (Source: Geofactsheet no 147)***

THINK ABOUT IT

As a class, discuss whether companies this big should be considered as countries. What would it mean if they were?

Table 4.7 The world's largest multinational corporations, 2006

Company	Revenue (US $m)
1 Wal-Mart Stores	351,139
2 Exxon Mobil	347,254
3 Royal Dutch Shell	318,845
4 BP	274,316
5 General Motors	207,349
6 Toyota Motor	204,746
7 Chevron	200,567
8 Daimler Chrysler	190,191
9 Conoco Phillips	172,451
10 Total	168,356
11 General Electric	168,307
12 Ford Motor	160,126

(Source: *Fortune*, 11 July 2007)

KEY TERMS

Globalisation – the increasing interconnectedness and interdependence of the world economically, culturally and politically.

New international division of labour (NIDL) – this divides production into different skills and tasks that are spread across regions and countries rather than within a single company.

Table 4.8 Aspects of globalisation

Dimension	Characteristics
Economic	World trade has expanded rapidly. MNCs have been the major force in increasing economic interdependence. The emergence of an increasing number of NICs has been the main success story of globalisation. However, many LEDCs feel excluded from the benefits of globalisation (Figure 4.51)
Urban	A hierarchy of global cities has emerged to lead the global economy. New York, London and Tokyo are at the highest level of this hierarchy.
Social/cultural	Western culture has diffused to all parts of the world through television, cinema, the internet, newspapers and magazines. The international interest in brand name clothes and shoes, fast food and branded soft drinks and beers, pop music and major sports stars has never been greater. However, cultural transmission is not a one-way process.
Linguistic	English has emerged as the working language of the 'global village'. In a number of countries there is great concern about the future of the native language.
Political	The power of nation states has been diminished in many parts of the world as more and more countries organise themselves into trade blocs. The EU is the most advanced model for this process of integration. On the other side of the coin is the growth of global terrorism.
Demographic	The movement of people across international borders has increased considerably in recent decades. More and more communities are becoming multicultural in nature.
Environmental	Increasingly, economic activity in one country has had an impact on the environment in other nations. The scale of the problems is so large that only coordinated international action can bring realistic solutions.

The reasons for globalisation

- Until the 1960s industrial production was mainly organised within individual countries. In the last 50 years a **new international division of labour (NIDL)** has emerged. Now many production processes are fragmented across national boundaries due to big differences in labour costs and skills.
- The increasing complexity of international trade as this process has developed.
- The lowering of trade barriers (tariffs, quotas, regulations).
- The development of an increasing number of NICs.
- The integration of the former Soviet Union and its Eastern European communist satellites into the capitalist system. No significant group of countries any longer stands outside the free market global system.
- The opening up of other economies, particularly those of China and India.
- The deregulation of world financial markets, allowing much freer trade in financial and other services.
- The 'transport and communications revolution' that has made possible the management of the complicated networks of production and trade that exist today.

THINK ABOUT IT

What is the evidence of the mixing of different cultures in the area in which you live? Is it mixing or is everyone becoming the same?

ACTIVITIES

1. What is an MNC?
2. Draw a bar graph to show the size of the revenue of the five largest MNCs in the world compared with the GDP of five LEDCs.
3. What is global shift?
4. Briefly describe the dimensions of globalisation.
5. Examine the reasons for globalisation.
6. Why are the economies of Brazil, Russia, India and China (BRIC, countries) considered to be so important for the future of globalisation?

RESEARCH LINK

Research 'shift happens' on the internet.

How MNCs affect employment opportunities and economic development

GET STARTED

Do any MNCs you have heard of have a good or a bad reputation?

The structure of MNCs

Large MNCs often have three organisational levels:

1 Headquarters
2 Research and development
3 Branch plants.

The headquarters of an MNC will generally be in the developed world city where the company was established. Research and development will most likely be located here too. It is the branch plants that are the first to be located abroad. However, some of the largest and most successful MNCs have divided their industrial empires into world regions, each with research and development facilities and a high level of decision making (Figure 4.52).

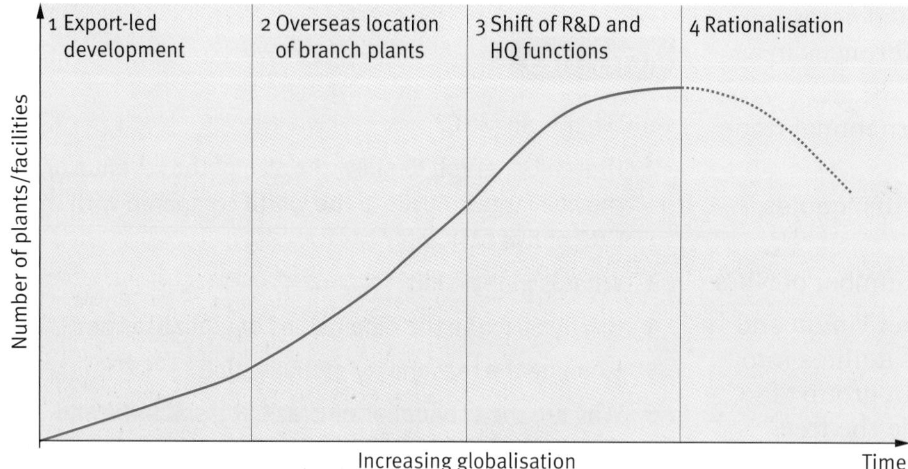

1 Export-led development
2 Overseas location of branch plants
3 Shift of R&D and HQ functions
4 Rationalisation
Number of plants/facilities
Increasing globalisation
Time
Activities concentrated in home country where labour and sourcing are established. However, exports may be subject to tariffs and other restrictions.
Incentives include cheaper labour, access to markets, and financial assistance from host governments.
New locations become semi-autonomous as products are more carefully tailored to new markets.
Increasing competition or recession necessitates concentrating activities in the best locations.

Figure 4.52 The development of MNCs.
(Source: *Advanced Geography: Concepts and Cases*, Hodder Education, 1999)

The employment benefits are greatest to those host countries that can attract as many different types of new jobs as possible.

MNCs are generally very successful because they can achieve **economies of scale**. To do this they operate very large factories, often employing many thousands of people.

The positive effects of MNC investment

Foreign direct investment by MNCs can bring a number of advantages to the countries in which they operate. This is why most countries encourage MNC investment. The location of a new MNC factory can create a large number of jobs. While wage levels vary, they are usually higher than those paid by local companies. If a number of MNCs locate in an NIC or LEDC the effect can be to (a) develop a much wider range of skills in the local population and (b) set off the multiplier effect. As most MNC production will be exported, this will benefit the host country's trading position.

The taxes paid by MNCs to host countries can be invested in health, education and infrastructure.

KEY TERMS

Deindustrialisation – the long-term absolute decline of employment in manufacturing.

Dependent – when an area is reliant on one company or type of company for the majority of its employment.

Economies of scale – a situation in which an increase in the scale at which a business operates will lead to a reduction in the average costs of production.

Reindustrialisation – the establishment of new industries in a country or region that has experienced considerable decline of traditional industries.

The potential disadvantages of MNC investment

A concern of many NICs and LEDCs is the speed with which MNCs can close factories in one country and open up in another. Because such factories may employ thousands of people, a closure can have a huge negative impact on the local economy. This is always a worry if a country or region relies on a small number of MNCs for employment. There is a danger that countries can become too **dependent** on MNCs. In some countries, MNCs may be so important to the economy that they can influence government decisions. Critics see this as being anti-democratic. The cultural impact may be of concern too. When MNCs come into a poorer country they bring aspects of 'Western' culture. Islamic countries in particular seem concerned about this.

The impact of MNC decisions on MEDCs

- Manufacturing employment in MEDCs has fallen substantially in recent decades. A major reason has been the global shift of industries to NICs.
- This process of **deindustrialisation** has caused significant economic and social problems in many traditional industrial regions.
- The greatest job losses have occurred in labour-intensive, low-technology industries.
- Manufacturing employment peaked in the UK in 1966 and in the USA in 1979.
- The manufacturing that has survived best in MEDCs has been innovative in the ways in which it operates.
- Some regions have been relatively successful in **reindustrialisation** while others have struggled to attract enough new industry.

Figure 4.53 Many large factories have had to close.

THINK ABOUT IT

As long as they provide jobs and pay taxes, should MNCs be able to do as they like in a country?

Tobacco MNCs

In the past, the World Health Organisation (WHO) has accused the tobacco MNCs of sabotaging efforts to control tobacco consumption through pressure tactics against the agency and other international organisations. A WHO report accused the tobacco industry of:

- using numerous third-party organisations such as trade unions to try to influence the WHO
- secretly funding 'independent' experts to conduct research and publish papers that would challenge WHO findings
- setting up press conferences to draw attention away from events organised by the WHO related to anti-smoking efforts.

PLENARY ACTIVITY

What would be the effects on towns in the UK of deindustrialisation? Devise indicators that could measure the decline.

ACTIVITIES

1 Write a definition of an MNC for a general readership.
2 What is meant by economies of scale? Give an example.
3 State three positive and three negative effects of MNC investment.
4 'MNCs are the enemy of sustainability.' Explain your opinions on this statement.
5 How have MNC decisions adversely affected MEDCs?
6 Would you want an MNC to invest in your town?

GET STARTED

Class activity: how many people in the class own something produced by Nike? What is the total for each type of product?

Figure 4.54 The Nike factory in Vietnam.

Nike is the world's leading supplier of sports footwear, apparel and equipment. It is one of the best-known global brands. It was founded in 1972. 'Nike' is the Greek word for 'victory'. Nike does not make any shoes or clothes itself, but contracts out production to South Korean and Taiwanese companies.

The subcontracted companies operate not only in their home countries but also in lower-wage Asian economies such as Vietnam, the Philippines and Indonesia. One hundred and fifty Asian factories employing 350,000 workers manufacture products for Nike. Nike's expertise is in design and marketing. The way the company operates is an example of the new international division of labour (NIDL).

RESEARCH LINK

The internet is full of opinion on this issue. Does Nike get a fair representation?

Fact file

- Nike employs 650,000 contract workers in 700 factories worldwide.
- The company list includes 124 plants in China, 73 in Thailand, 35 in South Korea and 34 in Vietnam.
- More than 75 per cent of the workforce is based in Asia.
- The majority of workers are women under the age of 25.

Figure 4.55 Vietnam.

Nike manufacturing in Vietnam

The manufacture of Nike products has expanded rapidly in Vietnam in recent years. The government has welcomed this process, as attracting a well known manufacturer brings a certain economic status. Nike's Vietnam HQ is in Ho Chi Minh City. About 75 million pairs of shoes are made for Nike in Vietnam each year. In 2007 and 2008 there were strikes in a number of 'Nike' factories over better pay and conditions.

Figure 4.56 *Ho Chi Minh City, the location of Nike's Vietnam headquarters.*

The potential benefits that Nike brings to Vietnam are:

- creates substantial employment in Vietnam
- pays higher wages than most local companies
- improves the skills base of the local population
- the success of a global brand has helped attract other MNCs to Vietnam, setting off the process of cumulative causation
- exports of Nike products are a positive contribution to the balance of payments
- sets new standards in efficient production for indigenous companies
- its contribution to local tax base helps pay for improvements to infrastructure.

However, critics of Nike's operations in Vietnam say that:

- the company image and its advertising, along with that of other MNCs, may help to undermine national culture
- there are concerns about the political influence of Nike and other large MNCs
- investment could be transferred quickly from Vietnam to lower cost locations in other countries.

Nike: the international context

The company aims to produce new shoes on a regular basis to cater for **niche markets**. To achieve this objective it utilises a just-in-time innovation structure, buying in necessary expertise at short notice. This involves short-term subcontracts, often allocated to firms based near to Nike's research and development headquarters near Beaverton in the state of Oregon, USA.

In response to years of 'sweatshop allegations', Nike produced a 108-page report in 2005 which gave the most comprehensive picture to date of the 700 factories producing its footwear and clothing.

The pressure group Human Rights First described the report as an important step forward in improving conditions for workers in poor countries. Phil Knight, the chairman of the company, said he hoped Nike could become a global leader in corporate responsibility.

KEY TERMS

Niche markets – small markets that deal in a specialised product.

ACTIVITIES

1. For which products is Nike famous?
2. How does Nike organise the manufacture of its products?
3. Where are Nike products manufactured?
4. Why can the way in which Nike operates be seen as an example of the new international division of labour?
5. Discuss the possible advantages and disadvantages to Vietnam of the manufacture of Nike products. Use a diagram to structure your ideas.

PLENARY ACTIVITY

'The operation of MNCs is at times very controversial and opinions may vary widely.' Discuss.

MNCs, globalisation and the future

GET STARTED

How do you think globalisation will change in the future? Will existing patterns continue or will new factors emerge? Who will benefit and who will not? What role does technology have in all of this?

Large MNCs are increasing their market share in many sectors of the global economy. It is becoming more and more difficult for small companies to start up or remain trading in many areas of business.

The evidence of this at a local level is the closure of small shops and other small businesses. For example, small food and clothes shops cannot compete on prices with large companies such as Tesco and Asda (Wal-Mart). Some parts of the UK have been described as 'food deserts' because of the absence of local food shops. These are frequently poor inner city areas where car ownership is low. However, people have no choice but to travel a long way to buy food. The large number of charity shops in a high street is often a good indication of the number of businesses that have had to close. It is likely that large MNCs will continue to increase their market share unless governments do more to protect smaller businesses.

Fact file

The 10 largest corporations in their field now control:

- 86 per cent of the telecommunications sector
- 85 per cent of the pesticides industry
- 70 per cent of the computer industry
- 35 per cent of the pharmaceuticals industry.

THINK ABOUT IT

Is your local shopping area affected by this trend? Does it have 'too many' charity shops? Why?

Surviving global shift in MEDCs

The decline in manufacturing concerns many people in MEDCs. This is mainly because so many jobs have been lost. The UK and other MEDCs have made big efforts to reindustrialise so that employment opportunities are not totally reliant on the tertiary sector. These new industries are usually very advanced in terms of technology and require a high level of investment.

The impact on specific groups of people

The widening wealth gap

The negative aspects of globalisation can be seen to have most impact on the poorest people, who are often from minority groups. This is the case in MEDCs, NICs and LEDCs. For

Figure 4.57 Small shops struggle to compete with large companies.

example, in the UK people with good qualifications and skills have generally benefited from changes in the global economy. Increasing globalisation has resulted in a big increase in business for the City of London and good salaries for the well-qualified people that work there – although these jobs are vulnerable to economic downturns too. Those with low levels of skills have found it difficult in the employment market because:

- so many traditional manufacturing jobs have moved from the UK to NICs
- increased levels of immigration have created more competition for lower-wage jobs.

The wealth gap between the rich and poor in the UK is thought to be at its widest since Victorian times.

The impact of pollution

It is the countries that are industrialising the fastest that now have the greatest pollution problems. And it is the poorest people in these countries that invariably live close to polluting factories. However, the problems caused by such high levels of pollution are global. This issue will be looked at in more detail in the next section.

Will Africa be brought into the global economy?

To a large extent Africa has been by-passed by the benefits of globalisation. An Oxfam report published in April 2002 stated that if Africa increased its share of world trade by just 1 per cent it would earn an additional £49 billion a year – five times the amount it receives in aid. The World Bank has acknowledged that the benefits of globalisation are barely being passed on to sub-Saharan Africa and may actually have accentuated many of its problems.

The rise of global civil society

A relatively recent trend has been the growing importance of international pressure groups such as Friends of the Earth and Amnesty International. These are elements of what has come to be known as **global civil society**. Their aim is to 'civilise' globalisation.

The combined actions of these organisations have resulted in a growing framework of international rules. These cover trade, the environment, human rights and other aspects of international relationships. The aim is to create a fairer world for ordinary people.

DISCUSSION POINT

What motivates people to invest so much in protesting against globalisation?

KEY TERMS

Global civil society – international groups, associations and movements that are not controlled by the state government.

Figure 4.58 ***Smoke billows from China's largest iron and steel works.***

ACTIVITIES

1. Give one drawback to the increasing control of large MNCs.
2. What is reindustrialisation?
3. Discuss the impact of globalisation on specific groups of people.
4. What is the global civil society? What role does it have to play in the future of globalisation?

PLENARY ACTIVITY

Whole-class discussion – what factors could halt or even reverse the process of globalisation?

A Decision Making Exercise – Should Ethiopia welcome multinational mining?

Multinational companies bring costs and benefits. In this exercise you will find out about the work of one of the largest mining companies in the world, Anglo American. One of the many products it mines is platinum, a valuable ore that is in high demand across the world. It has a long history of mining in the Limpopo region of South Africa. These pages give you details about the benefits it brings to the region. You will also read about *alleged* problems associated with its mining activities. In Ethiopia, rich reserves of platinum have been found. Should the Ethiopian government welcome inward investment from MNC mining companies? You decide.

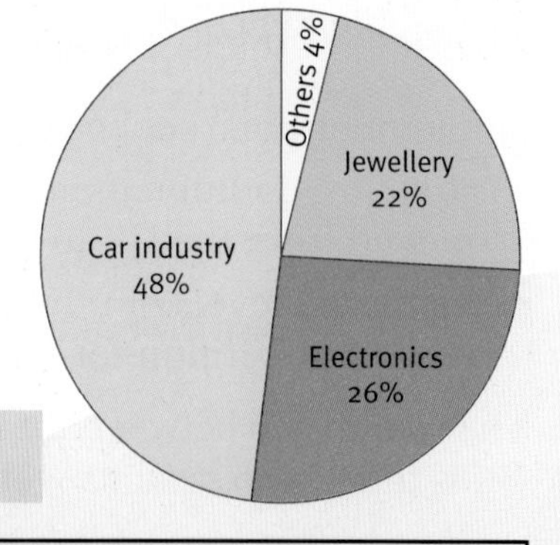

Figure 4.62 The market for platinum.

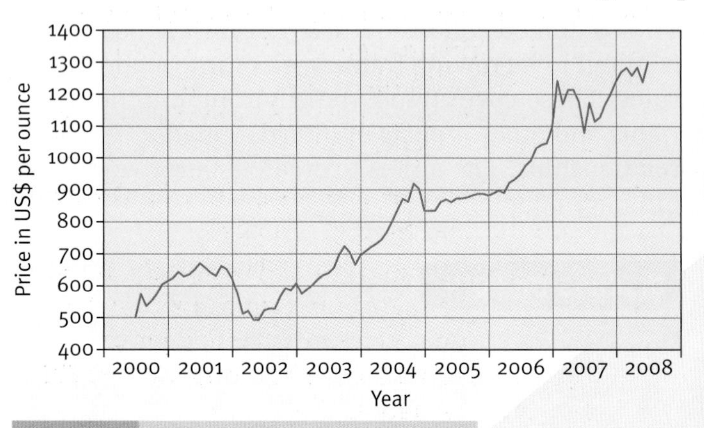

Figure 4.59 The price of platinum, 2002–2007.

Demand for platinum is greater than supply, hence the price increase. The world is desperate for platinum. In recent years, particularly, demand has come from the car industry (catalytic converters), and the electronics industry (PC hard drives and chips). Both are fast-growing industries.

Anglo American is one of the biggest mining companies in the world. It operates in every continent. Virtually all of its mining for platinum takes place in the Bushveld area of the Limpopo region of South Africa (see Figure 4.60). Eighty per cent of the world's platinum comes from the area around Mokopane, a town situated 100 km south-west of the capital Polokwane. This is a remote, undeveloped area. The Limpopo regional government welcomes investment from several mining companies. The mines are an important source of employment and companies such as Anglo American pay millions of dollars in tax to the regional government every year.

Figure 4.60 The Limpopo region, South Africa.

Figure 4.61 Mining for platinum.

Anglo American – the benefits to South Africa

The company made US $1.75 million profit after tax in 2007. It paid over US $1 billion in tax to the South African government. It donated huge sums to a range of social and economic projects in the country, particularly in the Limpopo region, including:

- rebuilding Portlake High School
- funding two mobile clinics to bring health care to remote areas
- education programmes to turn back the epidemic of HIV/AIDS
- rural skills programme: teaching plumbing, bricklaying and carpentry to the unemployed.

Anglo American employs over 150,000 workers in South Africa, most of them black South Africans.

These workers earn much more than the average worker in the country and have many fringe benefits, such as a health scheme.

Multinationals – all good news?

In a controversial report, ActionAid suggests that all is not well in the area surrounding the platinum mines. All of the issues raised in the report are strongly contested by Anglo American. Read the full report on the ActionAid website.

Figure 4.63 ***The environmental impact of mining.***

YOUR TASK

Platinum has been mined on a small scale in Welega in Ethiopia for several years. With the price of platinum reaching an all time high, many multinational companies want to open up new reserves in the area. Ethiopia is one of the world's poorest countries. Should it welcome inward investment based on platinum mining?

Consider the evidence on these two pages and the additional information on the CD. Write a report for the Ethiopian government outlining the likely costs and benefits that may result from a multi-million-dollar mine in the Welega region.

Would you recommend that the scheme should go ahead?

ActionAid is a charitable organisation 'fighting for a world without poverty', and fighting for 'justice for poor and marginalised people across the world'.

In March 2008, it published a report – *Precious Metal*. This followed an investigation into Anglo American operations in the Limpopo region of South Africa.

Extracts from the report:

- Near the village of Ga-Pila, 7000 residents were resettled to make way for the mining operations. Each family was paid just £356 in compensation.
- Near Ga-Molenke, sampling at a school and a community drinking tap found the water to be unfit to drink, with high levels of salt and nitrates. It is suggested that the probable cause was mining operations.
- Farmers lost 3600 hectares of grazing land when the mine was enclosed by fencing.
- In Ga-Puka, residents within 500 m of the mine have to evacuate their home every two weeks because of blasting noise and the danger of falling rock debris. Each person is given just £21 for the day's evacuation.

In a news release issued in April 2008, Anglo American challenged all of these allegations. It reminded all interested groups about the significant benefits that the mine brings. See its website for more information.

Figure 4.64 ***Ethiopia.***

How can economic activity affect the physical environment?

GET STARTED

How many types of pollution are there?

The primary and secondary sectors

Virtually all forms of economic activity have an impact on the environment. However, the greatest problems are usually caused by the primary and secondary sectors. Table 4.9 shows some of the effects that primary industry has on the environment.

Table 4.9 The impact of primary industry on the environment

Farming	High usage of chemical inputs for arable farming can reduce soil fertility and increase soil erosion. Destruction of hedgerows has an impact on wildlife. Excessive production of slurry from livestock can result in eutrophication in rivers.
Mining and quarrying	Can result in massive landscape scarring and pollution of groundwater. Vibration caused by blasting and large vehicles damages buildings and people.
Forestry	Large-scale felling in rainforests destroys the fragile forest ecosystem. Soil structure may deteriorate rapidly to leave barren landscapes.
Fishing	Overfishing has led to severe resource depletion in many fishing grounds. Natural food chains can be severely disturbed. Fishing vessels cause oil pollution.

The most serious polluters are the large-scale processing industries that can have a serious impact on land, air and water. These tend to form agglomerations as they have similar locational requirements. The impact of a large industrial agglomeration may spread a very long way and cross international borders. For example, pollution found in Alaska was traced back to the Ruhr industrial area in Germany.

Figure 4.65 The impact of deforestation.

DISCUSSION POINT

Should we be concerned about preserving primary industry in the UK if it is so damaging?

Tertiary and quaternary industries

The direct environmental impact of these industries is very limited compared with the primary and secondary sectors. However, the indirect effects are significant. Large out-of-town shopping malls and retail parks generate huge volumes of traffic. The issue of 'food miles' is near the top of the environmental agenda, with supermarkets stocking produce from all over the world (see page 99 for more on food miles). Many of the specialist materials used by the quaternary sector will have had an environmental impact earlier in the product chain. Computers have a significant impact on global warming through use of electricity.

THINK ABOUT IT

Should governments force out-of-town shops to charge their customers for parking?

Pollution control in MEDCs

Levels of pollution have generally fallen in MEDCs because:

- increasingly strict environmental legislation has been passed. This is the beginning of a process to make polluters pay for the cost of their actions themselves, rather than expecting society as a whole to pay the costs
- industry has spent increasing amounts on research and development to reduce pollution – the so-called 'greening of industry'
- the relocation of the most polluting activities, such as commodity processing and heavy manufacturing to NICs.

Figure 4.67 **Supermarkets stock produce from all over the world.**

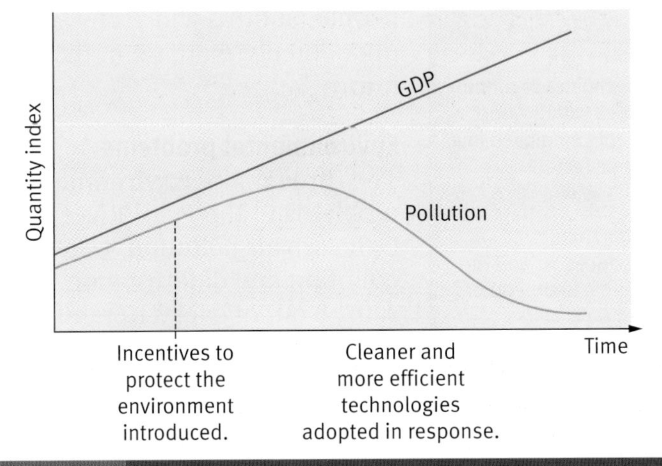

Figure 4.66 **The relationship between pollution and GDP per capita.**

So, after a certain stage of economic development in a country the level of pollution will usually decline (Figure 4.66). The 1990s witnessed the first signs of this. For example, in Germany the 1990 'take-back' law required car manufacturers to take responsibility for their vehicles at the end of their useful lives.

Where significant manufacturing pollution remains, it is the poor who are most directly affected. Middle and higher income groups have the money to avoid living close to polluting industry.

Pollution problems in LEDCs

However, as environmental legislation has tightened in the developed world many poorer countries have become dustbins for unwanted waste. This can generate much needed income in the short term, but cause long-term problems. As more and more LEDCs strive to industrialise, pollution levels rise. At this stage economic development is the main concern.

ACTIVITIES

1. With the help of Table 4.9, describe the environmental impact of one primary industry.
2. Explain the environmental effects that manufacturing industries can have on land, air and water.
3. How can tertiary industry have an impact on the environment?
4. What can be done to reduce the environmental impact of economic activity?
5. Do MEDCs have a moral responsibility to LEDCs to clear up the damage caused by 'exporting' pollution?

RESEARCH LINK

Where are the most polluted places in the world? Is there a pattern to their locations?

PLENARY ACTIVITY

Why have pollution problems steadily shifted from MEDCs to NICs and LEDCs?

CASE STUDY The balance between environmental concerns and economic development in China's Pearl River Delta

GET STARTED

How many products do you have with you now that were made in China?

China's Pearl River Delta

The Chinese economy is now so large it is being called the 'new workshop of the world'. This is a phrase first applied to the UK during the height of its industrial revolution in the 19th century. However, in the main industrial areas the environment has been put under a huge strain, leaving China with some of the worst pollution problems on the planet. One of China's main industrial regions is the Pearl River Delta. It faces the challenge of continuing to grow economically while trying to protect its environment.

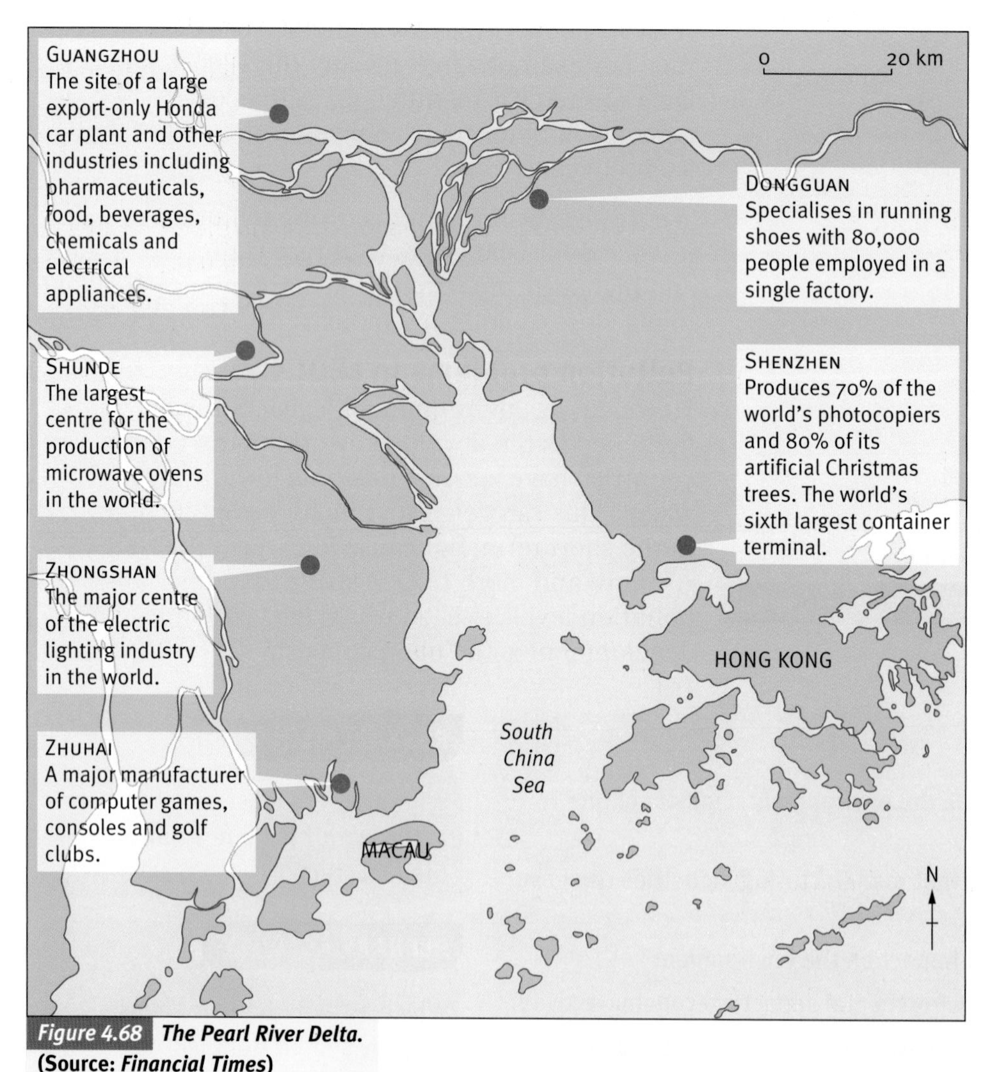

GUANGZHOU
The site of a large export-only Honda car plant and other industries including pharmaceuticals, food, beverages, chemicals and electrical appliances.
SHUNDE
The largest centre for the production of microwave ovens in the world.
ZHONGSHAN
The major centre of the electric lighting industry in the world.
ZHUHAI
A major manufacturer of computer games, consoles and golf clubs.
MACAU
South China Sea
HONG KONG
DONGGUAN
Specialises in running shoes with 80,000 people employed in a single factory.
SHENZHEN
Produces 70% of the world's photocopiers and 80% of its artificial Christmas trees. The world's sixth largest container terminal.
0 20 km
N

Figure 4.68 *The Pearl River Delta.*
(Source: *Financial Times*)

Location

The Pearl River Delta region is an area the size of Belgium in south-east China (Figure 4.68). It is the focal point of a massive wave of foreign investment into China. The Pearl River drains into the South China Sea. Hong Kong is located at the eastern extent of the delta, with Macau situated at the western entrance. The region's manufacturing industries already employ 30 million people, but this will undoubtedly increase in the future.

Environmental problems

The three major environmental problems in the Pearl River Delta are air pollution, water pollution and deforestation. In 2007, 8 out of every 10 rainfalls in Guangzhou were classified as **acid rain**. The high concentration of factories and power stations is the source of this problem, along with the growing number of cars in the province. The city has the worst acid rain problem in the province of Guangdong. The province's environmental protection bureau has reported that two-thirds of Guangdong's 21 cities were affected by acid rain in 2007. Overall, 45 per cent of the province's rainfall in 2007 was classified as acid rain.

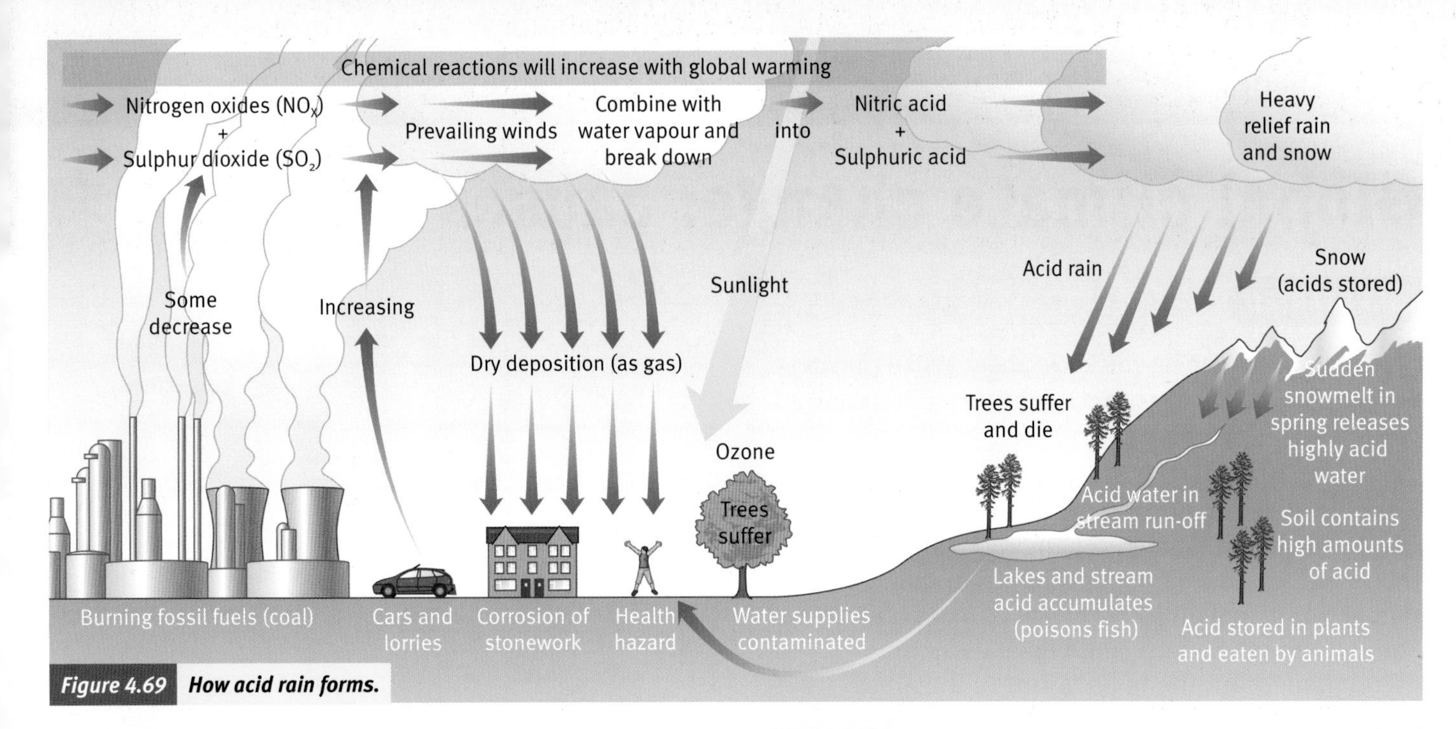

Figure 4.69 ***How acid rain forms.***

Half of the wastewater in Guangdong's urban areas is not treated before being dumped into rivers, compared with the national average of 40 per cent. Guangdong's government has pledged to reduce chemical pollution of water by 15 per cent by 2010 (from 2005 levels). It also aims to cut sulphur dioxide emissions by 15 per cent.

Almost all the urban areas have overexploited their neighbouring uplands, causing a considerable reduction in vegetation cover. This has resulted in serious erosion.

The Environmental Protection Bureau classifies the environmental situation as 'severe' and says the government is committed to taking the 'necessary measures' to reduce pollution. Among the measures used to tackle the problems are (a) higher sewage treatment charges, (b) stricter pollution regulations on factories and (c) tougher national regulations on vehicle emissions.

KEY TERMS

Acid rain – rainwater containing sulphuric acid, nitric acid and compounds of ammonia. These pollutants have been pumped into the atmosphere by manufacturing industry and vehicle emissions.

PLENARY ACTIVITY

The average income for people in the Pearl River Delta has greatly increased; was it worth it?

ACTIVITIES

1. Describe the location of the Pearl River Delta.
2. Name four industrial cities in the region. Which products does each city produce?
3. **a** What is acid rain?
   b How does acid rain affect an area?
4. On a sketch map of the area, explain the advantages to business of being in the Pearl River Delta area.
5. Suggest why industrial development has been so rapid in the Pearl River Delta.
6. **a** Summarise the main environmental problems in the region.
   b What is being done to try to reduce pollution in the region?
   c Why might the Chinese government drag its heels on cleaning up?
7. Draw a sketch of the Pearl River Delta and annotate it to show signs of economic activity and any possible environmental impacts.

Global climate change: causes

GET STARTED

Write down everything you know about global climate change. Divide your answers into causes, effects and solutions.

There is no doubt amongst geographers and scientists that the Earth's climate is changing. Many parts of the world are experiencing changes in their weather that are only explained by global climate change. Some of these changes could have disastrous consequences for the populations of the areas affected if they continue to get more severe.

The human impact on climate change

The global climate has always changed in temperature, up and down. However, the problem is that:

- the present rate of change is greater than anything that has happened in the past. In the 20th century, average global temperatures rose by 0.6°C. Most of this increase took place in the second half of the century
- most climate experts believe that this high rate of temperature change is due to human activity
- the predictions are for a further global average temperature increase of between 1.6°C and 4.2°C by 2100.

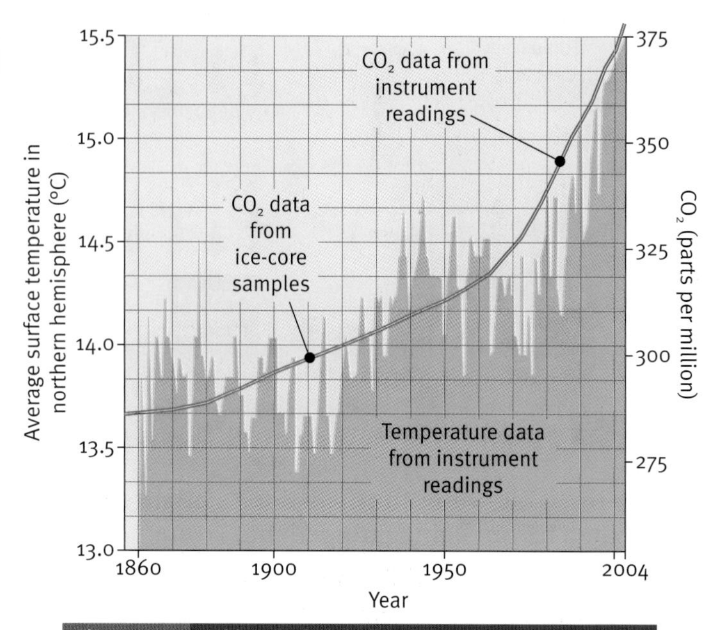

Average surface temperature in northern hemisphere (°C)
15.5
15.0
14.5
14.0
13.5
13.0
CO_2 data from instrument readings
CO_2 data from ice-core samples
Temperature data from instrument readings
375
350
325
300
275
CO_2 (parts per million)
1860
1900
1950
2004
Year

Figure 4.70 ***Evidence of global warming.*** **(Source: Scripps Institution of Oceanography, UC San Diego)**

KEY TERMS

Carbon footprint – the effect human activities have on the climate in terms of the total amount of greenhouse gas production.

Greenhouse effect – the property of the Earth's atmosphere by which long wavelength heat rays from the Earth's surface are trapped or reflected back by the atmosphere.

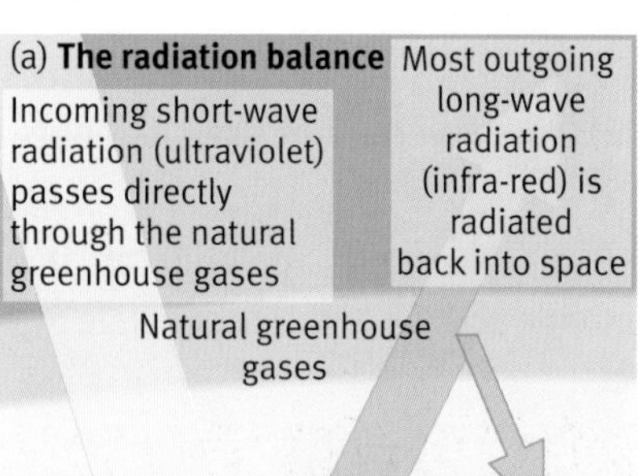

(a) The radiation balance
Incoming short-wave radiation (ultraviolet) passes directly through the natural greenhouse gases
Most outgoing long-wave radiation (infra-red) is radiated back into space
Natural greenhouse gases
Some outgoing radiation is absorbed by, or trapped beneath, the greenhouse gases
Previously a balance
Short-wave radiation is transformed into long-wave radiation (heat) on contact with the Earth's surface
(b) The greenhouse effect
Less heat escapes into space
Increase in greenhouse gases due to human activity
As more heat is trapped and retained, so the Earth's atmosphere will become warmer (global warming)

Figure 4.71 ***The greenhouse effect. The system acts like a garden greenhouse, where the glass allows the heat from the Sun to penetrate easily, but traps a certain amount of the outgoing heat.*** **(Source: Reproduced with the permission of Nelson Thornes Ltd from *Geography: An Integrated Approach* by David Waugh isbn 978-0-7487-7376-3 first published in 2000)**

The greenhouse effect

Global warming occurs because of the **greenhouse effect** of the Earth's atmosphere. Climatologists calculate that without these greenhouse gases, the average temperature of the earth would be 33°C lower than it actually is today.

Human activity has significantly increased the amount of greenhouse gases in the atmosphere and this has caused temperature to rise more rapidly than ever before. As the economies of China, India and other NICs expand even further, greenhouse gas emissions will continue to increase.

Figure 4.72 Urban smog in Shanghai.

The greenhouse gases are:

- **Carbon dioxide**: accounts for the largest share of greenhouse gas. It is produced by burning fossil fuels in power stations, factories and homes. Vehicle emissions are also a major source. CO_2 is also released into the atmosphere by deforestation and the burning of rainforests.
- **Methane**: released from decaying plant and animal remains and from farms (particularly from cattle and rice padi fields). Other sources include swamps, peat bogs and landfill sites.
- **Nitrous oxides**: from power stations, vehicle emissions and fertilisers.
- **Chlorofluorocarbons**: the main sources are aerosols, refrigerators, foam packaging and air conditioning.
- **Ozone**: from vehicle emissions.

Geographical sources of greenhouse gases

Figure 4.74 shows the countries most responsible for CO_2 emissions from 1840 to 2004. The USA has by far the highest total. If we were to look at very recent figures only, then the shares of China and India would be much higher than they are over the longer time period. This is because they have only industrialised on a large scale in recent decades.

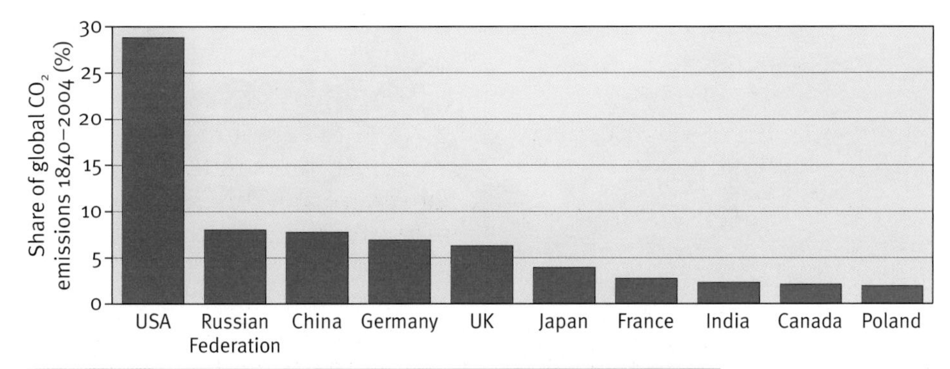

Figure 4.73 Countries most responsible for CO_2 emissions.
(Source: Human Development Report, 2007/2008, Palgrave Macmillan)

Fact file

Global climate changing timeline

1827: French scientist Jean-Baptiste Fourier compares the warming of the atmosphere to a greenhouse.

1979: The first World Climate Conference states the possibility of global warming.

1987: The warmest year on record to date.

1995: The warmest year on record to date.

1997: Kyoto Protocol agrees cuts in greenhouse gas emissions, but not all countries sign up.

1998: The warmest year on record in the warmest decade on record.

2005: The second warmest year on record. Kyoto Protocol comes into force. Stern Report on climate change published.

2006: The Intergovernmental Panel on Climate Change (IPCC) confirms the reality of global warming and that human activities are at least partly responsible.

ACTIVITIES

1. Define the greenhouse effect. Use a simple diagram to help you remember.
2. Describe the sources of two greenhouse gases.
3. With the help of an annotated diagram, explain the greenhouse effect.
4. Look at Figure 4.70. Describe and explain the geographical pattern of CO_2 emissions from 1860.
5. Which lifestyle choices do you think put the USA at the top of the polluters league? Why doesn't the UK have a similar record?

PLENARY ACTIVITY

What has your school done so far to reduce its **carbon footprint**? What else could it do?

Global climatic change: effects

GET STARTED

Think of one effect of global climate change you have heard of actually happening.

KEY TERMS

Thermal expansion – as sea and ocean temperatures increase, the water molecules near the surface expand and the sea level rises.

There are many potential consequences of global climate change. However, the global climate system is so complex that there is still much debate about exactly what could happen.

Rising sea levels due to thermal expansion: a global average sea level rise of 0.4 m from **thermal expansion** has been predicted by the end of this century.

Melting of ice caps and glaciers: satellite photographs show ice melting at its fastest rate ever. The area of sea ice in the Arctic Ocean has decreased by 15 per cent since 1960, while the thickness of the ice has fallen by 40 per cent. Sea temperature has risen by 3°C in the Arctic in recent decades. A report in the *Guardian* in March 2007 claimed that the Arctic Ocean may lose all its ice by 2040. This would disrupt global weather patterns. For example, it would bring intense winter storms and heavier rainfall to Western Europe.

In 2007, the sea ice around Antarctica had melted back to a record low. At the same time, the movement of glaciers towards the sea has speeded up. A satellite survey between 1996 and 2006 found that the net loss of ice rose by 75 per cent. Antarctica and Greenland are the world's two

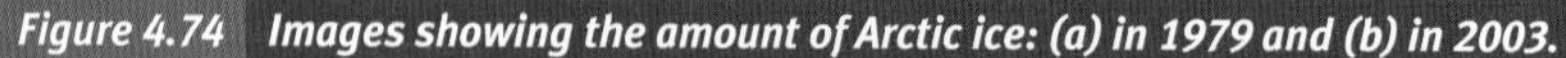

Figure 4.74 Images showing the amount of Arctic ice: (a) in 1979 and (b) in 2003.

(a)

(b)

major ice masses. Total melting of these ice masses (which is not predicted at present) could raise global sea levels by 70 m.

Research published by Chinese scientists in 2006 shows that glaciers in the Qinghai-Tibet plateau are melting faster than anyone previously thought. Warming has speeded up the shrinkage of more than 80 per cent of the 46,377 glaciers in this major plateau region. This could eventually result in water shortages in China and large parts of South Asia. The increasing rate of melting from the plateau has led to a rise in water run-off from the plateau, which has increased soil erosion and desertification. The UN has warned that Tibet's glaciers could disappear within 100 years.

Ice melting could cause sea levels to rise by a further 5 m (on top of thermal expansion). Hundreds of millions of people live in coastal areas within this range.

Growth of the tropical belt: a study published in 2007 warned that the Earth's tropical belt was expanding north and south. A further 207,199 km^2 of the Earth are experiencing a tropical climate compared with 1980. The poleward movement of subtropical dry belts could affect agriculture and water supplies over large areas of the Mediterranean, the south-western USA, northern Mexico, southern Australia, southern Africa and parts of South America. The extension of the tropical belt will put more people at risk from tropical diseases.

Changing patterns of rainfall: the amount and distribution of rainfall in many parts of the world could change considerably. Generally, regions that get plenty of rainfall are likely to receive even more and regions with low rainfall are likely to get less. This will include the poor arid and semi-arid countries of Africa. Experts predict that storms will hit the UK more frequently.

Declining crop yields: higher temperatures have already had an impact on global yields of wheat, corn and barley. A recent study revealed that crop yields fall between 3 and 5 per cent for every 0.5°C increase in average temperature. Food shortages could begin conflicts between different countries.

Impact on wildlife: many species of wildlife may be wiped out because they will not have a chance to adapt to rapid changes in their environments. The loss of Arctic ice will have a huge effect on polar bears and other species that live and hunt among the ice floes.

THINK ABOUT IT

What other disadvantages are there to the growth of the tropical belt?

Desertification is predicted to spread more widely in the future.

RESEARCH LINK

Study a selection of newspapers and magazines in your school library for a two-week period (or use a search engine that allows you to define the dates you are interested in) for any new details on global warming.

PLENARY ACTIVITY

Do the effects of global climate change actually matter to us in the UK?

ACTIVITIES

1. Describe why the sea level will rise using geographical terms.
2. Bubble map the effects of global climate change. Elaborate your explanations as much as possible.
3. On a blank world map, add labels to show the effects of climate change. Add more from your own research.
4. Write a 200-word report summarising the potential extent of flooding in the UK by 2080.
5. Read through these two pages and make a note of every effect of global climate change. Can you summarise these effects? Economically? By location?

Global climatic change: solutions

GET STARTED

There are lots of possible solutions to global climate change. Why hasn't anyone put them into action yet?

Human response to climate change must take two forms:

- acting to reduce the causes of greenhouse gas emissions
- preparing for the consequences of significant climatic change.

For both sets of actions it will be vitally important to manage the problem in a sustainable manner. If this does not happen, other major problems may be created in the process. In 2007 the IPCC warned that the world had until 2020 to avoid the most dangerous effects of climate change.

Figure 4.75 ***An example of an electric car.***

Reducing the causes of climate change

Table 4.10 summarises the views of the IPCC with regard to measures to combat climate change. The technology is already available for many important measures to lower emissions. Examples are nuclear power, renewable energy generation and measures promoting energy efficiency. However, there are two problems:

- it will need a very large investment to put some measures into widespread global use
- some measures are already considered to be unsustainable. For example, there is growing controversy about the increasing use of **biofuels**. This source of energy takes valuable land away from food production and uses a considerable amount of energy in the production process. Nuclear power is, as ever, a controversial issue.

RESEARCH LINK

Find out the pros and cons of the latest electric cars. Would you want your family to get one?

Table 4.10 Technologies and practices to combat climate change

Sector	Currently available	Available by 2030
Energy supply	Improved supply and distribution efficiency; switching from coal to gas; nuclear power; renewable energy	Carbon capture and storage for fossil fuel generating facilities; advanced nuclear and renewable energy
Transport	Higher fuel efficiency; hybrid vehicles; cleaner diesel vehicles; biofuels; shifts to rail, public transport and bicycles	Higher-efficiency biofuels; higher-efficiency aircraft; more powerful and reliable electric/hybrid vehicle batteries
Industry	Heat and power recovery; recycling and substitution of materials; control of emissions	Technological changes in the manufacture of cement, ammonia, iron and aluminium
Agriculture	Management of land to increase carbon stored in soil; dedicated energy crops to replace fossil fuel use	Improvement of crops yields; reductions in emissions from some agricultural practices
Forests	Increase in forested area; use of forestry products for bioenergy to reduce fossil fuel use	Tree species improvement to increase biomass productivity and therefore carbon capture

(Source: *Financial Times*, 5/6 May 2007)

Preparing for the consequences

For example, in the UK, improvement to the sewer network and drainage systems will be needed to cope with more severe storm events. Many homes will have to be made more flood resilient. An example of this is using stone, concrete or tiled floors so that if a house is flooded it can be hosed down and dried out more easily than a house with carpets. It costs about £40,000 to make an average house flood resilient.

In flood-prone areas in Japan, properties are being built on raised ground with sports fields and parkland used to hold flood water. In Germany, marshland has been used to construct 'reservoirs' to hold flood water.

More money is being directed into research to combat global warming. For example, agricultural scientists are trying to develop crops that can better withstand heat and drought.

Government legislation

In 2007, the British government stated its intention to become the first country in the world to set legally binding targets to reduce CO_2 emissions. The aim is to cut the UK's CO_2 emissions by 60 per cent by 2050. A new system of 5-year 'carbon budgets' will be introduced to set limits on total emissions. The limits will be set 15 years in advance. This will allow businesses and public organisations to plan ahead.

Many environmentalists want higher 'green taxes'. However, higher taxes are always unpopular with voters, especially if the economy slows down. In the transition to a **low-carbon economy** many proposals have been put forward, which include:

- a 10-year programme to convert all taxis to electric or hybrid power
- putting a halt on the expansion of airports
- huge investment in public transport
- free electric coaches operating on motorways in their own lanes
- introducing a personal climate allowance. Those who exceed it would be fined.

Many countries are looking at the concept of **community energy**. Energy is lost in transmission if the source of supply is a long distance away. Energy produced locally is much more efficient.

ACTIVITIES

1. What are the two general responses to climate change?
2. Look at Table 4.10. Suggest measures to combat climate change for one of the five economic sectors.
3. What measures does the UK need to take in its transition to a low-carbon economy?
4. What are the risks of leading the world in this project?
5. Do you think that the world can reduce emissions before catastrophic climate changes occur? Give reasons for your opinion.

Figure 4.76 **Wind power.**

THINK ABOUT IT

Why doesn't the government introduce personal carbon allowances?

KEY TERMS

Biofuels – any fuel derived from renewable biological sources such as plants or animal waste.

Community energy – energy produced close to the point of consumption.

Low-carbon economy – a country where significant measures have been taken to reduce carbon emissions in all sectors of the economy.

PLENARY ACTIVITY

Go back to the list you made at the beginning of this topic (page 200). How many of the things you listed were covered? Discuss whether you should cross out or research further the remaining statements.

GradeStudio

Higher:

Key Geographical Themes 8-mark case study question

Case study – The effects of economic development.

i Name and locate an economic activity.

ii How has the economic activity affected the natural environment? What has been done to minimise damage to the environment?

[8 marks]

Mark scheme

Level 1: Basic description of either effects on the environment or damage limitation – no development. Demonstrates limited relevant knowledge and information. Meaning may not be communicated very clearly because of mistakes in writing. **[1–3 marks]**

Level 2: Description of both effects on the environment and damage limitation, with limited development. Shows some relevant knowledge, based on a range of factual information and evidence. Meaning is communicated clearly. **[4–6 marks]**

Level 3: Thorough and developed description of both effects on the environment and damage limitation – with place-specific detail. Demonstrates thorough knowledge, based on a full range of relevant factual information and evidence. Meaning is communicated very clearly. **[7–8 marks]**

i Mining in the Amazon rainforest.

ii Mining causes deforestation in the Amazon. Ores mined include iron, bauxite, gold, nickel and copper. The main mining area is around Carajas. Large, open-cast mines have been created by foreign companies who are not bothered about the natural environment. Roads have been built through the forest so that the minerals can be exported. The Brazilian government has encouraged these mining companies because their activities bring in foreign currency to pay off debts.

The result of mining is large-scale deforestation. The loss of trees means that animal habitats are destroyed and many die. Some breeds are becoming extinct because they cannot reproduce. The nutrient cycle is broken when the forest is cleared, as the nutrients are soon washed out of the soil with nothing to replace them. Also, when the trees are cut down there is nothing to intercept the rain. This means that more rainfall reaches the ground and increases runoff. This can lead to rivers flooding and soil is washed off the ground and into rivers, which also increases the chance of flooding.

Not enough has been done to protect the environment in mining areas. The government is trying to control the development of new mines and roads, but there is still a lot of illegal mining and logging. The government needs to force companies to restore the forest after mining is complete.

Examiner's comment

Level 2 is the highest an answer can get without a valid named activity and location: this answer has done both of these well.

The introduction describes the mining activity and location, which is background information. The key effect on the natural environment is deforestation. This impact is developed into effects on habitats, animals, the nutrient cycle, runoff and flooding. These impacts are developed into a comprehensive answer.

The second requirement of the question is an evaluation of how damage can be minimised. The answer suggests government control of illegal logging is needed, but that it is not totally effective. It also suggests that the government needs to force companies to restore the forest on the completion of mining.

The answer is level 3, but could develop the idea of damage reduction. However, it is true to say that this issue has not been resolved in the Amazon.

End of unit activities

Thinking about geography – summarising information

Summarising is a thinking skill that helps us to refine and prioritise information. Examples in everyday life are when we tell friends about the interesting things we did on holiday, or watching *Match of The Day* on TV, which visually summarises the best of the football action.

By being able to summarise in geography, not only can we condense information and still understand its meaning, we can also *evaluate* how useful that information is, which further develops our *reasoning* capabilities.

Economic development – summarising using ICT

The challenge is to organise and summarise the text in the box opposite from over 200 words into fewer than 130 but still keep the meaning – this will help make learning and recalling facts easier.

ACTIVITIES

1 Draw a three-column table or 'solution frame' in a word document with the following headings:
   - Causes of the food crisis
   - Effects of the food crisis
   - Actions to address the problem

2 Read through the text and copy or paste each sentence into the appropriate column of your solution frame, depending on whether it is describing a *cause*, an *effect* or an *action*. It may help to put bullets in front of each sentence. Save your file as a first version.

Developing your summarising skills

3 Now that you have organised the sentences into three groups, start to narrow down the text by:
   - deleting unnecessary information or words without losing the overall sense
   - deleting repeated information or words
   - replacing or merging information with more general statements.

4 When you have finished summarising, save your file as a second version. Email this version to a classmate who will evaluate the quality of your summary based on: (a) information in the correct column, (b) quality of summarising and (c) solution frame presentation.

5 Decide whether to redraft your summary in the light of the comments made – go back to your first version if you need to – and produce your final version. Show this version to your teacher.

The Global Food Crisis

Food prices have risen 83 per cent since 2005, and in East Africa, millions are in urgent need of emergency food supplies. The global push for biofuel crops, which then take food crops out of production, is playing a big role in raising prices. NGOs are helping to stem the crisis by distributing meals through community kitchens and school canteens, handing out free seeds or fertilisers for farmers. On top of this, high oil prices have led to increases in the cost of fertilisers and other farm expenses, which in turn have a significant impact on food prices. Food prices have reached record levels and the number of hungry people has risen from 850 million to nearly 1 billion. Growing global demand for products such as meat and grain has made this situation worse. Farming policies that benefit rich countries are also having a huge impact. Increasingly unpredictable weather patterns mean that poor farmers are unable to grow as much. NGOs are pressuring governments to invest more in agriculture and to increase aid to those most at risk, get rid of subsidies that divert food production into fuel and increase fair trade deals. This is a serious crisis, with food riots happening across the world.

THINKING ABOUT SUMMARISING

- From your responses to activities 3 and 4, what was hard about narrowing the words down? Did you find it easy or hard to evaluate your classmate's summary? Explain why.
- Why is summarising important?
- In what other subjects do you summarise information? Offer some examples.
- How else could you present summarised information rather than just as text?

ExamCafé

Welcome

Welcome to Exam Café! These resources, here and on the CD that comes with this book, are designed to help you get ready for your exam. There are a range of revision tools and exam preparation activities that will help you get the most out of your revision time.

Revision

Revision checklist

On the CD you will find revision checklists for all four of the Key Geographical Themes and revision timetables that you can use or customise for your own plans. You can use the checklists to plan and structure your revision and to make sure you have covered everything you need to cover.

Theme 2: Population and settlement

b What are the causes and consequences of natural population change over time?	
Population change	• The growth of population on a global scale. • Birth rates and death rates vary between countries. • The rate of population change varies over time.
Overpopulation	How is the total number of people changing and where do they live? What do the following concepts mean? • Overpopulation • Optimum population • Sustainable living
Management	A case study to illustrate strategies to influence natural population change within a country. • China's one-child policy • Population management in Thailand Some strategies for population management are more sustainable than others.

Activities

1 Using a revision checklist from the CD, go over a key theme you need to revise. Identify areas that you are not confident about. How are you going to tackle these areas?

2 Use a revision checklist to make a set of revision flashcards, and either:

a use the cards with a partner to test each other's understanding; or

b with your classmates, try 'revision speed dating'. Each take a card and then half the class move round and sit with a new partner for two minutes to discuss the cards you have. By the end of the session you should have discussed every card!

Common mistakes

Josh
'It's really important to get your timing right in the exam: don't spend too long on the shorter questions or you won't have time to finish the longer ones. I always try to leave a bit of time at the end to look through what I've written too – just in case I spot any mistakes or see something I can add a bit to.'

Yasmin
'I make sure I always read the front of the exam paper really carefully, so I know how many questions I have to do and what sections I need to work through.'

Eden
'I make sure I read each question through twice before I start my answer. I check the command words really carefully so I am sure I know what the question wants me to do.'

Maisie
'I plan breaks in my revision so I might go for a walk round the block or something like that. My brain just switches off otherwise and nothing else goes in.'

Case studies can be tricky: you need to remember the details, but most of all make sure you only use what is relevant to the question. Just writing down everything you remember about a case study will waste your time and lose you marks if you aren't answering the question.

All exam papers are now marked on-screen. In the past examiners were sent the actual paper you write on in the exam, but now everything is scanned in and examiners work at their computers. This means two things:

- you need to write as clearly as you can
- you need to write inside the boxes provided for your answers – anything outside the boxes will not be scanned and the examiner will not see it. There are blank pages provided for your own notes.

Students often do not know key geographical terms well enough. Exam questions are usually based on testing your understanding of a key term or concept, so when students do not know exactly what they mean, a lot of marks get lost. For example, students often aren't sure what the term **distribution** means in questions like 'describe the distribution of …'. Do you know? Producing your own key word glossaries will develop your understanding and confidence. You can also highlight key geographic terms used in examination questions to focus your thinking.

You will need your geography skills in the exam: often students make errors and lose marks because they rush their map reading, or get confused about interpreting graphs or diagrams such as pie charts. Take your time reading maps, diagrams and graphs. Notice the title and any key or scales given to help your interpretation.

Seven secrets for revision success

1. When you revise, use whichever method, or methods, you feel most comfortable with. These could include:
   - using highlighter pens
   - creating lists and rhymes
   - making note cards
   - making up mnemonics (the first letter of words): for example, 'CASH' stands for corrosion, abrasion, solution and hydraulic impact types of erosion in a river or at the coast.
2. Take regular, varied breaks. It is difficult to concentrate for more than 40 minutes at a time. Have a 15-minute break after the first 40 minutes. After the next 40 minutes, take a longer break. Then after the next 40 minutes take a shorter break, and so on.
3. Test yourself. This could involve writing answers to past questions, drawing sketch maps, learning facts and figures, identifying symbols on an Ordnance Survey map, etc. Ask a teacher, parent or friend to assess you. If you ask a friend, then two of you are revising and helping each other's work.
4. Reward good revision by giving yourself a treat.
5. Do not work too late; it is important to get plenty of sleep.
6. Create a revision timetable and stick to it! Your timetable should take into account when your exams are in all other subjects.
7. Make sure you revise a topic more than once. Two or three revisions of a topic improve memory and recall considerably.

Exam preparation

There are lots of sample questions and answers on the CD that comes with this book: read through them and see what the examiner has written about how well the student has done in each case. You can also find some examples of student answers on the GradeStudio pages within this book. And there are plenty of exam-style questions throughout the book for you to practise with: remember, it is important to unpack each question before you attempt it.

Understanding exam language

Exam questions use very specific words to explain what they want you to do. There are lots of these **command** words and you'll find a list on the CD that comes with this book. Three common ones are:

- **name** – where you get marks for your knowledge and understanding of key geographical terms; also used as a command to name a case study example
- **describe** – where you use geographical knowledge and understanding to write about what something is like
- **explain** – where you apply all your geographical knowledge and understanding to give reasons for something.

A satellite image of a tropical storm in Asia.

Many questions will give you information for your answer, so you might be given a table, map or diagram and told to study that resource and then describe certain things about it. Other words might be used instead of name or describe, such as identify, or state (as in 'state two features' about ...), or list.

Let's look at a couple of examples: the first is from a higher-tier paper and the second from a foundation-tier paper.

Name and describe **two** processes of erosion which affect cliffs.

[4 marks]

1 Cliffs are eroded by undercutting at the bottom near the sea. A weakness in the rock is attacked by the force of the waves.

2 Corrasion is when the sea carries material such as sand and pebbles. When waves hit the cliff the material scrapes away at the rock.

Examiner says

The answer scores 4 marks. There is 1 mark for naming each erosion process and 1 mark for describing each process.

Study the satellite image of a tropical storm in Asia. State two features of a tropical storm shown in the satellite image.

[2 marks]

The area covered by cloud is large.
The cloud has a spiral pattern.

Examiner says

The candidate has successfully applied their knowledge of tropical storm features.
The candidate's answer is brief, clear and concise.

Here is another example from a foundation-tier paper:

Some MEDCs provide food aid to LEDCs.
Explain two problems associated with food aid to LEDCs. **[4 marks]**

The food might not get to those who are hungry – it could be stolen and sold.
If they rely on food aid they will suffer if and when the aid stops.

Examiner says

The candidate has given and explained two valid ideas about food aid problems. In the first example they have explained a possible cause of the problem, food aid being stolen. In the second example they have explained a consequence of the problem of relying on food aid. These are not the only ways of explaining the problems, but any more would not have scored any further marks. Notice that two ideas were needed and the use of the command word 'explain'. Four ideas without any explanation would have gained two marks.

And here is another example from a higher paper. This is worth the same marks as the foundation answer: can you see the extra **development** of the answer that is required for the higher tier? You'll see more examples of this on the CD that comes with this book.

Give **two** reasons why there is longer life expectancy in MEDCs than LEDCs. **[4 marks]**

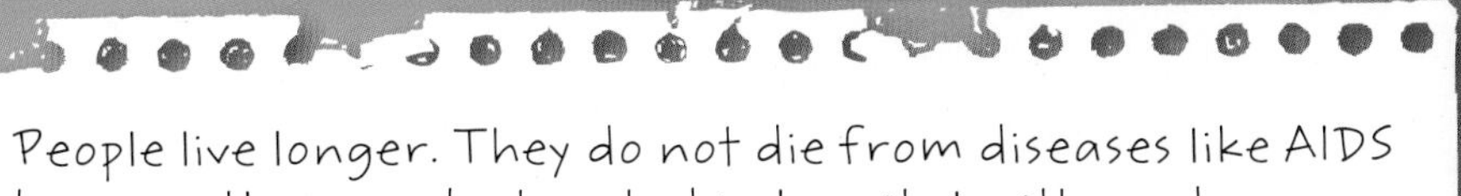

People live longer. They do not die from diseases like AIDS because they can be treated in hospital with modern drugs.
People in MEDCs take services such as sewers and toilets for granted. These are often not available in LEDCs. So people in MEDCs do not catch diseases such as typhoid and dysentery, which are caught from drinking dirty water.

Examiner says

This question requires two reasons to gain four marks, so each reason must be a developed explanation. The first basic idea refers to the availability of health care, which is developed by the idea of drugs being available for treatment. The second basic idea is that there is better water supply and sanitation in MEDCs. This is linked to the fact that fewer people will catch diseases. The answer scores 4 marks.

Chapter 5

Investigating geographical issues

What is a geographical issue?

A geographical issue is a current topical subject that affects people and the environment. An issue can affect one place or many places. It may be a local issue like the building of a bypass, or a global issue such as deforestation of the rainforest. People may have differing views about an issue, depending on how it affects them personally. We can learn about issues from a variety of sources, from the TV and internet to newspapers and magazines – and even school textbooks!

Issues may be reported in different ways and from varying viewpoints, and some reporting can be more biased than others. You need to use all your geographical enquiry skills, knowledge and understanding to be able to develop your own view on an issue.

For part of your controlled assessment you have to investigate a geographical issue. Your school will have the choice of nine issues.

Look at the article about sweatshops and the following enquiry questions. These are the geographical questions you need to consider when looking at an issue.

Exposed: high street clothes made in sweatshops that pay children just 60p a day

The UK's high street clothes are cheaper than ever, but the low-cost clothes come at a high price for young children toiling in the stores' Indian sweatshops.

An investigation revealed that children as young as 11 were working in squalid conditions, sewing tiny beads and sequins onto cheap T-shirts and embroidering dresses by candlelight.

QUESTIONS FOR INVESTIGATION

- What is the issue?
- Why is this an issue? Why may people hold different views on this issue?
- What places are affected?
- How are people affected?
- How am I connected to this issue?
- Can I do anything about this issue?
- When was this written? Has the situation changed since?
- This story comes from an article in a national newspaper. What is the purpose of the article?

What is the geographical investigation?

The geographical investigation is an important part of your GCSE exam, which you produce at school. You have to research a geographical issue and present a brief report based on your findings. You must use secondary data such as books, newspapers and information from websites. The report is worth 10 per cent of your final exam.

What is my investigation about?

There are nine possible topics (disease, trade, ecosystems, sport, tourism, energy, new technologies, crime and fashion). Each year, two questions will be set about each of the nine topics and your school will choose *one* to study. Your teachers will spend some lesson time introducing the main ideas about this issue.

How should I structure my investigation?

It is important that you present a well organised report and the following structure has been recommended:

- location of study/background information
- investigation question
- issues that arise from the investigation question
- research/data collected – evaluation of data and effects of the question investigated
- conclusion (including future scenarios).

What will my finished report look like?

You might decide (or be asked) to write a formal report, but there are many other ways in which this task can be done: for example, written pamphlet, website, posters, video, podcast or oral (spoken) report.

If you are writing a report, it should be about 800 words in length. You are encouraged to include information based on maps, tables, annotated diagrams and photographs, and none of these will count towards the total number of words. Nor will information included in footnotes or an appendix.

You should use information and ideas from books, magazines, newspapers or websites. However, it is important that you do not copy or make use of anybody else's work without acknowledging where it comes from. You must provide references to show where you got the information from. In the same way, if you want to quote what someone else has said you must make it clear it is a quotation, credit the author and give a reference (see Table 5.1). You could give your references in a list at the end of your work (in which case they should be in alphabetical order by surname) or you can insert footnotes. This can be done easily using the footnote tool on your word processing program.

Table 5.1 How to give references

A book	Miller, T. (2009) (ed.) *OCR GCSE Geography B*, Harlow: Pearson Education
An article in a journal or magazine	Womack, T. (2008) 'Protecting coral reefs', *Wideworld* 19 (3), 41–50
An article in a newspaper	Trapp, R. (2008) 'Heaven sent? The Divine Chocolate story', *Independent*, Tuesday 14 October
A website	http://en.wikipedia.org/wiki/Ecotourism [accessed 5 November 2008]
A quotation	'Nuclear power plants cost a lot to build but are then relatively cheap to operate. They are expensive to decommission (and no one knows the full cost of managing their long-lived wastes).' Friends of the Earth, June 2006

Figure 5.1 Geography student making a presentation.

Can I use IT?

You are encouraged to use IT for this task. You should use the internet to research information and you might also want to take digital photographs, or use photographs or video clips you find on the internet. Remember that you must acknowledge the sourccs for all your information, so the URLs must be noted.

You should use IT for your final report if possible, so if you are writing a formal report you should word process it. If you are presenting a poster or written pamphlet you may choose to use IT, and of course if you are presenting a podcast, video or designing a website then you should use IT for this. For an oral presentation you may decide to use PowerPoint®.

Can I have any help?

Your investigation is part of the controlled assessment and must be done under supervision. You are allowed to research your information in a library or homework club but *not* at home. When you put together your report, you must be under the direct supervision of your teacher. Your teacher will be able to discuss your ideas with you, but he or she will not be able to write any comments on your work and you are not allowed to submit a draft.

How will it be marked?

Your report will be marked in school by your teacher. Your tcacher will use the mark scheme set by OCR (your exam board). There are a total of 24 marks: 12 marks for application of knowledge and understanding, and 12 marks for analysis and evaluation. Some marks are allocated for writing concisely with good spelling, punctuation and grammar and for keeping within the word limit.

When teachers at your school have marked everyone's work, they will send a sample of work to the exam board to be moderated.

Top tips

- Use a variety of scales if appropriate (global, national, local).
- Organise your ideas into clear sections so your report is well structured.
- Use initiative in your research and acknowledge all your sources.
- Consider a variety of viewpoints from different groups of people.
- Have a clear conclusion, which answers the investigation question, and justify it.
- If you are writing a report, keep within the word limit and be careful about spelling, punctuation and grammar.
- Know the mark scheme that your teacher will use before you start writing.

Why is disease an issue?

A disease is an illness caused by an infection or a failure of health rather than by accident; it causes pain, distress and possibly death. Some diseases are infectious, and can be transmitted by coughs and sneezes, by hand to mouth, by bites of insects or other carriers of the disease, or from contaminated water or food. In addition, there are sexually transmitted diseases. Some diseases, such as some cancers and heart disease, are generally not caused by infection but are the result of lifestyle and are sometimes genetic.

How diseases are spread and the type of diseases people suffer from varies geographically. Life expectancy can fall due to problems such as famine, war, disease and poor health. On the other hand, improvements in health and welfare increase life expectancy. The map shows the global variation in life expectancy.

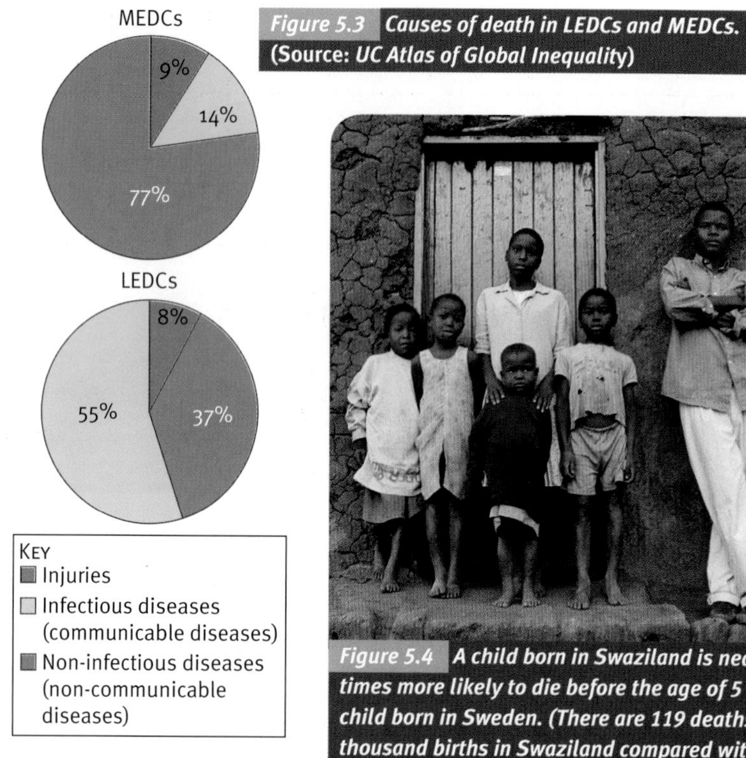

Figure 5.3 **Causes of death in LEDCs and MEDCs. (Source: *UC Atlas of Global Inequality*)**

Figure 5.4 ***A child born in Swaziland is nearly 30 times more likely to die before the age of 5 than a child born in Sweden. (There are 119 deaths per thousand births in Swaziland compared with four in Sweden.)***

Figure 5.2 ***Global variations in life expectancy.***

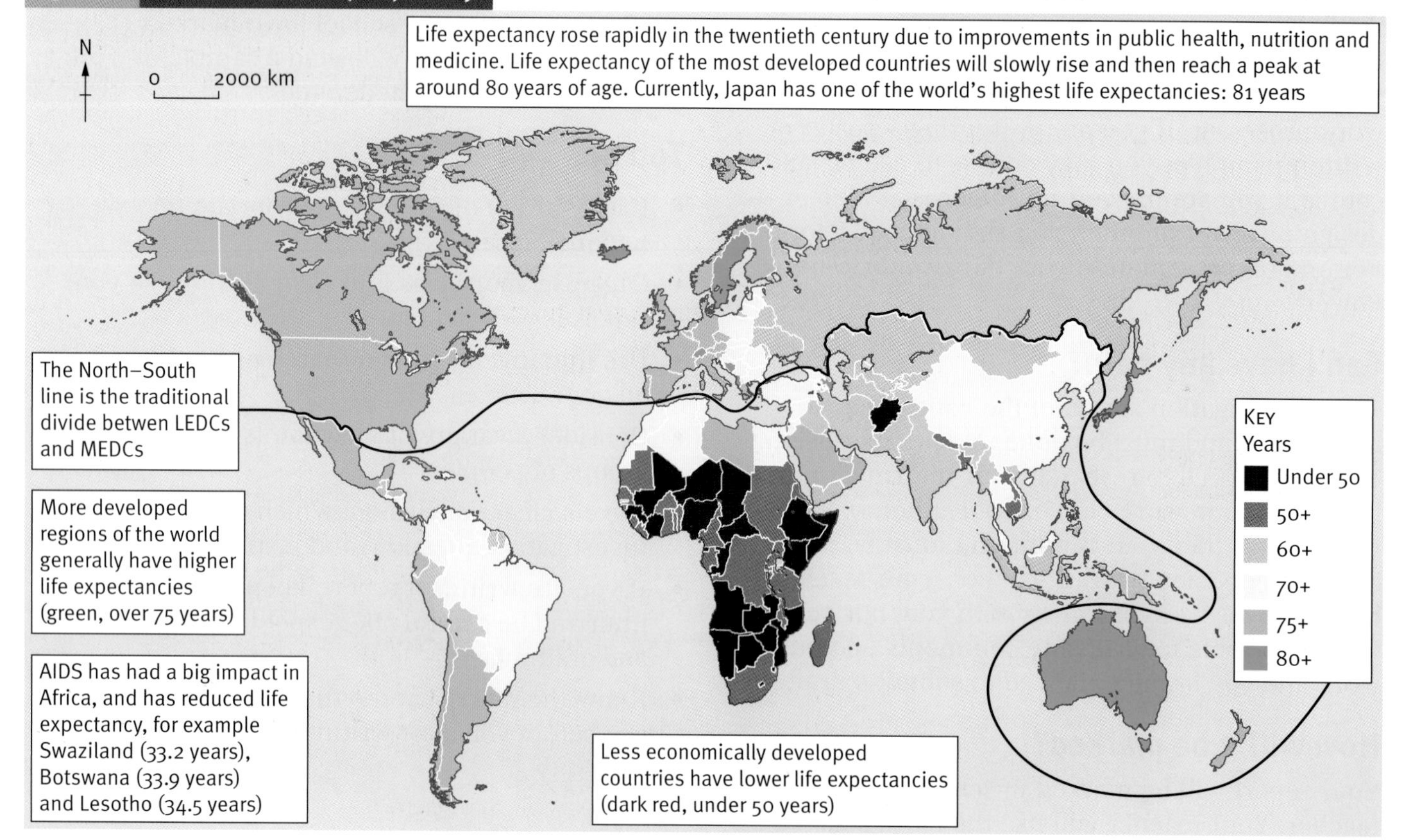

N
0 2000 km
Life expectancy rose rapidly in the twentieth century due to improvements in public health, nutrition and medicine. Life expectancy of the most developed countries will slowly rise and then reach a peak at around 80 years of age. Currently, Japan has one of the world's highest life expectancies: 81 years
The North–South line is the traditional divide betwen LEDCs and MEDCs
More developed regions of the world generally have higher life expectancies (green, over 75 years)
AIDS has had a big impact in Africa, and has reduced life expectancy, for example Swaziland (33.2 years), Botswana (33.9 years) and Lesotho (34.5 years)
Less economically developed countries have lower life expectancies (dark red, under 50 years)
KEY
Years
Under 50
50+
60+
70+
75+
80+

Table 5.2 Factors affecting health

	GDP per capita (US $)	Health spending (% GDP)	Adult literacy (%)	Life expectancy (years)	Infant mortality (per 1000 births)	Malnutrition (% under 5 years)	Stunted height (% under 5 years)	Immunised against measles (%)	HIV prevalence (% 15–49)	Access to improved water (%)
Malawi	800	12.0	64	48	76	18	53	85	14.1	73
Kenya	1700	4.5	74	52	77	18	41	77	6.1	61
Nepal	1000	5.3	49	61	64	39	49	85	0.5	90
India	2600	0	61	64	57	44	48	49	0.9	86
USA	45,800	15.9	99	78	7	1	0	93	0.6	100
UK	35,000	8.2	99	78	5	0	0	85	0.2	100

Why is there a variation in life expectancy?

- The type of disease that affects people varies significantly between LEDCs and MEDCs. In MEDCs, non-infectious conditions cause almost 80 per cent of deaths; these include cancers and heart disease. In LEDCs, over half of deaths are caused by infectious diseases, such as tuberculosis (TB), **HIV/AIDS**, malaria, diarrhoea, and measles. Look at Figure 5.3.
- Many of the diseases in LEDCs can be prevented by **immunisation**. There are programmes that have eradicated diseases such as measles and polio in some parts of the world.
- Global inequality in health care spending is large. Countries with very low life expectancy rates usually have extremely low numbers of doctors. Italy has 554 doctors per 100,000 people, whereas many countries in Africa have fewer than 25 doctors per 100,000 people.
- Environmental factors can play a role in death and disease, especially among young children. Chest infections, mainly pneumonia, are closely associated with exposure to indoor smoke from cooking with firewood. Diarrhoea accounts for 15–18 per cent of child deaths annually, and is largely caused by unsafe water, inadequate sanitation and poor hygiene.
- Inadequate food supply means that people who are malnourished are more susceptible to infectious diseases. Too much food, especially an unhealthy diet, causes **obesity**; people who are obese are susceptible to some non-infectious diseases such as heart disease and some cancers. This is more likely to affect people in MEDCs.

THINK ABOUT IT

Watch the video clip on the BBC website of Midge Ure talking about health problems in Africa. Why is it difficult for people in Sierra Leone to get good health care?

KEY TERMS

Communicable disease – an infectious or contagious disease that can be spread from one person to another.

Immunisation – being given a vaccine that helps the body's immune system fight infection.

Obesity – a condition in which a person has so much excess body fat that it affects their health.

RESEARCH LINK

Useful websites for data to show the contrasts between LEDCs and MEDCs are the UN Unicef website and the World Bank website.

ACTIVITIES

1. Choose one LEDC and one MEDC listed in Table 5.2. Describe what the table tells us about health in these two countries.
2. Suggest reasons why there is a large difference in life expectancy between Malawi (48 years) and the UK (78 years).
3. Draw a spider diagram to explain why countries with a high GDP such as the UK and USA generally have healthier populations than countries with a low GDP.
4. Why are some sections of Table 5.1 blank?

Does where you live determine how long you live – even in the UK?

The UK is an MEDC, one of the richest in the world with an average life expectancy of 78 years, yet this varies across the country. Look at the map of life expectancy in England, for example (Figure 5.5). People living in some areas have a higher chance of dying before they are 75 than others: for example, the life expectancy in East Dorset is 87 years.

The reasons for this can be partly explained by looking at the smoking rates and the obesity rates.

Smoking rates

If you smoke, your body is under increased risk of developing serious disease and health problems. All the major organs of your body are put under increased pressure by your habit, and smokers are more likely than non-smokers to develop a wide range of short- and long-term health problems such as cancer and lung diseases. This becomes a geographical issue when you see how deaths caused by smoking vary across the UK (see Figure 5.6, for England only).

Obesity

The number of people needing treatment for the consequences of clinical obesity – which can include heart problems, respiratory difficulties and damage to joints and bones – has also soared. In 1997–98, doctors recorded 23,951 such treatments. By 2006, that had risen to 85,302. The scale of the obesity crisis facing doctors in the UK has forced health ministers to draw up a strategy aimed at encouraging people to eat less and exercise more.

Disease and income

In MEDCs such as the UK, non-communicable conditions cause almost 80 per cent of deaths, including cancers and heart disease. Look at

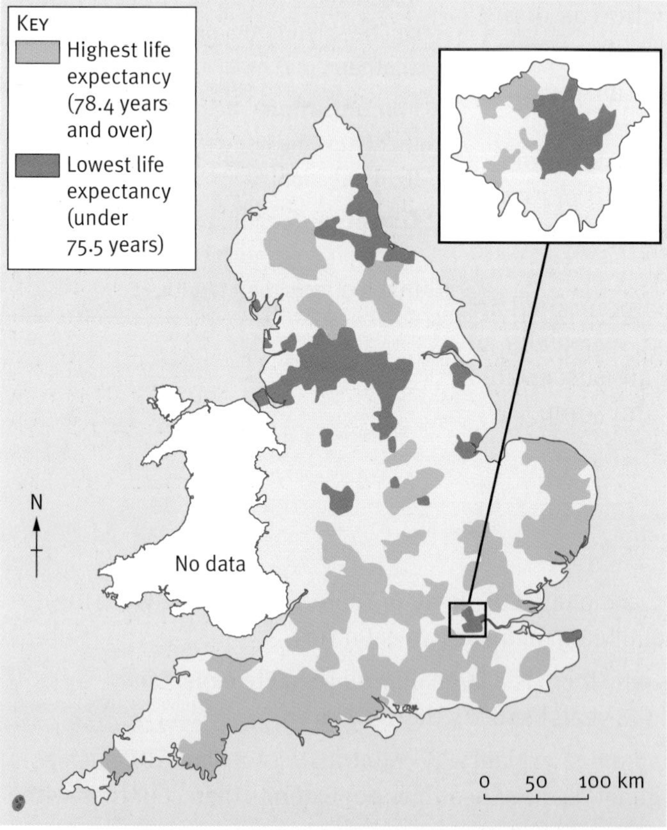

Figure 5.5 *Life expectancy at birth in England, males, 2002–2004.* **(Source: HMSO)**

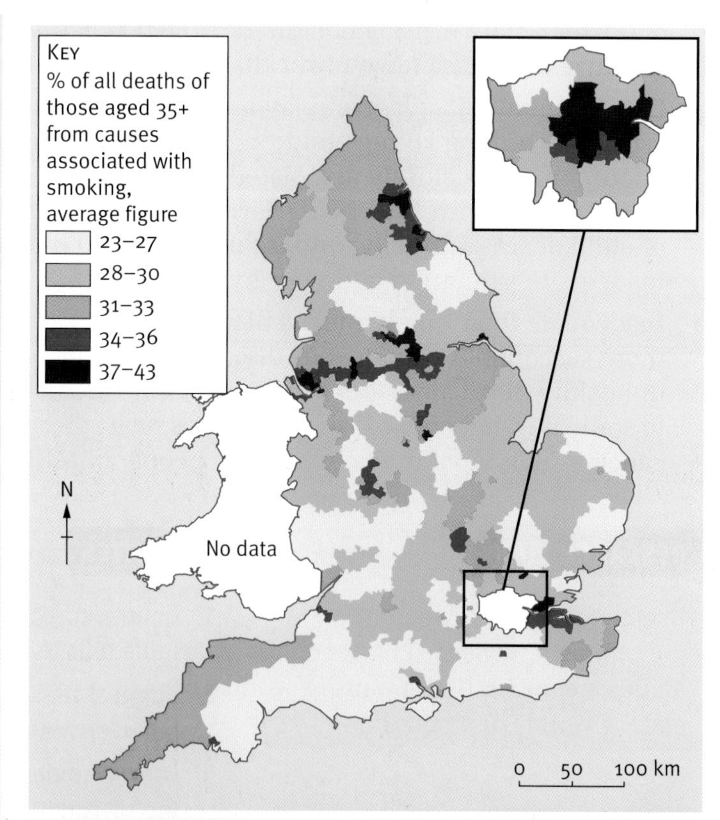

Figure 5.6 ***Deaths caused by smoking in England, 1998–2002.*** **(Source: Health Development Agency, *The smoking epidemic in England*, 2004, HDA-London, reproduced with permission)**

Figure 5.3 on page 216. These can be called 'diseases of affluence'. However, it is often not the affluent or well-off in these countries that suffer from them: poorer people are more likely to suffer. The pattern of disease reflects the pattern of other inequalities such as income and education. For example, 31 per cent of people in low-earning routine and manual jobs smoke, whereas only 19 per cent of people in high-earning managerial and professional jobs smoke. This can cause large contrasts in health over a small area. For example, life expectancy in Ecclesall, a suburb in the west of the Sheffield, is 88 years, whereas in the much less affluent Flowers estate in the east of the city, it is only 72 years.

Figure 5.8 A fitness camp in Rotherham.

What explains this large difference? Figure 5.7 shows how life expectancy is related to education.

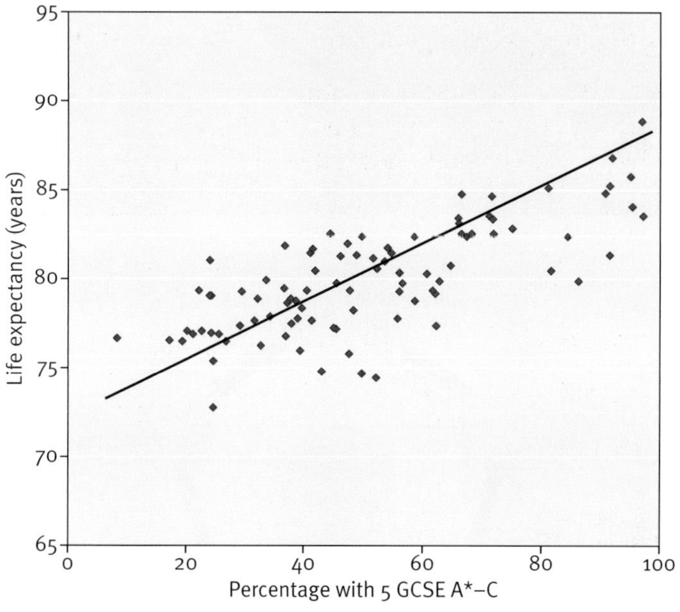

Figure 5.7 *The relationship between life expectancy and education in Sheffield areas.* (Source: NHS Across Sheffield)

Fight the fat in Rotherham

To combat the problem of obesity, the government in the UK is spending more than £1 million in one area of the country, on getting obese children in Rotherham to attend special residential fit camps where they'll spend 6 weeks losing weight and learning how to lead a healthier lifestyle.

It is estimated that 1 in 3 children in the town is overweight or obese, and 6 in 10 adults. Thirty-eight 8- to 16-year-olds are on a summer-long scheme, which is a last resort after other ways of losing weight have been tried and failed.

The parents of the children at the camp also have to get involved, and there are weekly check-up sessions after the residential course to make sure all the good things the children learn are actually put into practice at home. There are also healthy cooking classes for parents.

ACTIVITIES

1. It is said there is a North–South divide in life expectancy in the UK. Do you agree with this statement? Use the information on the maps in Figures 5.5, 5.6 and 5.7 to explain your ideas.
2. Look at the scattergraph for Sheffield wards in Figure 5.7. What does this scattergraph show? Why do you think that areas with higher levels of education also have a higher life expectancy?
3. What could the government do to try to increase the life expectancy in those parts of the UK where it is below average? Look at the Rotherham example and the video produced by the NHS's SMOKEFREE website.
4. Why does it matter if life expectancy differs by 16 years across one city?

Why does disease have such a big impact in LEDCs?

Figure 5.9 Every day 4000 children die of diarrhoea caused by dirty water.

What is the consequence of poor quality health care?

In LEDCs over half of deaths are caused by communicable diseases such as TB, HIV/AIDS, malaria, diarrhoea and measles. Much of this disease is preventable. Look at Figure 5.3 on page 216.

Disease can increase poverty too, because illness affects people's work and this damages economies. This is shown by the cycle of poverty diagram. People get trapped by poor health and are unable to escape from poverty.

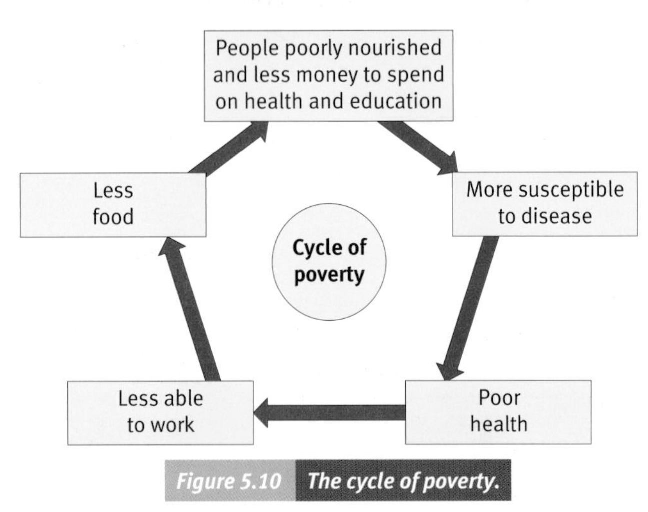

Figure 5.10 The cycle of poverty.

What are the big killers in LEDCs?

Malaria causes 1 in 5 of all childhood deaths in Africa. It is both preventable and treatable. Malaria is a disease that is contracted when a parasite is passed on from the bite of an anopheles mosquito. The symptoms of malaria are many and include fever, headaches, tiredness, nausea and diarrhoea. Infection can lead to death, although with anti-malarial drugs a full recovery can be made. The use of insecticide-treated nets can protect people from bites while they sleep and therefore prevent the disease.

In LEDCs every hour, 300 people die of an AIDS-related illness.

HIV/AIDS was first identified in 1985. While treatments are available that reduce its effects, there is no cure. It attacks the immune system and leaves people prone to infections that weaken and eventually kill them. It is transmitted by sex without the use of condoms, contaminated needles and from mother to baby during pregnancy. By the end of 2007 it was thought that 33.2 million people were living with the disease. Twenty-five million people have died from HIV/AIDS. Although the disease has affected both MEDCs and LEDCs, the highest number of people who are affected live in Sub-Saharan Africa (the countries south of the Sahara Desert).

The size of each country shows the proportion of all people aged 15–49 with the HIV virus living there. In 2003, the highest HIV prevalence was Swaziland, where 38%, or almost 4 in every 10 people aged 15 to 49 years, were HIV positive. All ten territories with highest prevalence of HIV are in central and south eastern Africa.

Figure 5.11 *The geography of AIDS.* **(Source: Copyright 2006 SASI Group (University of Sheffield) and Mark Newman (University of Michigan))**

What is the impact of AIDS?

HIV and AIDS cause huge problems in sub-Saharan Africa. They reduce life expectancy and slow development. The long incubation period between when a person becomes HIV positive and when a person develops AIDS means that the disease is spread more widely. It also makes containing and controlling the disease more difficult. Here are some of the impacts of the disease.

- Many children are left without parents, or left with very sick parents whom they have to care for.
- AIDS patients cannot get proper care or anti-viral drugs, which means that HIV/AIDS patients die much faster than in other countries.
- More people die as young adults. This may mean there are not enough knowledgeable and well-trained individuals. In Zambia, more teachers die of AIDS each year than are graduating as trained teachers.
- More money has to be diverted to health care, and away from other types of services that are also desperately needed, such as education.
- Household income declines while adults are sick and more money is spent by the family on health care. There is less food produced and less money to spend on food.

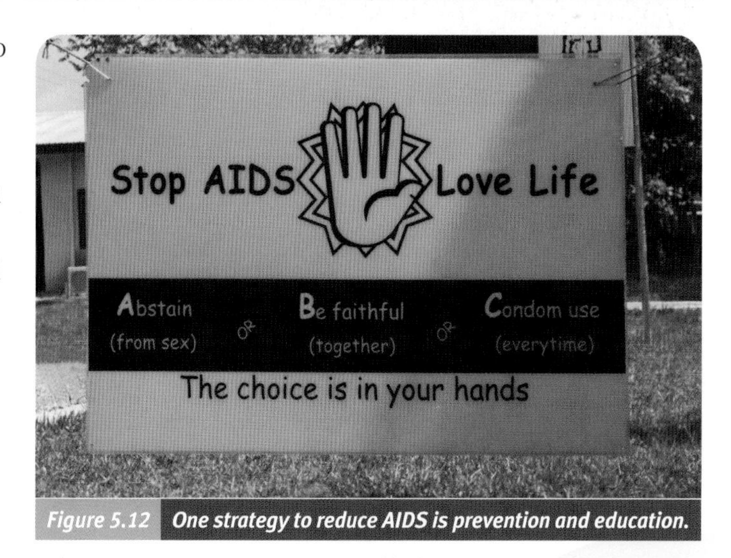

Figure 5.12 *One strategy to reduce AIDS is prevention and education.*

ACTIVITIES

1 Look at Figure 5.11. The world map is a strange shape, as the size of each country shows the proportion of all people aged 15–49 with the HIV virus.
   a Compare the size of the countries with the land area map. Which continent is the largest?
   b In your opinion, is this an effective way of showing information on a map? Justify your answer.
   c Suggest ways in which it could be shown in an even more useful way.

RESEARCH LINKS

Look at the Worldmapper website, where you can find similar maps for many diseases. You could select and annotate one for your controlled assessment.

- For more research on AIDS, use the UN AIDS website.
- For research on malaria use the Roll Back Malaria Partnership website.

On the BBC website watch the video clip 'AIDS awareness day is marked 1 December 2007' about AIDS orphans in South Africa. What problems has AIDS caused for the family on the video? What do you think could be some solutions for the family?

Why is trade an issue?

Trade is the movement of goods and services between countries. Countries sell products to make money in order to buy products they need or want. MEDCs have become rich mainly because they have made money trading with other countries. The pattern of world trade is very uneven. MEDCs produce 80 per cent of the world's **exports**.

LEDCs tend to trade in primary products, such as crops (sugar, tea, coffee) or raw materials (copper, iron ore). These are relatively low-value products and prices fluctuate depending on supply and demand. There is not a guaranteed price. LEDCs make little money from their exports and have to **import** expensive manufactured products. This means they often have a **trade deficit**.

MEDCs tend to have more trade in manufactured goods, such as cars and electronic goods. These are high value and prices are more stable. This means they usually have a good **balance of trade**, or **trade surplus**, with low-value imports and high-value exports at a profit.

This is one significant reason why MEDCs are rich and can afford to spend money on health care and education. LEDCs have much less money to spend on health and education because of their trade deficit.

Figure 5.13 Global exports in millions of dollars.

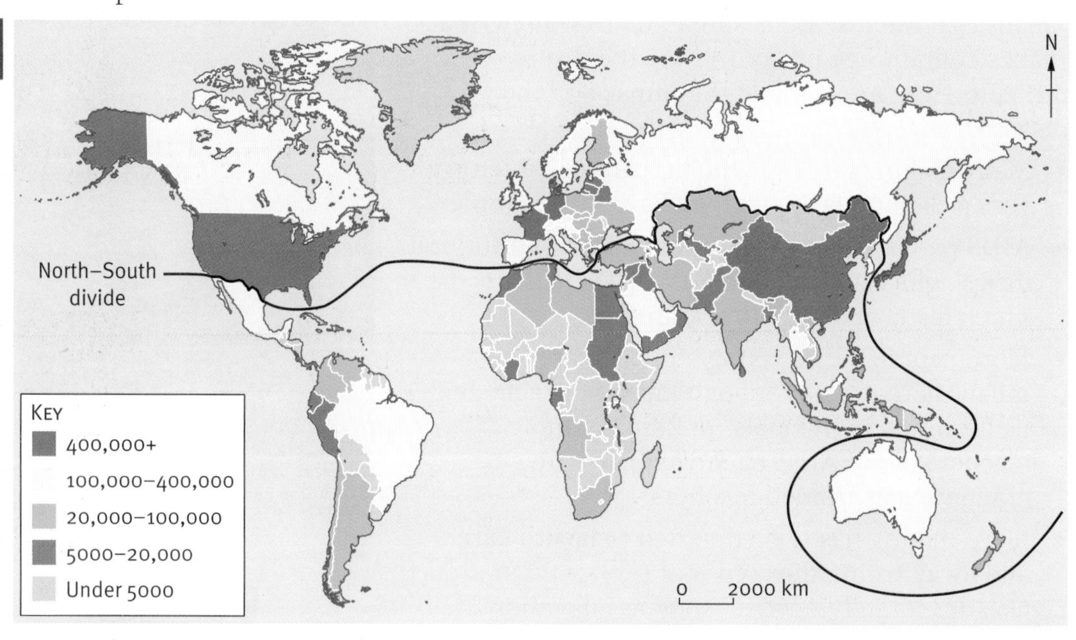
N
North–South divide
KEY
400,000+
100,000–400,000
20,000–100,000
5000–20,000
Under 5000
0 2000 km

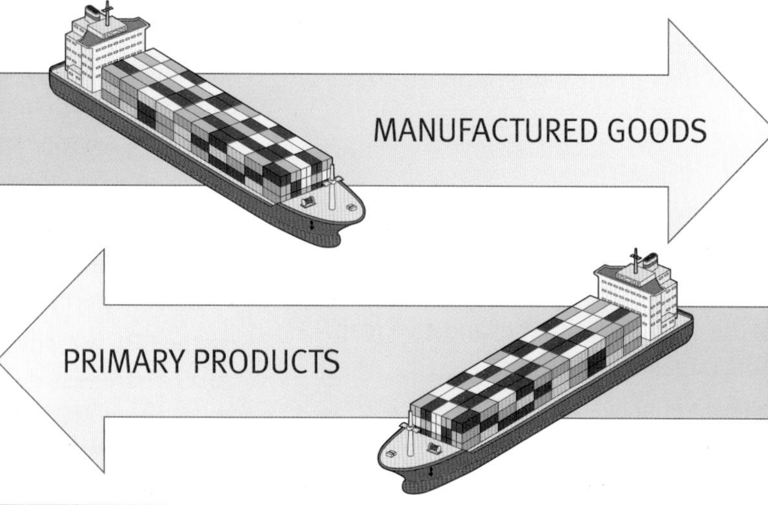
MEDCs
Produce manufactured goods from raw materials. These are worth a lot of money as value is added in manufacture. They are exported to LEDCs at high prices.
MANUFACTURED GOODS
PRIMARY PRODUCTS
LEDCs
Produce raw materials such as agricultural products e.g. tea, coffee. These are worth little money. Prices also fluctuate, depending on supply and demand. They are exported to MEDCs at low prices.

Figure 5.14 Trade between LEDCs and MEDCs.

Trade in Malawi

Malawi, a country in southern Africa, is one of the world's poorest countries.

It is the second largest tea producer in Africa, and was the first country in Africa to grow tea on a commercial scale. Today, large tea plantations account for 93 per cent of tea production, with the rest grown by some 6500 smallholders. Most of the tea is bought by major international companies from the UK and South Africa. Malawi is producing much better tea harvests than 10 years ago. However, the demand for tea is not keeping pace with the expanding supply, especially as increasing volumes of medium-grade tea is entering the market from Vietnam and China. This means that the value of Malawi's tea is getting less. Meanwhile the cost of imports, especially those from oil, are increasing.

Figure 5.15 ***Tea leaves from a plantation in Malawi are collected, sorted and processed.***

Table 5.3 Trade in Malawi

Malawi's trade	Products traded	Value of products (US $ millions)
Exports	Tobacco, tea, sugar, cotton, coffee, peanuts, wood products	657
Imports	Food, products made from oil, manufactured consumer goods, transportation equipment	892

THINK ABOUT IT

Malawi is typical of many of the poorest countries in the world. What is the impact of such a trade deficit on the people who live in countries like Malawi?

KEY TERMS

Balance of trade – the difference between the cost of imports and the cost of exports.

Exports – goods that are sold to other countries.

Gross National Income – total value produced within a country together with its income received from other countries.

Imports – goods that are bought from other countries.

Trade deficit – when the cost of imports is greater than the money made from exports.

Trade surplus – when the money made from exports is greater than the cost of imports.

ACTIVITIES

1 Study Figure 5.13 and name the five countries with the highest exports.
2 Where are the countries with the lowest-value exports? Give some reasons why a country may have low-value imports.
3 How does the pattern of high-value and low-value exports relate to the position of the North–South divide? Why is China an exception?
4 Look at Malawi's trade figures in Table 5.3. What is its balance of trade?
5 Why is Malawi's tea production unlikely to improve its balance of trade in the future?
6 How do you think that Malawi could improve its balance of trade?

Is trade unfair?

Why can't Grace sell her chickens at the market?

Grace Kumajor goes to the market in Kumasi, Ghana, but is finding it more difficult to sell her chickens. The reason is clear. 'Walk into any of the supermarkets and you will find that they are bulging with imported frozen chicken,' she says. 'People don't want to buy local chicken because imported chicken is much cheaper.'

For the last few years, the Ghanaian market has been flooded with cheap imported chicken from the European Union (EU). EU farmers receive generous subsidies for their products and so there are more chickens produced than are needed. European customers prefer the fillet to the chicken legs because of the bones, so there is no demand and no market for the rest of the chicken parts. The EU then sell it to LEDCs at prices that are so low, they ruin local markets. This phenomenon is known as 'dumping'.

Imported chicken also has its health hazards, as it tends to thaw out between freezing several times in transit from the EU to Africa. The Ghanaian health service does not have the resources to detect and prevent an outbreak of salmonella, which might accompany imported chicken.

In 1992, domestic poultry farmers supplied 95 per cent of the Ghanaian market, but by 2001 their market share had shrunk to just 11 per cent, threatening the livelihoods of over 400,000 poultry farmers like Grace and her family.

The current world trading pattern often seems to work against people like Grace in Ghana and other producers in LEDCs, and does not operate on the principles of **free trade**. There are many reasons for this.

Colonial past

Many LEDCs used to be colonies and were governed by another country. For example, many African countries were colonies of European countries, such as Ghana, which was ruled by the UK. Colonies provided their rulers with a cheap source of raw materials which they used in their own manufacturing. Few colonies developed industries of their own, and so although they are now independent they still rely on MEDCs for manufactured goods. In the 1970s and 1980s many LEDCs took out loans from banks and MEDCs, and since interest rates have risen they have been left with huge debts to repay.

THINK ABOUT IT

What are all the reasons why Grace can't sell her chickens? How is the global trade system unfair to Grace and the chicken producers of Ghana? How do people in the UK affect Grace?

Figure 5.16 **Grace has a small farm. She grows crops to feed her family and also raises chickens, which she sells to bring in some extra money.**

Trading blocs

Trading blocs are groups of countries that have trade agreements. They encourage trade between themselves without charging any **tariffs**. But they then charge tariffs on any imports from outside the trading bloc, making these goods more expensive. The EU is an example of a trading bloc, so is the North American Free Trade Association (NAFTA).

Subsidies

Within the trade blocs such as the EU, farmers are paid subsidies to encourage them to increase production and make their products cheaper on the world market. This makes it even harder for LEDCs to compete and often food produced by MEDCs is cheaper than locally produced food in LEDCs, such as chicken in Ghana.

The European Union

The EU is the most powerful trading bloc in the world, with about 30 per cent of the world's GDP. It has 500 million citizens in 27 countries. The EU has developed a single market, guaranteeing the freedom of movement of people, goods, services and capital between its members. Fifteen countries have adopted a common currency, the Euro. Through its Common Agricultural Policy the EU gives subsidies to its farmers. Tariffs encourage trade between EU countries and make it more difficult for countries outside the EU to trade with its members.

Fact file

The World Trade Organisation (WTO) is an international body whose purpose is to promote free trade by persuading countries to abolish import tariffs and other barriers. The WTO is the only international agency overseeing the rules of international trade. The organisation frequently has talks to discuss tariffs and subsidies and mechanisms to protect poor farmers, but these have made little progress.

Figure 5.17 ***The European Union, the most powerful trading bloc in the world.***

KEY TERMS

Free trade – trade between countries when the prices paid are determined by what the producer wants to be paid and what the consumer is prepared to pay and there are no barriers.

Tariff – a tax put on goods when they are moved from one country to another.

Figure 5.18 ***The EU logo.***

RESEARCH LINK

To find out more about the WTO watch the video 'World trade talks collapse' on the BBC website. The WTO also has its own website.

ACTIVITIES

1. Draw a spider diagram showing all the reasons why trade is unfair, using the information from these pages.
2. Is world trade free trade? How could the WTO help trade be more just for LEDCs?
3. Who in the EU would be prepared to spend less on supporting its farmers? Who wouldn't? Explain your answer.

Can trade be fair?

Bananas are the UK's most popular fruit. Until recently, for every £1 spent on bananas in a supermarket, only 4p went to the producer and plantation worker. However, this is changing. In 2000, the first Fairtrade bananas were available in UK supermarkets and now they are available in all supermakets.

Fairtrade products began with raw foods such as bananas and cocoa, to ensure that farmers were being given a fair price for the food that they grew. This is why the FAIRTRADE Mark can be seen mostly in supermarkets on the actual food products.

Figure 5.19 **The FAIRTRADE Mark. Any product that shows this logo has been fairly traded.**

There are now increasing numbers of products such as chocolates and cotton and clothes such as T-shirts in Marks & Spencer. There are now even sport balls carrying the FAIRTRADE Mark.

The demand for Fairtrade-certified foods has grown rapidly; in 2003, there were approximately 150 certified Fairtrade products and by 2007 they were over 3000.

Divine tasting chocolate

Life is hard for cocoa farmers – they are some of the poorest people in the world and on average earn about £50 a year. In Ghana, West Africa, there are about 2 million cocoa farmers. They depend on selling their beans to pay for the essential things in life, such as school fees, doctors' bills, farm tools and wellington boots to protect their feet from the scorpions that live among the cocoa trees. However, world prices for cocoa are now so low many farmers often cannot even afford the basics.

In 1993, a small group of Ghanaian cocoa framers set up their own co-operative business called Kuapa Kokoo. In 1994, Comic Relief gave a grant so that Kuapa could develop their business. Now Kuapa has over 35,000 members in over 650 villages and produces 1per cent of the world's total cocoa crop!

Fact file

What does Fairtrade mean?

- Focuses on poor producers to help them develop a sustainable livelihood through trade.
- Pays a fair price that covers costs and a living wage.
- Pays extra money for social development work.
- Encourages fair treatment of all workers.

In 1998, Kuapa Kokoo joined forces with The Day Chocolate Company. This meant that for the first time, cocoa growers began to benefit from selling chocolate as well as growing cocoa beans and they launched their first chocolate bar – Divine milk chocolate – and later developed the Dubble chocolate bar.

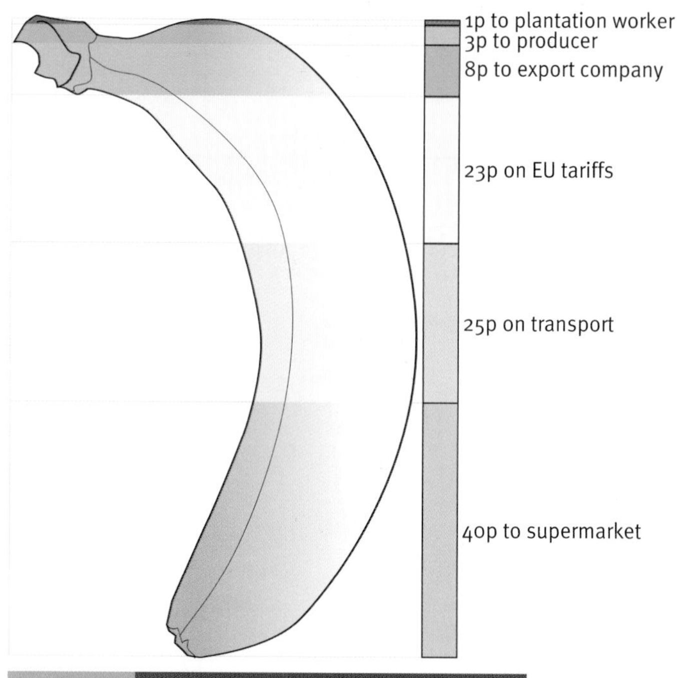

Figure 5.20 **Who gets the money spent on bananas?**

RESEARCH LINK

There are many stories on the internet of how buying Fairtrade products is making a difference to farmers. Look at the Divine Chocolate or Dubble websites for more information. The Fairtrade Foundation website also has information on how Fairtrade works.

Figure 5.21 Fairtrade products.

The cocoa farmers benefit as:

- they earn more money per tonne for beans
- money from the Fairtrade is spent on community projects, such as clean water and better toilets
- a share of the profits at the end of each season is for personal and community use
- they have training in quality control, record keeping and management.

Some people think that the Fairtrade movement is not the whole answer, but that the trade rules need to change. The Trade Justice Movement campaigns for trade justice, with the rules weighted to benefit poor people and the environment. They think that the trade rules benefit the rich and powerful countries, putting profits before the needs of people and the planet.

RESEARCH LINK

Look at the Trade Justice Movement website to see how it is trying to change trading rules to benefit poor producers.

ACTIVITIES

1. How does Fairtrade work?
2. What are the benefits to the producer?
3. Can Fairtrade products help LEDCs develop?
4. Why is it more beneficial to LEDCs if the Fairtrade products are not just raw materials, but also manufactured products such as shoes, clothes and footballs?
5. Are there any benefits to the consumer?

THINK ABOUT IT

Watch the video clip of the 'Bishop's chocolate challenge' on the BBC website. Do you think this is a good idea? How would you encourage people to buy Fairtrade products?

What are ecosystems?

Tundra
Climate:
Very low rainfall (less than 250 mm), short cool summers (6–10°C), long cold winters well below freezing.
Vegetation:
It is too cold for trees to grow. Grasses, flowers, mosses and lichens.

Coniferous forest
Climate:
Low rainfall all year (less than 500 mm), cool summers and below freezing for the winter months.
Vegetation:
Trees are evergreen with needles.

Desert
Climate:
Very dry all year (less than 250 mm/yr rainfall), very hot temperatures in summer (over 40°C), can be very cold at night.
Vegetation:
Very few plants can survive, only those specially adapted such as cacti, and those with very deep roots.

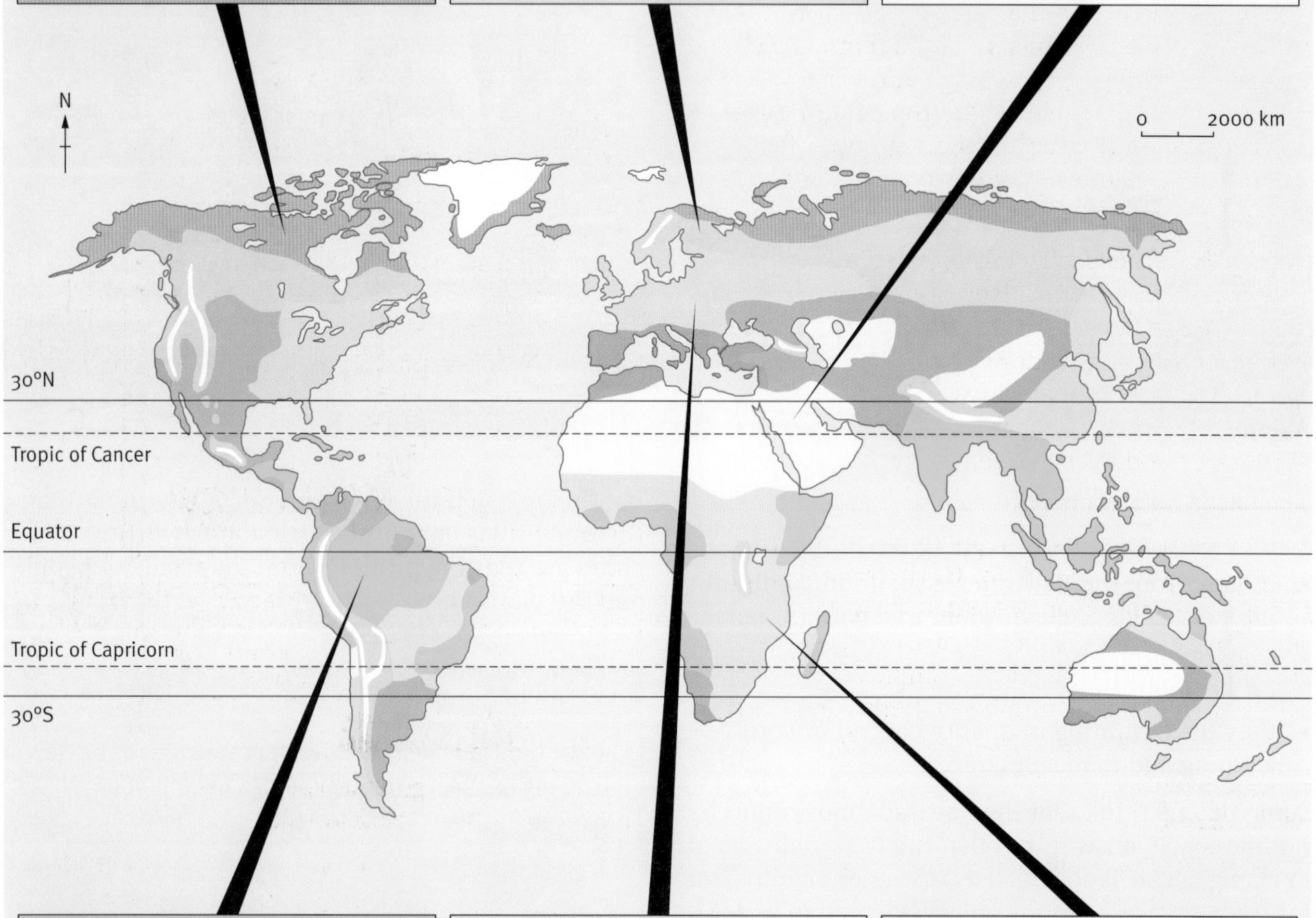

N
0 2000 km
30°N
Tropic of Cancer
Equator
Tropic of Capricorn
30°S

Tropical rainforest
Climate:
High rainfall all year (over 2000 mm/yr), constant high temperature all year (26–27°C).
Vegetation:
Very tall trees, high density and great variety, layers of forest with continuous canopy.

Temperate deciduous forest
Climate:
Rain all year (500–1500 mm), low teperature range, cool summers (15–20°C), mild winters above freezing.
Vegetation:
Deciduous trees that lose their leaves in winter, shrubs and grasses.

Savanna
Climate:
A dry season with several months of very little rain and a wet season with high rainfall, hot in the summer (30°C), warm winters (20°C).
Vegetation:
Deciduous trees that lose their leaves in winter, shrubs and grasses.

Key
Tropical rainforest
Savanna
Desert
Mediterranean
Temperate deciduous forest
Coniferous forest
Temperate grassland
Tundra
Polar and high-mountain ice

Figure 5.22 ***The world's major ecosystems.***

An ecosystem is a natural system in which the living parts (animals and plants) and non-living parts (water, air, sunlight, soil and rock) interact and interrelate.

Ecosystems vary in size from a small pond to vast grass plains. They also vary in their type, from barren deserts to complex tropical rainforests. The climate is a major influence on the plants and animals that live in an ecosystem. Rain provides water and sun provides heat, which are both essential for photosynthesis and therefore for plant growth.

Energy flow

The sun provides energy which, along with water and CO_2, is made into new plant material by the process of **photosynthesis**. Where there is plenty of rainfall and sunlight, ecosystems are very productive, for example in the tropical rainforest. However, when they are in short supply – for example, where the conditions are very dry or very cold – there is little plant growth.

Photosynthesis can only take place when the temperature is over 6°C. However, plants can adapt to survive in these harsh conditions. For example, in desert seeds can lie dormant or inactive until there is a sudden rainstorm and then come to life and quickly flower. Baobab trees in the savanna store water in their massive trunks to last them through the dry season.

When new plant growth happens, the energy is passed along the food chain. **Herbivores** eat green plants and **carnivores** or predators eat the herbivores or other, smaller carnivores. In reality this happens through quite a complex food web, as plants are eaten by a number of herbivores and carnivores eat several different animals.

Nutrient cycling

Nutrients are chemicals that are needed by plants and animals. Rocks are broken down by weathering to form the basis of soil, and this provides some nutrients. Others are added from the breakdown of dead animals and plants by decomposers: fungi, bacteria, worms and other insects. Nutrients dissolved in rainwater are taken up by plants, through roots. The nutrients then become part of the living plant tissue and eventually pass into animals along the food chain. To complete the cycle, when the plants and animals die their nutrients return to the soil by decomposition (Figure 5.23).

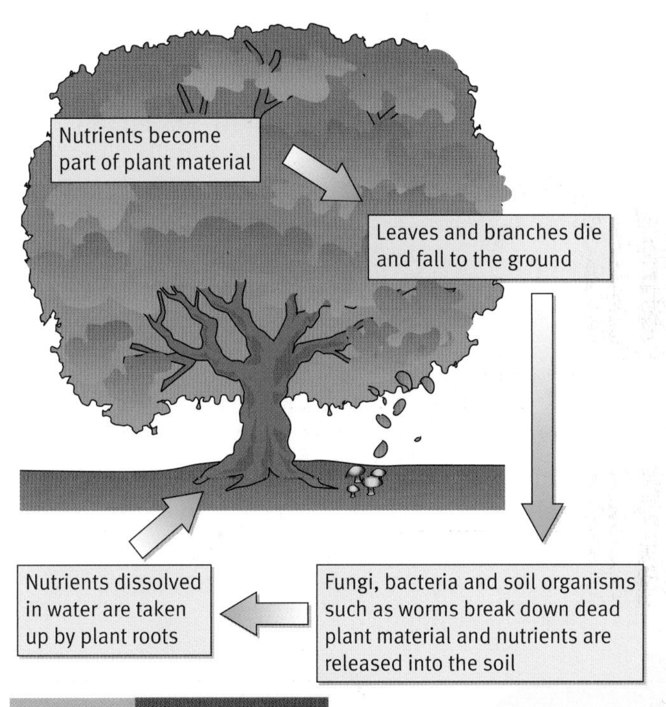

Nutrients become part of plant material
Leaves and branches die and fall to the ground
Fungi, bacteria and soil organisms such as worms break down dead plant material and nutrients are released into the soil
Nutrients dissolved in water are taken up by plant roots

Figure 5.23 The nutrient cycle.

ACTIVITIES

1 Write a sentence about how these parts of an ecosystem are linked:
   a climate and plants
   b animals and plants
   c animals and decomposers
   d soil and plants
   e climate and animals.
2 Plants are specially adapted to their environment. Find out how one plant is ideally suited to each of the six major ecosystems in Figure 5.22.
3 Find examples of two herbivores and two carnivores that live in each of the world's major ecosystems.
4 What is the major ecosystem in the UK? Are there any examples of this near where you live?

KEY TERMS

Carnivore – an animal that eats other animals, also known as a predator.

Herbivore – an animal that eats only plants.

Nutrients – chemicals used in plant growth.

Photosynthesis – a process by which green plants turn sunlight into plant growth.

Why are ecosystems an issue?

Ecosystems have been affected by human activity for thousands of years, as people have destroyed the natural vegetation in order to use the land for other activities. In the last 100 years, as the human population has increased, the demand for the Earth's resources (and the technology to access them) has also increased, and most of the world's ecosystems have become under threat.

Figure 5.25 *Swallow's Wood contains many species of trees, including beech, birch, oak, and lime, flowers such as orchids and iris, animals such as badgers, voles, bats and hares, and birds such as jay, kingfisher and cuckoo.*

Britain's ecosystem: what has changed?

In 7000 BC, 80 per cent of the UK was covered by a temperate woodland ecosystem: woodland made up of the native species of trees such as oak, ash, elm, beech, lime and alder. Now only 6 per cent of the ancient woodland remains, as woodland has been cleared for farming, industry, roads and settlements. There are still threats to the natural woodland today.

Building a by-pass

The A57/A628 connects Manchester with Sheffield and it passes through the villages of Mottram and Hollingworth. The high volume of traffic, including large numbers of heavy goods vehicles, causes long delays and congestion for traffic during peak hours. There is noise, visual intrusion and an unpleasant environment for residents, as well as a high accident rate.

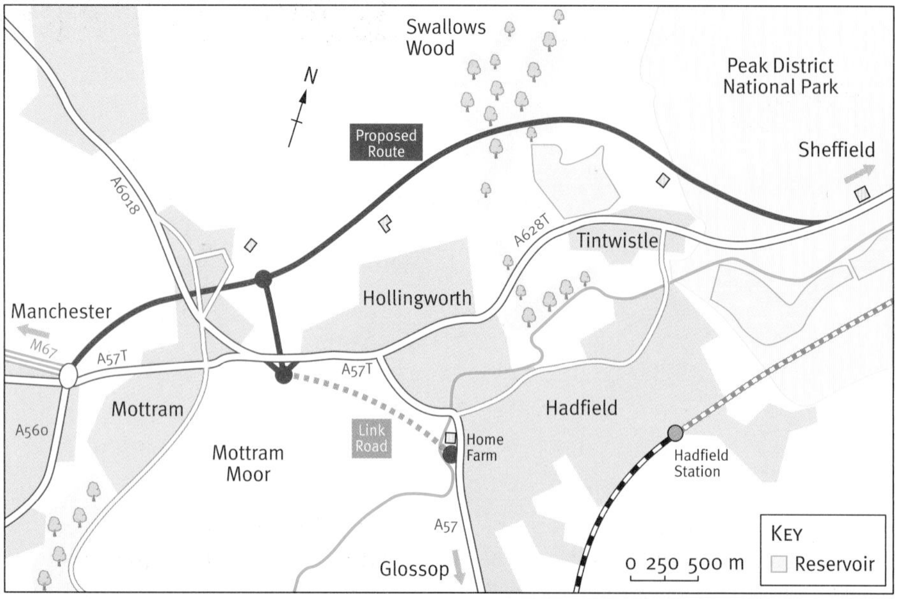

Figure 5.24 *The proposed route of the by-pass. 1.3 km of the road would pass through the Peak District National Park and also pass through green belt.*

RESEARCH LINK

Suggest some alternative strategies to building the by-pass that could result in solving the traffic problems of the Mottram and Tintwistle area. The Campaign to Protect Rural England (CPRE) has suggested some ideas, and you could also see the Save Swallowswood campaign website.

A 5.6-km by-pass has been proposed that would go around the villages and link the M67 to the A628 Woodhead Pass (Figure 5.24). The route would pass through Swallow's Wood. Swallow's Wood is a nature reserve containing woodlands, meadows, ponds and marsh areas. There has been a long campaign against building the by-pass. However, many people are very keen for it to be built.

The Tundra ecosystem: what's changing?

The Arctic tundra is a very inhospitable environment, but its fragile ecosystem is under threat as people are demanding more resources that are found in these areas. Look at Figure 5.22 on page 228 to locate this ecosystem.

Figure 5.26 *Twenty-four hours a day, 7 days a week, the chimneys of the nickel ore smelter at Norilsk pump out a toxic cocktail of pollutants, causing acid rain and smog.*

Mining in Norilsk

Norilsk, in Siberia, is one of the northernmost cities on the planet. The average temperature is approximately −10°C, and can be as low as −50°C. The city is covered with snow for over 250 days a year. The polar night lasts from December through mid-January, so that the people of Norilsk do not see the Sun at all for about 6 weeks.

Norilsk is situated on some of the largest mineral deposits on Earth and has become the centre of a region where nickel, copper, cobalt, platinum, palladium and coal are mined and smelted. Norilsk is now one of the 10 most polluted cities in the world. A major pollutant is sulphur dioxide, which combines with water in the atmosphere to form acid precipitation, which often falls as yellow acid snow that has been recorded at pH3. As a result there is not a single living tree within 48 km of the nickel smelter.

Much of the reindeer pastureland has been lost; this is because the mosses and the lichens have become poisonous. Some areas are like deserts; the vegetation has gone and the soil is badly eroded. The lakes in the surrounding area are becoming polluted by streams contaminated by the smelters. People are also affected; the life expectancy of a worker in Norilsk is 10 years less than for the average Russian.

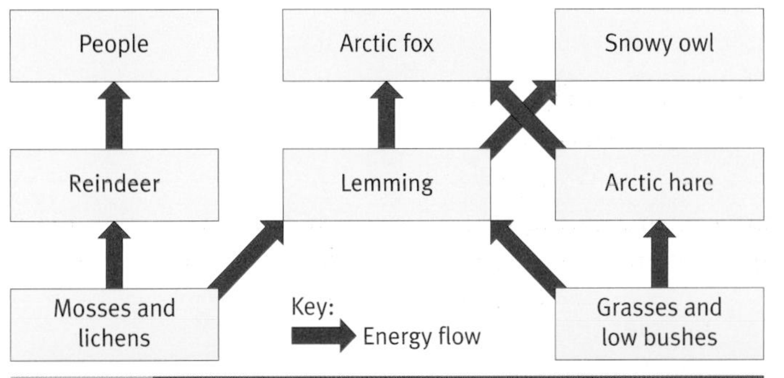

Figure 5.27 *The tundra food web is quite simple and very fragile. It is a difficult environment for plants and animals to survive in.*

THINK ABOUT IT

Norilsk is a very remote area. The world needs metals such as nickel and the Russian people need to earn money. Does it matter what happens to the tundra? How could the industry in Norilsk reduce its impact on the ecosystem?

RESEARCH LINK

Watch the film about Norilsk Nickel and study its environmental policy on the company website. Does that change your opinion about nickel smelting in Norlisk?

ACTIVITIES

1. What are the advantages of building the by-pass in Figure 5.24? What groups of people would support the building of the by-pass?
2. What are the disadvantages of the proposed by-pass? What groups of people would be against the proposal?
3. Do you think it matters if woodland areas such as Swallow's Wood have roads through them?
4. Describe the tundra food web using the terms 'plants', 'herbivores' and 'carnivores'. How does pollution pass along the food chain? Give some examples.
5. Find a map that shows the location of Norilsk, and annotate it to show the characteristics of the ecosystem and how this is being damaged by the industry of Norilsk.
6. Why does the Russian government tolerate the smelter at Norilsk?

How can ecosystems be managed for the future?

People can damage ecosystems but can also implement strategies to manage ecosystems for a sustainable future.

Why can't Mr Bundani's farm feed his family?

Desertification is causing the destruction of savanna ecosystems as they are changed into desert. In the last 50 years, the Sahara desert has spread southwards to cover an extra 65 million hectares. It leads to a reduction in food and an increase in poverty, and often results in people leaving the rural areas to go to towns and cities. When desertification is combined with drought it can lead to famine, malnutrition and death. Desertified land can be returned to savanna if the land is managed well.

Figure 5.28 ***Mr Bundani is a substance farmer in Burkina Faso. He tells what has happened to his land.***

Our land used to be good land, and we could feed our families from the crops we grew and our animals provided us with milk and meat. The rains came earlier and lasted longer. But times have changed and we can hardly survive. Our land is now desert.

There were more people living in our village, which meant we had to grow more food. So instead of leaving some fields empty each year to give them time to rest and recover, we planted crops every year. The soil little by little became exhausted and we have had to abandon some fields. Our cattle have over-eaten the grass and less grows back each year. We needed more and more wood for fuel and the trees over years were chopped down and never re-grew. The land was exposed to the wind and the topsoil was blown away, and the ground became as hard as concrete and the rains were no longer able to soak into it.

Fact file

Strategies that can help return desert back to productive savanna:

- small barriers are built in the soil to retain rain water and allow it to seep into the ground, instead of draining away
- hedges of trees or shrubs are planted that ultimately serve as protection against the wind. These plants also provide firewood, fodder and fruit. Belts of trees and grass can be used to reduce wind velocity and stop sand spreading
- leguminous plants, such as beans, which 'fix' nitrogen from the air, can be used to help restore fertility in already damaged areas
- the ground is covered with straw in the dry season to keep it more moist
- local people are trained to manage the scheme themselves
- fuel-efficient stoves and solar cookers help reduce the use of fuelwood.

RESEARCH LINK

Desertification happens in many parts of the world. The Gobi Desert in China is advancing on agricultural land. The Chinese solution is a series of measures known as the 'Green Wall of China'. What is the 'green wall'? How does it combat desertification?

Can forestry be sustainable?

The tropical rainforest is home to over half the world's plant and animal species. It is highly bio-diverse. In the 1940s, 15 per cent of the world's land surface was covered in tropical rainforest. Today less than half of that is left. Some estimates claim that 40 hectares are being destroyed every minute.

The Solomon Islands are a chain of 900 islands in the eastern Pacific Ocean. They are covered in tropical rainforest and fringed by coral reefs. However, the Solomon Islands have been devastated by Australian and Asian logging companies.

Figure 5.29 *This tropical rainforest in the Solomon Islands contains 4500 different species of tree and is home to 163 types of birds and many mammals that are found nowhere else in the world.*

In the Marovo Lagoon area, an eco-forestry project is now encouraging environmentally sustainable forest management. The timber harvesting can continue indefinitely at the same rate as replacement timber growth. Single trees are selected and felled, the logs are sawn into timber where they fall with a small portable sawmill, and the timber is carried out by hand. There is minimal damage and the forest regenerates. The timber is marketed as 'eco-timber' or 'good wood' to New Zealand and Australia.

Other small-scale cash earning schemes, such as fishing, tourism, carving and other crafts provide additional income for the local people.

Figure 5.30 *The Forestry Stewardship Council (FSC) logo. Products with the FSC logo are certified to assure consumers that they come from forests that are managed to meet the social, economic and ecological needs of present and future generations. B&Q has increased the number of wood products with this logo and is top of Greenpeace's garden furniture league.*

ACTIVITIES

1. Desertification has both natural and human causes. Write two lists of causes of desertification from the quote from Mr Bundani.
2. Desertification is likely to get worse in the future as a result of climate change. Why is this?
3. Look at the Fact file showing ways of reversing desertification. Choose two methods and explain how they work and how they can be considered sustainable.
4. How can we encourage sustainable forestry in the tropical rainforest? What role does the Forestry Stewardship Council play?

RESEARCH LINK

Look at the FSC website and watch the video 'Buyer be fair'. Do you think the FSC can make a difference to the world's forest ecosystems?

Look at the Greenpeace and WWF websites to see how these organisations are working for a sustainable future for the Earth's ecosystems.

Why is sport an issue?

People all over the world are interested in sport, and many people participate either individually or as part of a team. Many more attend sporting events or watch on television.

Sport is a global industry and generates huge amounts of money. The most successful sportsmen and women earn huge sums. Tiger Woods is the world's wealthiest sports star, earning US $115 million between July 2007 and July 2008. David Beckham, the UK's wealthiest sports star, earned US $50 million in the same period. About 80 per cent of their earnings come from sponsorship and endorsements. Large companies such as Nike, Pepsi and Gillette are prepared to pay high-profile sports stars to endorse their products, as this helps to increase sales worldwide and in turn make more profit.

Sports clubs involved with popular sports also earn vast sums. Manchester United is well known for the huge profits it generates every year. In 2007, this was about £375 million. Manchester United is now a huge multinational company, and has interests well beyond the football club, the merchandising and the corporate entertainment that goes with it. Many American football, baseball and basketball teams in North America also have huge followings and make vast profits, although they may not be so well known worldwide.

Only eight members of the 2008–9 Chelsea squad were British. The rest of the team came from all over the world, including France, Germany, Brazil, Italy, Portugal, Czech Republic, Serbia, Ivory Coast, Ghana, Slovakia, Argentina and French Guyana, showing that this major Premier League club is truly a global concern.

Where does the money come from?

Television has enabled sport to become a global industry. Broadcasting companies pay vast sums for the rights to show major sporting events because of the huge audiences these attract. The popularity of sport helps broadcasting companies to increase their audiences, and in turn this attracts more advertisers. In the case of commercial broadcasters such as Sky, they can increase the number of people who pay a monthly subscription to watch their programmes. For example, in 2007 Setanta bought Premier League football rights, and within a year the number of subscribers had risen from 200,000 to 1.1 million. Sky paid £300 million for the exclusive rights to broadcast Test Match cricket for four years from 2010 to 2013, which they hope will attract more people to pay to watch Sky Sports.

Although people all over the world enjoy sport it is not equally available to all. People in MEDCs are

Figure 5.31 Chelsea Football Club team at the start of the 2008–09 season.

more likely to participate in sport, either by playing or watching, than people in LEDCs. In all countries people with higher incomes have greater access to sport than people who earn less.

Figure 5.32 shows the distribution of Formula One (F1) motor racing Grand Prix circuits. The map shows the circuits used for Grand Prix events in 2008 and those planned to be used until 2011. The global distribution is clearly unequal. Most F1 races are held in Europe, where F1 racing started, but in recent years new circuits have been introduced, for example in Bahrain, China, Malaysia, Turkey and Singapore, and future races are planned in Moscow, India and Abu Dhabi. The sport attracts huge television audiences, which means it can attract sponsors who are prepared to invest vast sums so their product can be associated with an exciting sport and be seen by at least 55 million people worldwide each time an F1 race is broadcast.

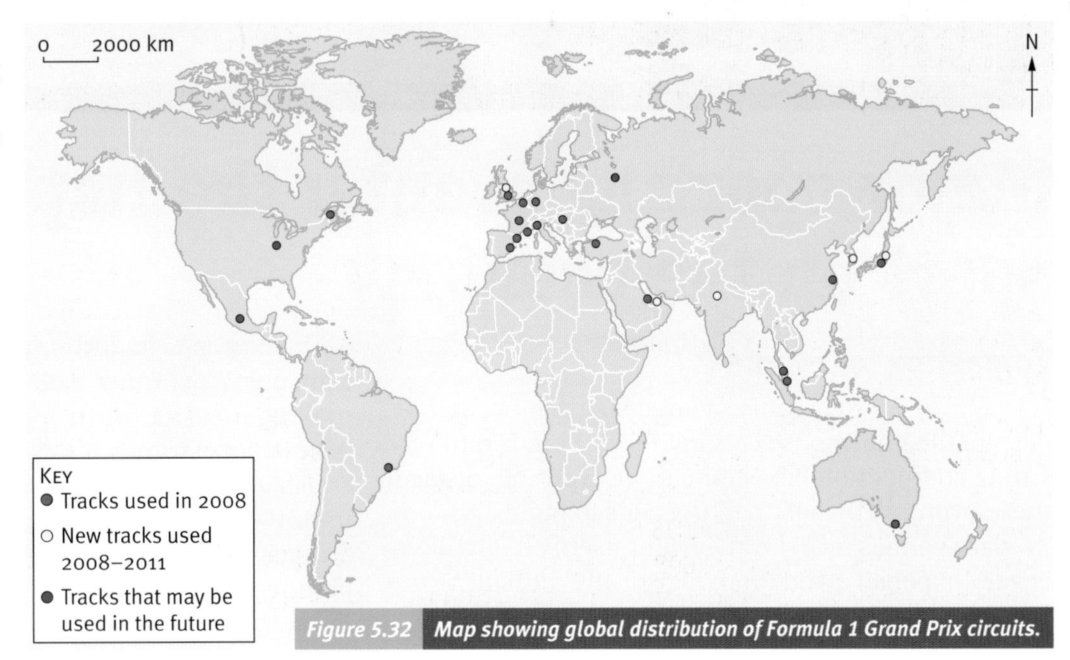

Figure 5.32 Map showing global distribution of Formula 1 Grand Prix circuits.

We should not be surprised that richer countries in the world are more likely to win major sporting events, because they can invest in better facilities and can provide funding for athletes to train. At the Beijing Olympics, Team GB were in fourth place in the final medal table – the country's highest placing since 1912, with China in top spot. However, if other factors are taken into account, such as population of the country and its GDP, then other countries are at the top of the leader board.

Fact file

- The opening ceremony of the Beijing Olympics was watched by an estimated 4 billion people worldwide, almost certainly the biggest audience ever for a sporting event.
- The FIFA World Cup Final in 2006, between Italy and France, was watched by a live global audience of 260 million people, with more than 600 million watching at least some part of the match.

RESEARCH LINK

Channel 4's alternative medal table will show you what happens when other factors are taken into account for an Olympic medal table: look at the Channel 4 website for more details.

ACTIVITIES

1. Describe the global distribution of F1 Grand Prix circuits. Name continents and countries in your answer and refer to the North–South divide.
2. Suggest reasons to explain the distribution.
3. Where are the newest race circuits? Why would a country want to host a Grand Prix?
4. Choose your favourite football club or any other sporting team. Use the internet to research where its players come from. Label all these countries on a world map to show the global distribution of players.

Figure 5.33 A Formula 1 Grand Prix car.

Major sporting events: the impact of sport on the economy

Competing for events

It is a huge job to organise a large sporting event, but despite the enormous costs involved there is often strong competition between nations wanting to stage these events. The host nation needs to build and prepare venues and run the event itself, all of which is very expensive and often logistically difficult.

So why do countries want to host these large events? Does it really help the country, city or region during the event itself? And what are the long-term impacts once it is over? Hosting big sporting events can be very profitable and raise the status of the cities and countries staging them. On the other hand, it can be very expensive and fail to make any real difference once the event is over.

The Athens Games, 2004

Athens hosted the 2004 Olympic Games at a cost of approximately £9 billion. There have been benefits to the city, particularly the much-improved public transport (including a new airport and metro), making it easier for people to travel around Athens. A new pedestrian walkway has also been constructed round the Acropolis, which has improved visitors' experience of Athens' most famous ancient site. These changes had been planned for years, but it took the Olympics to turn them to reality.

Elsewhere in the city, the benefits are less clear. Twenty-one out of 22 Olympic venues are now abandoned and some of the magnificent stadiums are overrun with rubbish, weeds and graffiti. It is costing the Greek authorities millions each year to maintain and protect the stadiums whilst they struggle to find buyers. Hosting the Olympics does not seem to have revitalised the Greek economy as was predicted and there must be questions as to whether the massive cost was money well spent.

Figure 5.34 *This abandoned stadium in Faliron Bay, Athens used for volleyball, basketball and judo in the 2004 Olympics shows that hosting a major sporting event is not always a long-term success.*

The Beijing Games, 2008

China was able to boost its international standing during the 2008 Olympics. There was a lot of positive publicity during the Games, and China hopes to attract more tourists and greater business investment as a result. The Chinese built or renovated 31 venues in and around Beijing, including the construction of the six main sporting facilities. These include the Bird's Nest stadium and the Water Cube, which cost more than US $20 billion. As well as the impressive venues, the city built a new airport terminal, new subway lines and a new light railway. It also tightened laws on industrial and vehicle emissions. Whether the massive investment will provide long-term benefits for the people of Beijing, or just the government, remains to be seen.

Figure 5.35 This logo helped to promote Beijing to a global audience during the 2008 Olympic Games.

The Commonwealth Games, Manchester 2002

The Manchester 2002 Commonwealth Games is a good example of a very successful event. From the outset, Manchester aimed to host a world-class event in order to showcase itself as a city. However, it also aimed to create a lasting legacy for the city, using it as a catalyst for major regeneration of rundown east Manchester. This area had suffered large-scale deindustrialisation, resulting in 30 years of underemployment, but the Games and the investment helped present a new image of the area. After the Games, Manchester City football club took over the stadium and draws up to 40,000 spectators to home matches, helping the local economy. Concerts and events at other centres, including the Velodrome, are also important in drawing people into the area. By 2008, about £600 million had been invested, creating about 20,000 jobs in a new business park and large retail centre. There is still work to be done, but the benefits brought by the Games are clear.

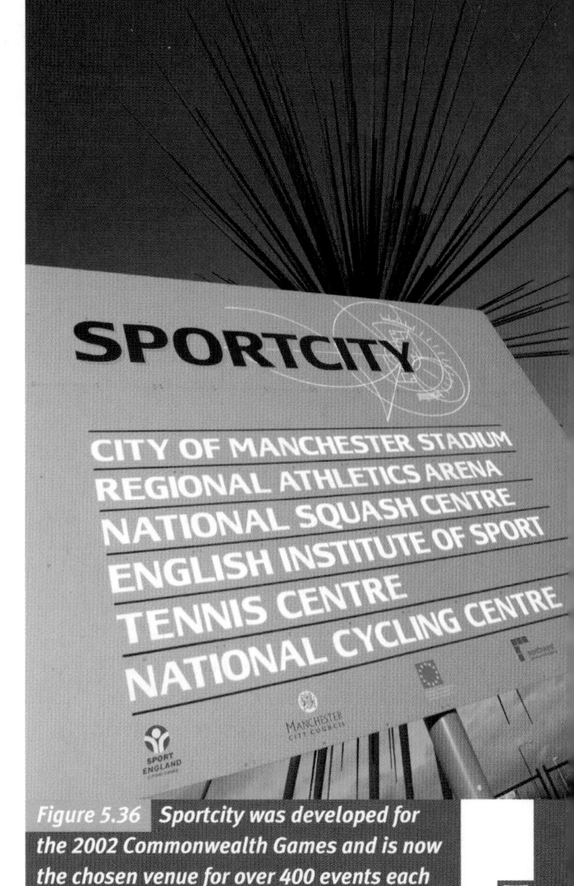

Figure 5.36 Sportcity was developed for the 2002 Commonwealth Games and is now the chosen venue for over 400 events each year. It receives over 4,500,000 visits annually. It was built on previously derelict land and the site has undergone major environmental improvements and created new employment and other opportunities for local people.

ACTIVITIES

1. Draw a spider diagram to show the types of jobs that could be generated by the hosting of a major sporting event.
2. List five benefits or opportunities that result from hosting a major sporting event such as the Olympic Games.
3. Give reasons why some people might object to their country or city staging a big sporting event, both in the short term and in the long term.
4. Choose a recent major sporting event. Use the internet to research both the positive and negative impacts of this event.

RESEARCH LINK

Find out about the London 2012 Olympics. How much is this event expected to cost? How will this money be found? How will east London benefit? What problems might there be?

Decide whether you think the Olympic Games in London will be money well spent or not and explain your reasons carefully. Go to the London 2012 and Games Monitor websites for more information.

The impact of sport on the environment

Many people not only want to watch sport but also to take part. Some of the most popular participation sports in the UK are swimming, walking, cycling, fishing and football. Governments try to encourage more people to get involved in sport because being active benefits people's health and well-being, both physically and mentally. However, in some cases increased numbers of people participating in sport can cause problems for the environment and therefore strategies need to be adopted to manage this impact.

Figure 5.37 Les Arcs ski resort.

Environmental impacts of winter sports

Skiing and snowboarding are very popular activities in both Europe and North America. The Alps attract over 120 million visitors each year. In summer, visitors climb, hike or admire the stunning scenery, but most visit in winter to enjoy some of the best skiing in the world. The large-scale development of ski resorts has brought prosperity to many Alpine areas but has also had a major impact on the environment.

Economic benefits include:

- jobs are created in construction as new roads, buildings and ski facilities are built
- once open, resorts need thousands of people to work on the slopes and in hotels, shops, restaurants and bars in the resort
- local people find winter employment as ski guides, lift operators and in behind-the-scenes organisation, or providing accommodation in their houses
- the local economy benefits from employment as people spend money in local businesses
- young people are more likely to stay in the area rather than move to larger towns and cities to find employment
- once-remote areas become more accessible as new roads and improved services are developed for the visitors.

Despite these benefits, many people feel that winter sports are causing excessive environmental damage. Small villages have mushroomed into large resorts, with new roads linked to the motorway network bringing much more traffic along with noise, congestion and air pollution, especially at weekends. Development of huge numbers of new buildings has changed the landscape for ever, and some modern high-rise apartments and hotels are very different in appearance to traditional chalets, which of course some people do not like.

Perhaps the greatest problem is the damage being done by the ski installations themselves. Huge pylons and cables are erected for the ski lifts, and trees may be cut down to make space for the lifts and the ski runs, leading to soil erosion on the steep slopes. Heavy use of the pistes can damage fragile vegetation below, and during the summer the main ski runs become huge scars on the landscape.

As more areas are used for winter sports, the alpine ecosystem is threatened and habitats of birds and animals are being lost. If resorts are developed at higher altitudes, either as a response to global warming or to satisfy the growing demand for skiing, these fragile environments will be put under even more pressure.

> The Les Arcs ski resort is committed to act to protect the environment.
>
> Without a massive increase in awareness, our environment is threatened. The professionals who work in the mountains are taking the following steps:
>
> - collection and recycling of ski cards
> - reforestation and seeding of new ski run development
> - reduction of the number of pylons where possible when building new lifts
> - use of organic oils for machine maintenance
> - protection of birds (black grouse) around cables, in conjunction with the Vanoise National Park
> - distribution of pocket ash trays
> - installation of many dustbins, which are emptied daily
> - collection of garbage under the chairlifts at the end of the season
> - protecting natural environments for the future.
>
> (Source: the website of the Les Arcs resort)

Visitor impacts in the Lake District

National parks have been created in many countries to protect the beautiful scenery, but they are also attractive to visitors, many of whom visit in order to take part in sports such as hiking, camping, climbing and mountain biking. It is important that steps are taken to protect the natural environments for the future.

Figure 5.38 *Helicopters are the only way of transporting materials into the hills, and this is one of the reasons why footpath repair is so expensive.*

The Lake District National Park in England attracts over 12 million visitors a year, and it has been estimated that over 10 million people use footpaths and bridleways leading to the fells. This number of people using the tracks for hiking or mountain biking causes a lot of damage and some have become ugly scars visible from miles away, unpleasant and even dangerous to use. In 2002, the Upland Path Landscape Restoration Project (UPLRP) was formed in the Lake District, starting a 10-year strategy to repair 180 eroded footpaths. This is expected to cost over £5 million; some coming from Heritage Lottery Funding, some from organisations such as the National Trust and some from donations. The project also benefits from volunteer labour. Footpath repair is very expensive, but it is essential to protect the landscape for the future.

Figure 5.39 *Footpath repair in the Lake District.*

ACTIVITIES

1. Construct a table or matrix showing the economic benefits and the environmental costs of mountain biking, hiking, skiing and snowboarding.
2. Choose a sport that you enjoy. Describe any economic benefits and/or environmental losses that result from people taking part in your chosen sport.

Why is tourism an issue?

Why has tourism increased so rapidly?

Tourism is the largest and fastest-growing industry in the world. In 2006, there were 846 million international tourist arrivals worldwide. In fact, tourist numbers have increased every year, rising from only 25 million international tourists in 1950, to an estimated 1.6 billion by 2020. At present, Europe is the largest source of tourists and is also the most popular destination, but other areas are expected to catch up as people travel further afield for their holidays.

There are several reasons why tourism has increased so rapidly, and is expected to continue rising.

- People, especially those in MEDCs, have more money to spend than in the past. Wages are higher and people have money to spend on themselves after paying their living expenses, so **disposable income** is higher. Also many people have more days of paid leave from work so they are able to take more frequent, longer holidays and travel further.
- Transport, especially road and air transport, has developed rapidly in the last 50 years so travel is quicker and easier. The increase in low-cost airlines has also had an impact, encouraging people to take more holidays abroad.
- The number of retired people has increased and many have enough money to take frequent holidays; many choose to travel extensively while they are still able to do so.

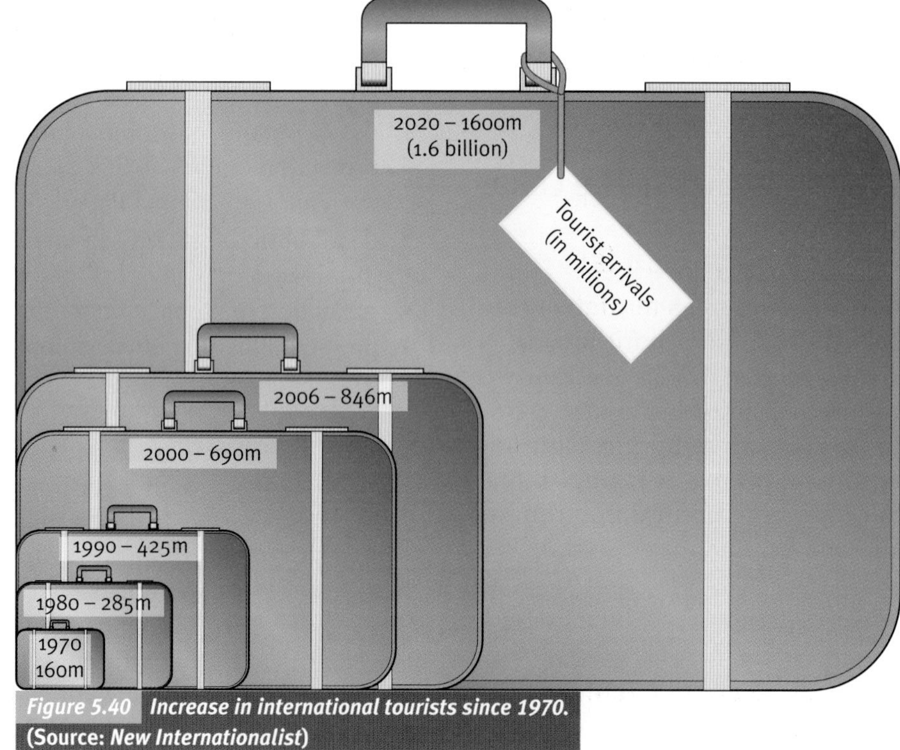

Figure 5.40 ***Increase in international tourists since 1970.*** **(Source: *New Internationalist*)**

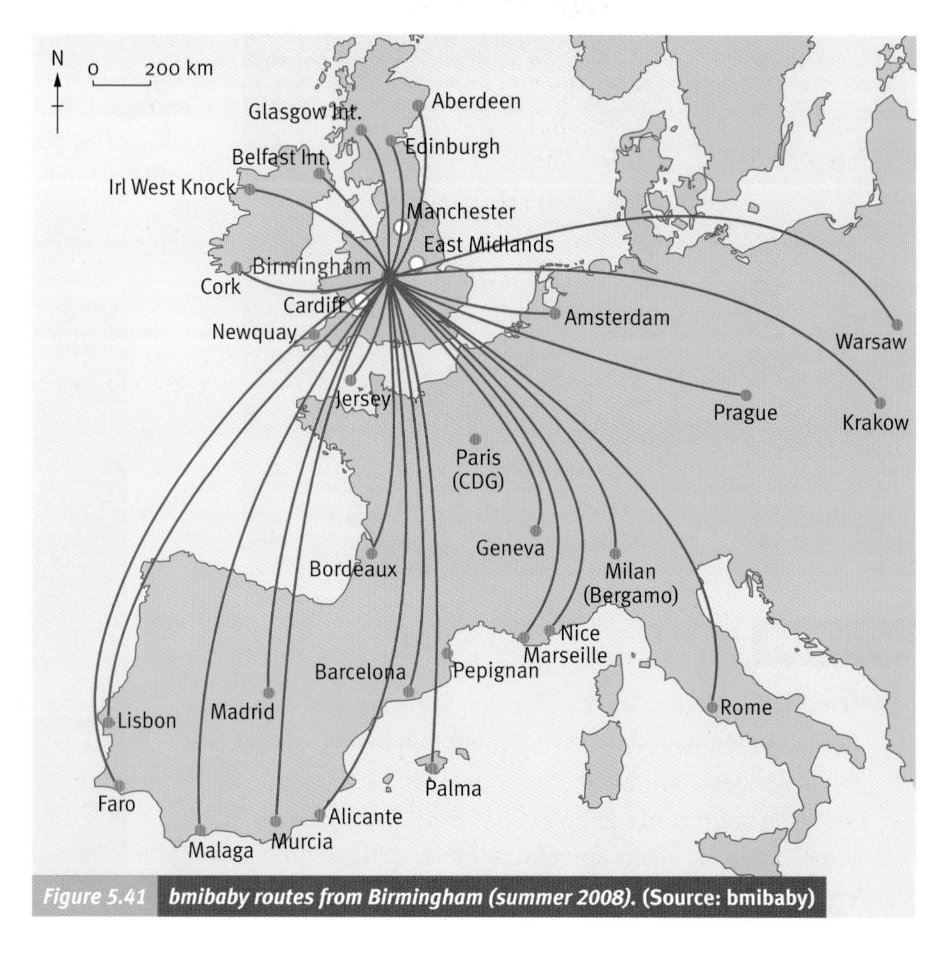

Figure 5.41 ***bmibaby routes from Birmingham (summer 2008).*** **(Source: bmibaby)**

- Travel is made easier by travel companies providing **package holidays**. Easier access to the internet also means people can book their own holiday accommodation and make travel arrangements such as booking flights and car hire online.
- Education, television, films and the internet mean there is more information about different places in the world, which encourages people to travel more widely for their holidays.

Why is tourism an issue?

There are many beautiful places in the world, so how do tourists choose where to visit? What do you think are the most important factors: the weather, scenery, accommodation or the cost? How can places compete with each other and market themselves to ensure that tourists choose them as a destination?

Fact file

- Global tourism is worth US $733 billion (2006).
- Tourism employs 8 per cent of the global workforce, according to Tourism Concern.
- In 2006, 75 countries earned more than US $1 billion from international tourism, according to *New Internationalist*.
- Every year, an area about half the size of Paris is cleared for golf courses, each of which needs up to 2.3 million litres of water every day, according to *Earth Report*.

KEY TERMS

Disposable income – the amount of income a person has left after paying for all necessities.

Package holiday – travel and accommodation sold together by a tour operator.

RESEARCH LINK

Which were the first bmibaby routes to be cut in the credit crunch of 2008? Why? Several low-cost airlines have been offered money to start flying to new cities. Who offers the money and why?

Figure 5.42 ***In the summer, Benidorm's population rises from about 65,000 to half a million people. The huge developments you can see in this photo have completely changed the area, which used to be a small fishing village.***

ACTIVITIES

1. Suggest reasons why Europe is the largest source of tourists and also the most popular destination. Explain why this might change in the future.
2. Give three reasons why low-cost airlines such as Easyjet, Ryanair and bmibaby were so successful in the first few years of the 21st century.
3. bmibaby recently added three new routes to Milan, Warsaw and Krakow (shown on the map in Figure 5.41).
   a. Use the internet to find out which countries these three places are in and why they are attractive to tourists.
   b. What other reason could explain why airports in Poland have been added to the routes provided by bmibaby and other low-cost airlines?
4. How do you think the map showing routes might change in the future? Explain why you think these changes will take place.
5. Look at the photograph in Figure 5.42 of Benidorm, one of Spain's most popular resorts. Why should people choose Benidorm when there are hundreds of other exciting destinations to choose from? What impact would there be on the resort and the people who live there if visitors stopped coming?

Tourism: good or bad?

Tourism has significant impacts on both people and the environment. It can generate a great deal of money for the area, but it can also divide communities, threaten local culture and destroy fragile ecosystems.

Impact on people

The most obvious benefit for tourist destinations is the money spent in the area. Travel companies invest millions on developments that benefit the whole area, such as roads, better water supplies and sewage systems, and new buildings. In addition, tourists spend money on accommodation, food and drink, transport, excursions, entertainment and souvenirs. Tourism enables local people to set up their own businesses: for example, making and selling souvenirs or providing activities such as boat trips, sightseeing or entertainment. There will initially be jobs in construction and, as tourism develops, local farmers and fishermen can benefit by supplying their produce to hotels. Thus both the local and national economy benefits through the **multiplier effect**.

However, it is not all good news. Many jobs in tourism are unskilled and poorly paid: for example, jobs in hotels and restaurants, shops and bars. These jobs may also be **seasonal**, meaning people only have a job during the time of the year when most tourists visit. Higher-level, better-paid jobs are often given to people from outside the area who may be better qualified or have more experience.

Figure 5.43 ***How the money from tourism leaks away: where each pound goes that is spent on a safari holiday in Kenya.*** **(Source: Leeds Development Education Centre)**

Another problem is that often only a small percentage of the money that tourists spend benefits local people and businesses, with most going abroad to the tour operators, airlines and hotel chains. This phenomenon is called 'leakage' (Figure 5.41).

Impact on environments

Successful tourism often relies on an attractive natural environment, so it is essential that the environment is protected or tourists will stop going to the area and find an alternative. Resorts want to attract more visitors to increase their income, but as numbers increase it becomes more and more difficult to protect natural environments.

Eighty per cent of all tourism takes place in coastal areas, with beaches and coral reefs among the most popular destinations. These areas may be environmentally fragile and large numbers of visitors can have a huge impact, especially when tourism is concentrated in just a few centres.

- Local infrastructure, water supplies and sewage systems may not be able to cope and there can be problems removing waste.
- Sea life such as whales, dolphins and sea birds are disturbed by increased numbers of boats, and by people approaching too closely. On some beaches, such as on the Greek island of Zakynthos, nesting sites for endangered marine turtles have been destroyed.
- Removing corals and shells from the sea, either by individual tourists or local people who then sell them as souvenirs, damages marine ecosystems, such as the Great Barrier Reef.
- Litter, especially on the beach, is unsightly and can damage sea life. Plastic bags are particularly dangerous and take years to degrade. Turtles can be killed by bags they have mistaken for jellyfish.

Figure 5.44 Australia's Great Barrier Reef has about 2 million visitors a year. This puts huge pressure on local infrastructure as well as potentially damaging the fragile coral reef ecosystem.

Protecting natural environments for the future

National parks have been created in many countries as one way to try to protect important natural environments. Yellowstone Park in Wyoming, USA was established in 1872 and was the first national park in the world, but since then many more areas have been protected, including 12 areas in England and Wales.

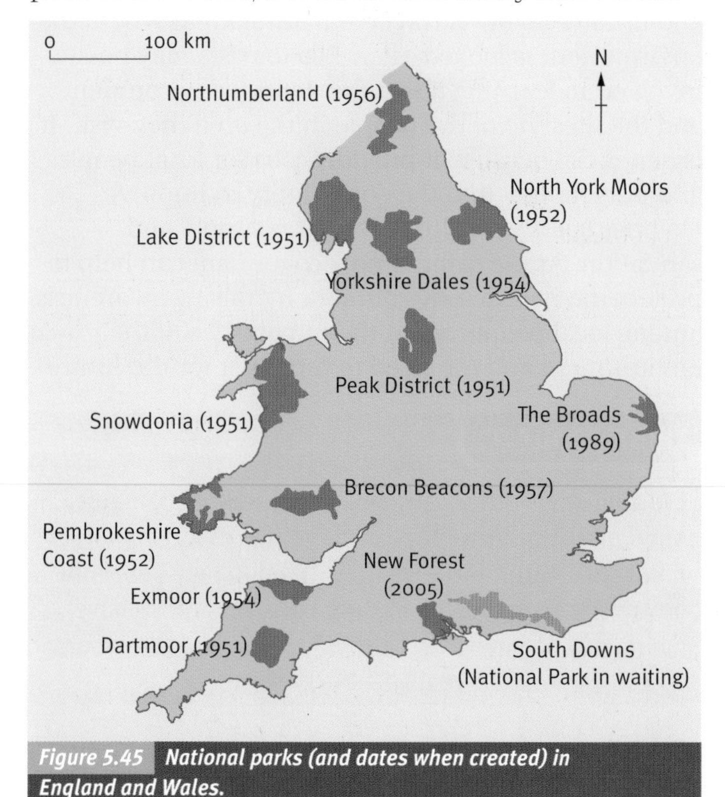

Figure 5.45 National parks (and dates when created) in England and Wales.

Places that are particularly popular with visitors are referred to as **honeypot sites**. In national parks, small, attractive settlements, well known beauty spots and popular walks are all potential honeypots and it is important to look after these areas. However, this needs careful planning and usually considerable investment.

KEY TERMS

Honeypot sites – particular locations that attract a lot of visitors.

Multiplier effect – the process whereby a significant increase in economic growth can lead to even more growth as more money circulates in the economy.

Seasonal employment – temporary employment for part of the year only.

ACTIVITIES

1 Construct a table to summarise the positive and negative impacts of tourism on people and the environment.

2 Use the internet to find out more about coral reefs. What other activities in addition to tourism are threatening these fragile ecosystems, and what steps are being put into place to protect coral reefs?

Tourism: is there another way?

Tourism is a major world industry and many LEDCs see it as an important way for them to develop in the future. However, international tourism relies on air transport and this is a big contributor to climate change. For this reason, some people consider it is unrealistic to think that tourism can contribute to genuinely *sustainable* development for the future. It may be more useful to think about 'responsible tourism'. This encourages individuals, organisations and businesses to take responsibility for their actions and the impacts of their actions, in this way reducing the damage being done by tourism, particularly mass tourism.

There are a number of NGOs campaigning for more responsible tourism. One example is WWF, well known because of its panda logo. WWF works with large holiday companies and tour operators to encourage them to ensure that both the natural environment and local people are looked after. WWF also want tourists themselves to act more responsibly.

This is a list of some of the ways you can make a difference when you go on holiday:

- use facilities and trips run by local people whenever possible
- dispose of rubbish responsibly – it can be hazardous to wildlife – and take all rubbish home from the beach
- turn off lights, taps and air conditioning when you leave hotel rooms
- reuse towels and participate in any green schemes run by hotels
- do not be tempted to touch wildlife or disturb habitats on land, at the coast or under water
- do not eat shark's fin soup or any dishes you suspect of containing endangered species
- be careful what you choose to bring home as a souvenir. Some species have become endangered because of this, including coral, elephants and alligators
- boats and jet-skis create noise and chemical pollution that disturbs wildlife; do not keep the engine running unnecessarily
- if you are sailing, surfing or windsurfing, keep a distance of at least 100 m from seal resting and bird nesting sites to avoid disturbing them.

KEY TERMS

Ecotourism – environmentally responsible travel to natural areas in a way that has a low impact on the natural environment and benefits the local economy.

THINK ABOUT IT

Although around 80 per cent of UK package holidaymakers believe that it is important that their holidays do not damage the environment, they are ultimately motivated by cost when choosing a holiday. In your opinion, what, if anything, could make people more concerned about the environment and less about cost when deciding on their holiday plans?

Eco-tourism is an example of sustainable tourism. It is small scale, improves the well-being of local people and usually involves travel to natural areas where the environment is looked after. Eco-tourists may be involved in learning about the natural environment and the lifestyle of the people in the area they visit. It is beneficial because it provides jobs for local people and puts money into the community to improve facilities such as health care, water supply and sanitation. At the same time, eco-tourism can help to protect the natural environment by raising awareness among local people about the importance of their own environment and the need to protect it for the future.

THINK ABOUT IT

Eco-tourism is essentially small scale and sustainable, but if places become popular there is pressure for larger hotels and better facilities. New developments may not be environmentally friendly and not all countries have the money or expertise to develop facilities sustainably. So can eco-tourism really be the answer?

The 3 Rivers Eco Lodge and Sustainable Living Centre, Dominica

Dominica is a small, mountainous island covered by tropical rainforest. It was chosen as the film location for *Pirates of the Caribbean* because of its natural beauty. In the past, the economy depended on agriculture, particularly bananas. Now, tourism is important, although compared with other Caribbean islands tourism is underdeveloped with only 65,000 visitors a year. This is partly because there are no extensive sandy beaches, no large hotel chains, no international airport and only one main road. The Dominican government encourages development of tourism, particularly eco-tourism.

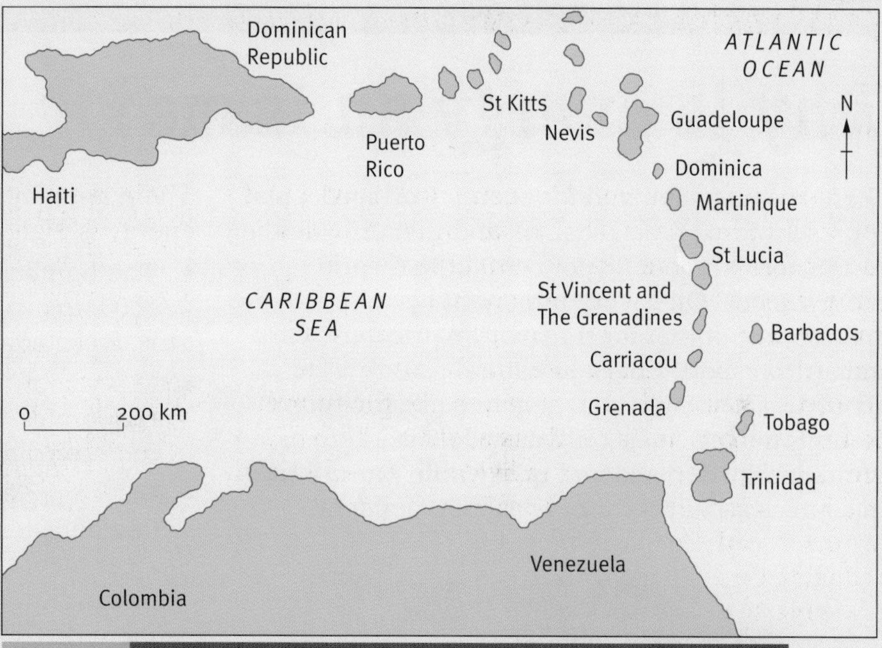

Figure 5.46 ***Dominica: one of the tiny Windward Islands in the Caribbean.***

The 3 Rivers Eco Lodge has won awards for sustainable tourism. It aims to cause as little harm to the environment possible; the sun powers the entire property, including a solar-powered hot water system and a solar-powered pump to extract water from the river, which works in silence to avoid disturbing animals. All water is treated and reused in the garden to minimise consumption and the lodge composts kitchen waste, using the compost to grow organic food. Even the pick-up truck has been modified to run on used vegetable oil as well as diesel, reducing harmful emissions by over 90 per cent and providing a use for cooking oil from local restaurants that would otherwise be discarded. The hotel avoids purchasing packaged goods, shops locally, recycles and uses biodegradable products.

Figure 5.47 ***The natural beauty of Dominica.***

Equally importantly, the lodge works with the local community. All 14 employees come from local villages and visitors are encouraged to take part in activities such as helping at the village primary school, learning traditional farming methods and even joining in with a reggae band. This helps local people to earn money and also to realise the importance of the natural environment and their own culture and the need to look after it for the future.

ACTIVITIES

1. Summarise the qualities of an eco-tourism holiday.
2. In what way are such holidays sustainable?
3. Do you think there is such a thing as a truly eco-friendly holiday? Why?

RESEARCH LINK

Find out about the *New Internationalist* from its website. Is this a website you can trust or does it have a bias in the stories? What are the signs of bias, if any? Would this mean it was of no use to you?

Investigate eco-holidays in the UK. Where do they happen? What are they like? Are you tempted? Explain your answer.

Why is energy an issue?

There is a huge demand for energy worldwide, and at present fossil fuels (coal, oil and gas) provide the vast majority (about 88 per cent) of the world's energy needs. Oil is very important because it is the main source of fuel for transport, particularly cars and air transport. Coal and natural gas are used mainly for heating and to generate electricity. At present, nuclear fuels only provide 6 per cent of world energy demand and **renewable energy** from all sources accounts for only about 6 per cent.

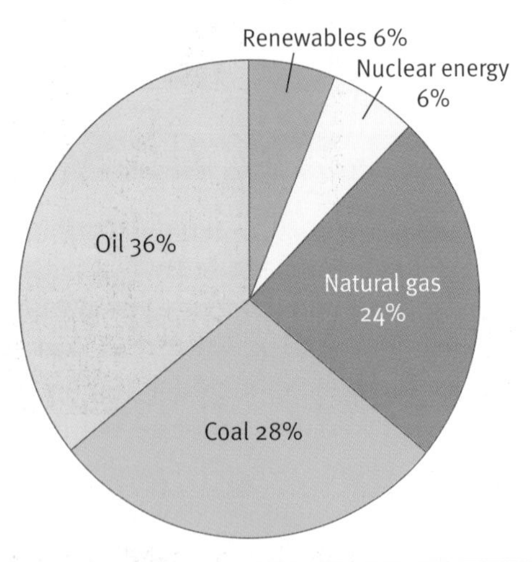

Figure 5.48 **World energy consumption 2006.** **(Source: *BP Statistical Review of World Energy 2008*, BP plc)**

KEY TERMS

Energy gap – the gap between demand for energy and the energy that can actually be supplied.

Non-renewable energy – energy from sources that are finite and will eventually run out, such as fossil fuels.

Renewable energy – energy generated from sources that are naturally replenished, such as sunlight, wind, tides and geothermal heat.

There are a number of reasons why demand for energy is high and continuing to rise.

- The number of people in the world is rising. In 2008, the world population was 6.6 billion and it is expected to reach 9 billion before 2050.
- Many countries have rapidly developing economies, and as industries grow there is increasing demand for energy.
- With increasing wealth, people have greater demands for energy. They may be able to afford a car, new electrical appliances for their homes or travel more in their spare time. All this increases demand for energy.
- Recent decades have seen huge advances in technology, particularly the rise in the use of computers and the internet, and this increases demand for energy.

Generating electricity

Fossil fuels are still the most important fuels in the UK because they are used to generate much of our electricity. The graph shows that we use very little oil to generate electricity, but still rely heavily on coal and natural gas. Over 20 per cent of our electricity is generated at nuclear power stations and only 4 per cent from renewable sources, including hydro-electric power.

Supplies of coal and natural gas in the UK are now running low and both are increasingly expensive. For this reason, both coal and natural gas are imported, and within the next 10 years we could be importing three-quarters of the fuel we need as our own runs out.

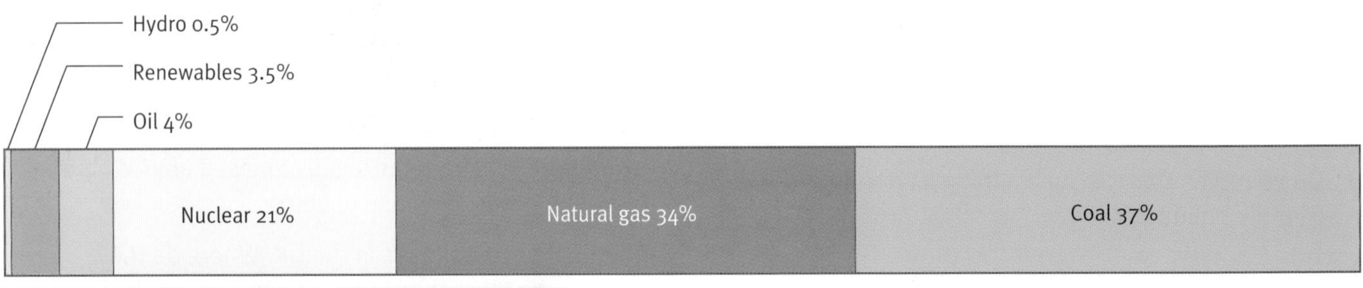

Figure 5.49 **Fuels used in UK electricity generation 2004.** **(Source: *The Oil Drum: Europe*, Crown copyright)**

The energy gap

The demand for energy is rising at an estimated 2 per cent a year. MEDCs currently use most energy, but demand is rising quickly in LEDCs as their industries expand. Fossil fuels such as coal, gas and oil are the main sources of energy at present, but these **non-renewable** sources, particularly coal and oil, are running out. When demand for energy cannot be met there is an **energy gap**. Alternative sources of energy such as wind, water and solar power need to be developed in order to fill the 'gap', but it may not be possible to do this quickly enough to keep pace with the decline in fossil fuels.

Figure 5.50 ***Ferrybridge power station cooling towers in Yorkshire. This power station uses 800 tonnes of coal and 218 million litres of water per hour. The station has two 198-m-high chimneys and eight cooling towers, which are the largest in Europe.***

Environmental pollution

Using fossil fuels for energy has major environmental impacts. Burning oil, coal and gas causes acid rain, which damages lakes and forests, often many miles from the source of the pollution. Burning fossil fuels also releases CO_2 and other greenhouse gases that contribute to global warming. CO_2 accounts for the greatest proportion of greenhouse gas emissions in the UK. The government is committed to reducing CO_2 emissions, it is important to change to cleaner methods of generating electricity, particularly from renewable sources of energy.

The share of electricity generated from renewable energy (including hydro) in the UK is only 4 per cent. The government hopes that by 2020 we will be generating 20 per cent of our electricity from renewables. However, this still means that the remaining 80 per cent will have to come from other sources. Nuclear power is unpopular with many people, but it is one way of generating electricity without releasing greenhouse gases, particularly CO_2.

ACTIVITIES

1 The UK government has set a target of reducing CO_2 emissions by 20 per cent of 1990 levels by 2010.
  - a How does it hope to do this?
  - b Explain why this is likely to be difficult to achieve.

RESEARCH LINKS

How has the global price of oil changed over the past 12 months? Has this hindered or helped the development of renewables?

How well is the UK government able to meet its CO_2 targets? Why?

THINK ABOUT IT

China uses more coal than the USA, the EU and Japan combined. Every week, a new coal-fired power plant opens somewhere in China that is big enough to provide electricity for over a million people.

- Will the UK's attempts to replace coal with other fuels and reduce carbon emissions make any difference in the face of China's expansion?
- Does China have the right to generate electricity this way?

Nuclear energy

The UK has 10 nuclear power stations, which generate just over 20 per cent of our electricity. These power stations are old, and all but one are due to be decommissioned (closed) in the next 15 years; by 2023, all but one will have shut. No new nuclear power stations have been built since Sizewell B, which opened in 1995.

However, the rising price of fossil fuels and concerns about their contribution to global warming have led to changing opinions about nuclear power. It is now seen as an important way to generate electricity in the future, so much so that in January 2008 the government announced plans to build a new generation of nuclear power stations on the same sites as existing plants. At least eight new power stations will be built and it is hoped that the first will be on-stream by 2017. A number of European countries are already building new power stations, including France, Finland, Bulgaria, Romania, and Slovakia.

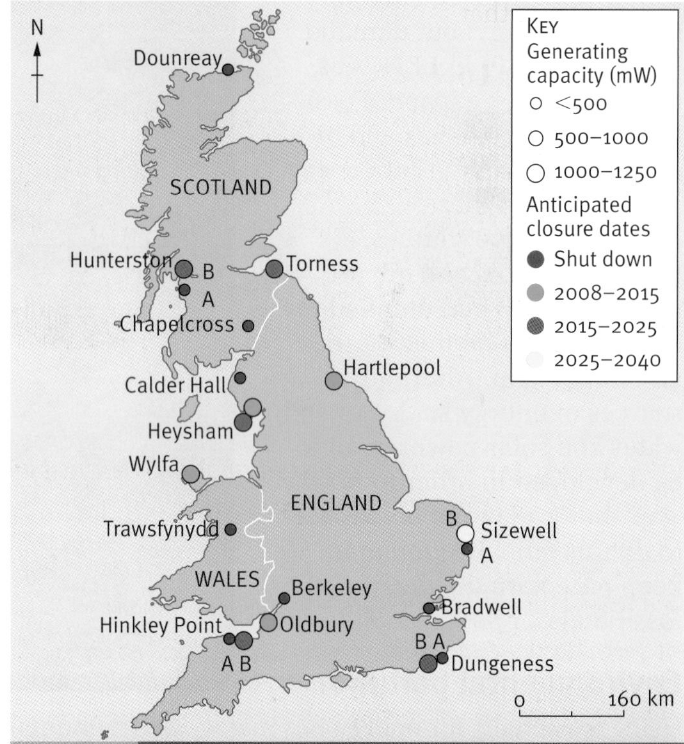

Figure 5.51 **UK nuclear power stations. The current nuclear power stations are all sited away from large centres of population and it is likely the eight proposed new stations will be on some of the same sites.**

Figure 5.52 **Sizewell B was the last nuclear power station to be opened in the UK (1995). It supplies about 3 per cent of the country's electricity. It is due to be decommissioned in 2035.**

Opinion is divided about increasing the amount of nuclear power generated in the UK. Here are a selection of viewpoints from individuals and organisations: read them carefully and then answer the questions that follow.

James Lovelock, Environmentalist

'Only nuclear power can halt global warming. It is the only realistic alternative to fossil fuels that is able to supply our energy needs and also reduce greenhouse emissions. I entreat my friends in the environmental movement to drop their wrongheaded objection to nuclear energy.'

Professor Ian Fells, Chairman of the New and Renewable Energy Centre

'Britain cannot do without nuclear power; it's like a slow-motion train crash. If we don't do anything about this next year, and the next, nothing will happen. We will find soon that our electricity supply becomes fragile and we get power cuts and it will get worse.'

Greenpeace

'Research shows that even 10 new reactors would only cut the UK's carbon emissions by 4 per cent sometime after 2025. Most of the gas we use is for heating and hot water and for industrial purposes and nuclear power cannot replace that energy. And oil is virtually all used for transport – nuclear power can't take its place.'

Friends of the Earth

'There is no need for nuclear power; the UK's vast renewable resources combined with simple energy-saving tactics would provide a safer, cleaner and more sensible solution.

Renewable sources could generate more than half our current electricity needs by 2025. Many renewable technologies can be implemented within 3 years, but we'll be waiting at least 10 for nuclear.'

John Hutton MP, Business and Enterprise Secretary, 2008

'Nuclear energy is safe, clean and good for the environment. Britain should become a gateway for the new wave of nuclear technology and commit itself to much greater use of atomic energy. Reactors have been proved safe and any delay in using them more widely could be disastrous for the environment.'

Michael Meacher MP, former Environment Secretary

'There is still no practical method of dealing with radioactive waste from nuclear power stations. There is already 10,000 tonnes of high and intermediate level radioactive waste, 90 per cent of which is being stored at Sellafield in Cumbria. This will grow to half a million tonnes by the end of this century even without any new build. Do we really want to generate more nuclear reactors producing even more waste when we don't know what to do with all the waste that is building up?'

ACTIVITIES

1 Put together a list of reasons in favour of nuclear power and reasons against it.

2 Study the list of arguments in favour of nuclear power. Rank them in the order you think is most important. Explain why you have decided on this order.

3 Imagine you live close to the site chosen for a new nuclear power station. Draw a speech bubble and write your thoughts about nuclear power and how it might affect you.

Fact file

- France has 59 nuclear power plants, which generate about 88 per cent of the country's electricity.
- They generate more electricity than they need and export it to neighbouring countries, including the UK.
- In the future, France may help to train a new nuclear workforce and co-operate with the UK in building long-term waste disposal facilities.

Renewable energy

The supply of renewable energy needs to increase at more than 5 per cent per year to keep pace with increasing demand for energy as the supply of fossil fuels declines. This is a tall order.

Figure 5.53 Highland and coastal areas, where the wind is strongest, are the best areas for wind farms, but there are already several offshore wind farms in operation and several more are planned.

Types of renewable energy

Renewable technologies include **biomass**, geothermal, hydro-electric, solar, tidal, wave and wind power. At present hydro-electric power (HEP) is by far the most significant, providing about 6 per cent of global energy demand but up to 80 per cent of electricity in some countries such as Paraguay, Norway and Brazil. However, not every country has the large rivers or steep mountains needed for HEP. Some countries have potential for geothermal energy, particularly volcanic areas such as New Zealand, Iceland and Japan. In countries with plenty of sunshine, for example Spain and Portugal, solar power is a real possibility. For the UK, the best options for significant energy production from renewables are wind, wave and tidal power.

Challenges

For all types of renewable energy, the problems are that technology is not well developed and it is expensive to build the new installations. Another issue is that some technologies, particularly those based on the weather, are intermittent and cannot be relied upon for a constant supply of electricity. Furthermore, large, new installations such as wind farms and tidal barrages are criticised for being unsightly and spoiling the landscape, especially when built in areas of countryside. On the other hand, using renewable sources means that there are no carbon emissions, and because they will not run out this is the only sustainable way of supplying our energy demands in the long term.

Figure 5.54 The biggest solar park of Europe: Alentejo, Portugal.

In the UK renewables are expanding fast, but they cannot meet all our energy requirements and currently less than 5 per cent of UK electricity is generated from renewables. The government has set itself the target of increasing this to 10 per cent by 2010, and 20 per cent by 2020. It is a challenging target and many experts feel we are not making progress fast enough to achieve this.

Wind power

Wind power is growing globally at a rate of about 30 per cent each year. Many wind farms have been set up in the UK, particularly in Wales and Scotland. However, progress has been faster in Germany, Spain, India, China and the USA. According to the British Wind Energy Association, 8 out of 10 people are in favour of wind energy, and costs have fallen in recent years, making wind power more competitive, particularly when compared with increases in fossil fuel prices.

One of the problems in the UK is that local objections to wind farms influence planners, who may block applications for new wind farms.

To the editor, Sheffield Telegraph

I write to express my grave concern about the proposed wind farm at Sheephouse Heights on the outskirts of Sheffield. This beautiful hillside, which is designated as green belt land and is currently grazing land, should surely be protected from any sort of development, including wind farms. If you have ever seen a wind farm at close quarters you will know the impact these giant concrete and steel structures, over 100 metres tall, have on the landscape. In my view, the only place for these huge industrial installations is in already industrialised areas, and it is the duty of planners and politicians to prevent wind farm developers from blighting our precious countryside.

Yours,

Chris Singleton, Sheffield

Solar electricity generation accounts for a tiny amount of production worldwide but, like wind, it is growing very rapidly, and it has been developed on a large scale in the USA and Spain. Electricity is generated using photovoltaic panels; these have been expensive to construct, but new technologies are now bringing the cost down.

Saving energy

Every year, we waste more than eight times the amount of energy supplied by all of the UK's nuclear power stations combined. According to government research, we could save up to 30 per cent of all energy we use, and of course this would save everyone money on their bills.

We can all make changes that would save energy – for example, by:

- walking, cycling, or using public transport
- reducing the number of flights taken, especially short flights
- using smaller, more energy-efficient cars
- switching off lights, phone chargers and TVs when not in use
- recycling and reusing as much as possible
- using energy-efficient light bulbs and rechargeable batteries
- insulating roofs, blocking draughts, and using double glazing.

KEY TERMS

Biomass – biological material used to generate energy, for example plants grown for biofuel.

Fact file

According to Greenpeace

- Traditional-style light bulbs waste as much as 95 per cent of the energy they use.
- If all retailers in the UK only stocked energy-efficient light bulbs, we could save over 5 million tonnes of CO_2 emissions a year.
- That's the equivalent of the output of two nuclear power stations, or more than the CO_2 emissions of the 26 lowest-emitting countries combined.

The Energy Saving Trust says

- Almost £875 million worth of electricity is wasted each year in the UK by leaving gadgets and appliances on standby.

ACTIVITIES

1. Write a letter to the *Sheffield Telegraph* replying to the one from Chris Singleton. In your letter, try to argue the case for allowing the new wind farm at Sheephouse Heights on the outskirts of Sheffield.
2. Look at the video called 'Harnessing off-shore wind' at the BBC website. In your opinion, should more wind farms, both onshore and offshore, be set up in the UK? Give your reasons.
3. Thinking about the way your family uses energy, how could you reduce the amount of energy you all use?

Acid rain – rainwater containing sulphuric acid, nitric acid and compounds of ammonia. These pollutants have been pumped into the atmosphere by manufacturing industry and vehicle emissions.
Aftershocks – smaller earthquakes that occur in the same general area during the days to years following a larger event.
Amenity value – used for recreational and leisure activities.
Appropriate technology – aid supplied by a donor country whereby the level of technology and the skills required to service it are properly suited to the conditions in the receiving country.
Asylum – to claim safety in another country.

Back offices – offices of a company handling high-volume communications by telephone, electronic transaction or letter. The location of such low-to-medium level functions can be relatively flexible and they have been increasingly decentralised to locations where space, labour and other costs are relatively low.
Backwash – the movement of water down a beach by the action of gravity.
Balance of trade – the difference between the cost of imports and the cost of exports.
Beach replenishment – building up the beach by pumping sand or shingle onto it.
Bedload – larger particles moved along a river bed.
Biofuels – any fuel derived from renewable biological sources such as plants or animal waste.
Biomass – biological material used to generate energy, for example plants grown for biofuel.
Birth rate – the number of children born in a year for every 1000 people in a population.
Blind fault – one where the fault line does not reach the surface.
Brownfield land – land on which there have already been buildings.

Canalisation – making a river more like an artificially built canal.
Carbon footprint – the effect human activities have on the climate in terms of the total amount of greenhouse gas production.
Carbon neutral – not adding to the net amount of CO_2 in the atmosphere.
Carnivore – an animal that eats other animals, also known as a predator.
Catchment area – the area from which people come to a shop or a shopping centre.
Catchment Flood Management Plans (CFMPs) – plans setting out strategies to reduce flood risks in a river basin over 50–100 years.
Central Business District (CBD) – the main shopping and business centre of a city.
Charity shops – shops that raise money for charities, often selling goods donated by the public.
Communicable disease – an infectious or contagious disease that can be spread from one person to another.
Community energy – energy produced close to the point of consumption.
Commuter range – the distance people will travel from their homes to places of work.
Conservation areas – parts of a town with historic buildings that are protected.
Consolidated rock – rock with strong structure that is hard to break down.
Constructive waves – waves that build up beach material to create landforms.
Consumer goods – things that people buy.
Conurbation – a large urban area formed when cities and towns merge as they grow towards each other.
Convenience goods – goods that people buy frequently for everyday needs such as groceries.
Counterurbanisation – movement of people away from a city or town.
Cumulative causation – the process whereby a significant increase in economic growth can lead to even more growth as more money circulates in the economy.
Cyclone – a system of winds rotating inwards to an area of low pressure.

Death rate – the number of people who die in a year for every 1000 people in a population.
Deindustrialisation – the long-term absolute decline of employment in manufacturing.
Delta – where a river breaks into many distributaries before it reaches the sea.
Demographic transition – model showing how the population in a country changes over time as birth and death rates fall.
Demography – the study of population.
Dependency ratio – the ratio between the economically active population and those who are dependent on them.
Dependent – when an area is reliant on one company or type of company for the majority of its employment.
Desertification – the degradation of land in arid and semi-arid areas resulting primarily from human activities and influenced by climatic variations.
Destructive waves – waves that erode coastlines.
Development – the use of resources and the application of available technology to improve the standard of living within a country.
Discharge – the volume of water in a river passing a point in a given time, measured in cumecs (cubic metres per second).
Disposable income – the amount of income a person has left after paying for all necessities.
Dredging – taking sediment from river or sea bed.

Eco-homes – houses designed in ways that conserve resources and energy.
Economic core region – the most highly developed region in a country with advanced systems of infrastructure and high levels of investment, resulting in high average income.
Economic migrants – people who move to another country to get a better job and improve their standard of living.
Economies of scale – a situation in which an increase in the scale at which a business operates will lead to a reduction in the average costs of production.
Ecotourism – environmentally responsible travel to natural areas in a way that has a low impact on the natural environment and benefits the local economy.
Eco-towns – towns designed to be sustainable and that do not cause environmental problems.

El Niño – large climatic disturbances in the southern Pacific Ocean that occur every 3–7 years.
Emigrant – a person who leaves a country to go to live in another country for longer than a year.
Energy gap – the gap between demand for energy and the energy that can actually be supplied.
Environmental impact assessment – a document required by law in many countries, detailing all the impacts on the environment of a project above a certain size.
Epicentre – the point on the Earth's surface directly above the focus of an earthquake.
Erosion – the wearing away and removal of rocks by the action of water, wind or ice.
Estuary – tidal part of a river mouth.
e-tailing – buying and selling goods and services online through the internet.
Evaporation – water turning into water vapour.
Evapotranspiration – the sum of evaporation from the Earth's surface together with the transpiration from plants.
Exponential growth – a rate of increase that quickly doubles.
Exports – goods that are sold to other countries.
Extensive farming – when one type of farming and large farms dominate a very large area. Inputs per hectare are low compared with intensive farming, such as market gardening.

Farmers' market – a place where farmers sell produce direct to customers, usually from stalls on one day a week.
Fault – a fracture in the Earth's crust that shows signs of movement.
Fertility rate – the average number of children to which each woman gives birth.
Flood plain – flat area next to a river that is liable to flood.
Food miles – the distance that food has been transported before it is sold.
Foreign direct investment – overseas investments in physical capital by MNCs.
Free trade – trade between countries when the prices paid are determined by what the producer wants to be paid and what the consumer is prepared to pay and there are no barriers.
Freeze–thaw – the continued freezing and thawing of moisture in rocks that will eventually cause them to break.

Gross Domestic Product (GDP) – total value of goods and services produced by a country in a year.
GDP at purchasing power parity (PPP$) – the GDP of a country converted into US dollars on the basis of the purchasing power parity of the country's currency. It is assessed by calculating the number of units of a currency required to purchase the same representative basket of goods and services that a US dollar would buy in the USA.
Geology – the nature and structure of rocks.
Geothermal – energy generated by heat stored deep in the Earth.
Global city – major world city supplying financial, business and other significant services to all parts of the world. The world's major stock markets and the headquarters of large MNCs are located in global cities.
Global civil society – international groups, associations and movements that are not controlled by the state government.
Globalisation – the increasing interconnectedness and interdependence of the world economically, culturally and politically.
GPS (Global Positioning System) – a group of satellites that allow people to find out their exact location on the Earth's surface.
Greenfield land – land on which there has not been any previous building.
Greenhouse effect – the property of the Earth's atmosphere by which long wavelength heat rays from the Earth's surface are trapped or reflected back by the atmosphere.
Gross National Income – total value produced within a country together with its income received from other countries.
Groundwater flow – movement of water underground through rocks.
Groyne – wooden or concrete construction built across a beach.

Hard engineering – use of concrete barriers to control water.
Herbivore – an animal that eats only plants.
Hierarchy – an arrangement in order with one at the top and an increasing number at lower levels.
High-technology cluster – where high-tech companies group together in a region because their location factors are similar and such companies benefit from being in close proximity.
HIV/AIDS – Acquired Immune Deficiency Syndrome (AIDS) is a set of symptoms and infections resulting from the damage to the human immune system caused by the Human Immunodeficiency Virus (HIV).
Honeypot sites – particular locations that attract a lot of visitors.
Housing Association – an organisation that manages the building of houses for local people.
Hurricane – a violent tropical storm in the Caribbean region.
Hydro-electricity – electricity produced by flowing water.

Immigrant – a person who moves to live in another country for longer than a year.
Immigration – the movement of people into a country, to live there.
Immunisation – being given a vaccine that helps the body's immune system fight infection.
Impermeable – a surface that does not allow water to pass through it.
Imports – goods that are bought from other countries.
Infiltration – seeping of water into soil.
Informal sector – jobs that are often without regular hours and payment.
Infrastructure – the basic amenities that people need in a city, such as roads, sewerage, electricity and water supplies.
Interception – collection of water by vegetation.
International aid – the giving of resources (money, food, goods, technology, etc.) by one country or organisation to another poorer country. The primary objective is to improve the economy and quality of life in the poorer country.
Irrigation – artificial watering of the land.

Lahar – a type of mudflow composed of pyroclastic material and water that flows down from a volcano, typically along a river valley.

Land use – the different ways that land is used, for example for industry or recreation.
Lateral erosion – erosion of the sides of a valley.
Less Economically Developed Countries (LEDCs) – countries with low economic output per person, often measured by Gross National Product (GNP).
Life expectancy – the average number of years a person may expect to live when born, assuming past trends continue.
Literacy rate – the percentage of people who have basic reading skills.
Longshore drift – the movement of material along a coastline by the action of the waves.
Low-carbon economy – a country where significant measures have been taken to reduce carbon emissions in all sectors of the economy.

Managed retreat – allowing the sea to flood lowland areas.
Mangroves – evergreen trees and shrubs growing in tropical coastal areas, whose roots trap sediment and aid beach development.
Marine erosion – the wearing away of rocks by the action of the sea.
Meanders – large bends in a river.
Migration – to move from one country to live in another.
Monsoon – annual period of heavy rainfall in Asia.
More Economically Developed Countries (MEDCs) – countries where there is a high level of economic activity and where there is generally a good standard of living for most people.
Multinational companies (MNCs) – firms that produce goods in more than one country.
Multiplier effect – the process whereby a significant increase in economic growth can lead to even more growth as more money circulates in the economy.

Natural increase – an increase in population when there are more births than deaths in a year.
Nature reserve – area set aside to preserve plants and animals.
Net migration – the difference between people moving into a country and those who move out.
New international division of labour (NIDL) – this divides production into different skills and tasks that are spread across regions and countries rather than within a single company.
Newly Industrialised Countries (NICs) – countries that have moved rapidly from having limited economic development to having many new industries producing goods for both the home and export market.
Non-Governmental Organisations (NGOs) – private organisations that work on big issues affecting humanity. They may get money from governments and work with governments, but they are private organisations, not government-controlled.
Niche markets – small markets that deal in a specialised product.
Non-renewable energy – energy from sources that are finite and will eventually run out, such as fossil fuels.
North–South divide – the simple division between MEDCs (mainly in the north) and LEDCs (mainly in the south).
Nuées ardentes – highly destructive, fast moving clouds of dust and ash.
Nutrients – chemicals used in plant growth.

Obesity – a condition in which a person has so much excess body fat that it affects their health.
Optimum location – the location that best satisfies the objectives of the company. For most, this will be the maximum profit location.
Optimum population total (OPT) – the number of people that an area can support in a way that allows them to have a sustainable standard of living.
Outsourcing – where a company contracts out some of its work to another company. This usually happens because a company can save money. However, it can also happen if a company lacks certain skills.
Overpopulated – the idea that there can be too many people in an area for its resources to sustain.

Package holiday – travel and accommodation sold together by a tour operator.
Pandemic – a disease that affects a very large area, often crossing between continents.
Periphery – the parts of a country outside the economic core region. The level of economic development in the periphery is significantly below that of the core.
Permeable – a surface that allows water to pass through it.
Photosynthesis – a process by which green plants turn sunlight into plant growth.
Planning permission – all changes to how land is used need to be given planning permission by a local council.
Plate tectonics – the study of the distribution and movement of the Earth's crustal plates.
Population pyramid – a graph to show a country's population structure.
Population structure – the balance of people of different ages and genders in a country's population.
Precipitation – moisture that falls from the atmosphere in any form.
Prevailing wind – most frequent wind direction.
Product chain – the full sequence of activities needed to turn raw materials into a finished product.
Public Inquiry – a public meeting held to reach decisions about difficult planning proposals.
Pyroclastic flows – avalanches of hot volcanic debris.

Quality of life – this term sums up all the factors that affect a person's general well-being and happiness.

Rapid Transit System – public trains or trams with short waiting times and regular stops.
Reclaimed land – land that has been drained so that it can be used for development.
Refugees – people who have fled from their homes in one country, usually against their will, to seek a more secure life elsewhere.
Region – a large area in a country that has some common features, for example of landscape and the type of economy.
Reindustrialisation – the establishment of new industries in a country or region that has experienced considerable decline of traditional industries.
Renewable energy – energy generated from sources that are naturally replenished, such as sunlight, wind, tides and geothermal heat.

Replacement rate – the number of children that need to be born to replace the present population.
Residential – an area with houses.
Retirement village – a small settlement designed for elderly people.
Rurbanisation – the process of bringing features of a city to a rural area.

Salt marsh – an area of tidal mudflats, partially flooded at high tide.
Seasonal employment – temporary employment for part of the year only.
Seismic waves – vibration generated by an earthquake.
Shanty towns – areas of housing that people build themselves, often on the edges of big cities in LEDCs, on land they do not own.
Shield volcano – volcano that covers a large area with very gently sloping sides.
Shopping mall – an undercover area with a variety of shops.
Silt – very fine sediment deposited by flowing water.
Sink estate – a housing area with a poor reputation for its living conditions.
Slums – areas of poor, crowded housing in cities.
Squatters – people who occupy land illegally and may build houses there.
Storm surge – extreme storm waves created by strong winds and low pressure, leading to higher sea levels and flooding.
Strato volcano – tall, conical volcano composed of many layers.
Sub-aeriel processes – processes active on the face and top of cliffs.
Subduction zone – the area of a destructive plate boundary where one plate descends beneath another.
Subsidies and grants – money that a government can give or loan to a business.
Subsistence farming – the most basic form of agriculture where the produce is consumed entirely or mainly by the family who work the land or tend the livestock. If a small surplus is produced it is sold or traded.
Surface runoff – all water flowing on the Earth's surface.
Sustainable – capable of existing in the long term.
Sustainable development – development that meets the needs of the present without harming the ability of future generations to meet their own needs.
Sustainable management – long-term management that does not harm people or environments.
Swash – the force of breaking waves moving up a beach.

Tariff – a tax put on goods when they are moved from one country to another.
Thermal expansion – as sea and ocean temperatures increase, the water molecules near the surface expand and the sea level rises.
Through-flow – movement of water through the soil.
Trade deficit – when the cost of imports is greater than the money made from exports.
Trade surplus – when the money made from exports is greater than the cost of imports.
Transpiration – loss of moisture from plants.
Tsunami – tidal wave caused by an underwater earthquake.
Typhoon – a violent tropical storm over the Indian and Pacific oceans.

Upgrading – making conditions better.
Urban regeneration – rebuilding an old part of a city to improve it.
Urbanisation – the process through which an increasing percentage of a country's population lives in urban areas compared with rural areas.

Vertical erosion – downward erosion of a river bed.

Waterfront sites – land alongside or near rivers or old docks, usually in a city.
Water table – the upper level of underground water.
Weathering – breaking up of rocks by the action of weather, plants, animals and chemical processes.

Zero carbon – not releasing any CO_2 into the atmosphere.
Zero growth – when the birth rate and death rate in a country are about the same and the population is not increasing.
Zone – an area with mainly one type of land use.

Index

Single User Licence Agreement: OCR GCSE Geography B ActiveBook CD-ROM

Warning:
This is a legally binding agreement between You (the user or purchasing institution) and Pearson Education Limited of Edinburgh Gate, Harlow, Essex, CM20 2JE, United Kingdom ('PEL').

By retaining this Licence, any software media or accompanying written materials or carrying out any of the permitted activities You are agreeing to be bound by the terms and conditions of this Licence. If You do not agree to the terms and conditions of this Licence, do not continue to use the OCR GCSE Geography B ActiveBook CD-ROM and promptly return the entire publication (this Licence and all software, written materials, packaging and any other component received with it) with Your sales receipt to Your supplier for a full refund.

Intellectual Property Rights:
This OCR GCSE Geography B ActiveBook CD-ROM consists of copyright software and data. All intellectual property rights, including the copyright is owned by PEL or its licensors and shall remain vested in them at all times. You only own the disk on which the software is supplied. If You do not continue to do only what You are allowed to do as contained in this Licence you will be in breach of the Licence and PEL shall have the right to terminate this Licence by written notice and take action to recover from you any damages suffered by PEL as a result of your breach.

The PEL name, PEL logo and all other trademarks appearing on the software and OCR GCSE Geography B ActiveBook CD-ROM are trademarks of PEL. You shall not utilise any such trademarks for any purpose whatsoever other than as they appear on the software and OCR GCSE Geography B ActiveBook CD-ROM.

Yes, You can:
1 use this OCR GCSE Geography B ActiveBook CD-ROM on Your own personal computer as a single individual user. You may make a copy of the OCR GCSE Geography B ActiveBook CD-ROM in machine readable form for backup purposes only. The backup copy must include all copyright information contained in the original.

No, You cannot:
1 copy this OCR GCSE Geography B ActiveBook CD-ROM (other than making one copy for back-up purposes as set out in the Yes, You can table above);
2 alter, disassemble, or modify this OCR GCSE Geography B ActiveBook CD-ROM, or in any way reverse engineer, decompile or create a derivative product from the contents of the database or any software included in it:
3 include any materials or software data from the OCR GCSE Geography B ActiveBook CD-ROM in any other product or software materials;
4 rent, hire, lend, sub-licence or sell the OCR GCSE Geography B ActiveBook CD-ROM;
5 copy any part of the documentation except where specifically indicated otherwise;
6 use the software in any way not specified above without the prior written consent of PEL;
7 Subject the software, OCR GCSE Geography B ActiveBook CD-ROM or any PEL content to any derogatory treatment or use them in such a way that would bring PEL into disrepute or cause PEL to incur liability to any third party.

Grant of Licence:
PEL grants You, provided You only do what is allowed under the 'Yes, You can' table above, and do nothing under the 'No, You cannot' table above, a non-exclusive, non-transferable Licence to use this OCR GCSE Geography B ActiveBook CD-ROM.

The terms and conditions of this Licence become operative when using this OCR GCSE Geography B ActiveBook CD-ROM.

Limited Warranty:
PEL warrants that the disk or CD-ROM on which the software is supplied is free from defects in material and workmanship in normal use for ninety (90) days from the date You receive it. This warranty is limited to You and is not transferable.

This limited warranty is void if any damage has resulted from accident, abuse, misapplication, service or modification by someone other than PEL. In no event shall PEL be liable for any damages whatsoever arising out of installation of the software, even if advised of the possibility of such damages. PEL will not be liable for any loss or damage of any nature suffered by any party as a result of reliance upon or reproduction of any errors in the content of the publication.

PEL does not warrant that the functions of the software meet Your requirements or that the media is compatible with any computer system on which it is used or that the operation of the software will be unlimited or error free. You assume responsibility for selecting the software to achieve Your intended results and for the installation of, the use of and the results obtained from the software.

PEL shall not be liable for any loss or damage of any kind (except for personal injury or death) arising from the use of this OCR GCSE Geography B ActiveBook CD-ROM or from errors, deficiencies or faults therein, whether such loss or damage is caused by negligence or otherwise.

The entire liability of PEL and your only remedy shall be replacement free of charge of the components that do not meet this warranty.

No information or advice (oral, written or otherwise) given by PEL or PEL's agents shall create a warranty or in any way increase the scope of this warranty.

To the extent the law permits, PEL disclaims all other warranties, either express or implied, including by way of example and not limitation, warranties of merchantability and fitness for a particular purpose in respect of this OCR GCSE Geography B ActiveBook CD-ROM.

Termination:
This Licence shall automatically terminate without notice from PEL if You fail to comply with any of its provisions or the purchasing institution becomes insolvent or subject to receivership, liquidation or similar external administration. PEL may also terminate this Licence by notice in writing. Upon termination for whatever reason You agree to destroy the OCR GCSE Geography B ActiveBook CD-ROM and any back-up copies and delete any part of the OCR GCSE Geography B ActiveBook CD-ROM stored on your computer.

Governing Law:
This Licence will be governed by and construed in accordance with English law.